Creating the Built Environment

Creating the Built Environment

An introduction to the practicalities of designing, constructing and owning buildings

Leslie Holes

E & FN SPON
An Imprint of Chapman & Hall
London · Weinheim · New York · Tokyo · Melbourne · Madras

Published by E & FN Spon, an imprint of
Chapman & Hall, 2–6 Boundary Row, London SE1 8HN, UK

Chapman & Hall, 2–6 Boundary Row, London SE1 8HN, UK

Chapman & Hall GmbH, Pappelallee 3, 69469 Weinheim, Germany

Chapman & Hall USA, 115 Fifth Avenue, New York, NY 10003, USA

Chapman & Hall Japan, ITP-Japan, Kyowa Building, 3F, 2-2-1 Hirakawacho,
Chiyoda-ku, Tokyo 102, Japan

Chapman & Hall Australia, 102 Dodds Street, South Melbourne,
Victoria 3205, Australia

Chapman & Hall India, R. Seshadri, 32 Second Main Road, CIT East,
Madras 600 035, India

First edition 1997

Typeset in 10/12pt Palatino by Saxon Graphics Ltd, Derby
Printed in Great Britain by St. Edmundsbury Press, Bury St. Edmunds, Suffolk

ISBN 0 419 20820 8

A catalogue record for this book is available from the British Library

Library of Congress Catalog Card Number: 96-70573

Printed on acid-free paper, manufactured in accordance with ANSI/NISO Z39.48-1992 and ANSI/NISO Z39.48-1984 (Permanence of Paper).

Contents

Preface

About 7 per cent of the working population of the UK earn their living from buildings and other constructions, working mainly in small specialist groups. Some are involved in design, and some with construction, maintenance and renewal. Others will be concerned with the extraction, transportation and processing of raw materials, and in the manufacture and distribution of products. Yet others will be employed in buying and selling, or managing existing buildings.

Every new building project is unique, and the individuals who will work on it must learn about it first. Thus, a timely flow of information on its design and construction is vital. While those responsible for its management are sometimes described as a 'team', they may not have worked together before, and their relations with each other are not always harmonious.

All are likely to have been trained in exclusive groups by those with that specialism. Their concentration on specifics will have obscured the fact that much of the knowledge they use in their work is also used by other specialists. They are not likely to have been encouraged to take an interest in the complexities of buildings and the building process as a whole, nor in the wider and possibly more financial viewpoints of employers and building owners.

Practitioners, educationalists, students, funders, government bodies, members of school and health trusts, industrialists, would-be owners, occupiers – in fact, all of us – have financial and other interests in buildings. We spend most of our lives within them. Our homes may absorb up to half our incomes. Yet the construction and care of buildings and the economics of home ownership are ignored in school curricula.

It is commonly expected that, in the future, individuals will change their vocation from time to time during their working lives, and will need structured information on how things are done in their new situation, and why. Educational and training institutions are increasingly seeking to provide general introductory courses for large numbers of students. Students on such courses will wonder about the opportunities this unknown industry might offer and where they might find job satisfaction.

My aims have been to:

- explain what buildings are and how they come about;
- provide lists of things to think about and do at each stage in the procedure, while avoiding specialist details;
- identify some of the common knowledge, practices and terminology which enable those involved to communicate with one another and to work together;
- express this knowledge in as timeless and universal a way as I could, while conveying the sense of an ever-changing industry applying an ever-developing technology;
- offer it all in a sequence that moves from the descriptive and qualitative to the more quantitative, and from the physical to the financial;
- encourage the reader to appraise and learn from existing buildings;
- foster the development of the information specialist.

The book is intended to be a kind of mirror in which this great and important industry with its many ramifications is reflected. I hope this introduction to the business of designing, constructing and managing buildings will help readers to place their own interests in a wider context and to appreciate the work and problems of their colleagues.

Leslie Holes
1996

Acknowledgements

I gratefully acknowledge the courtesy of the following in permitting the reproduction of certain material.

The American Institute of Architects for permission to quote sections from *The Handbook of Professional Practice*.

The British Standards Institution for permission to quote extracts from British Standards. Complete copies can be obtained by post from BSI Customer Services, 389 Chiswick High Road, London W4 4AL.

The Building Research Establishment and Alan Stevens for the quotation regarding the time spent in making a building grow.

The Royal Institution of Chartered Surveyors and the Building Cost Information Service for permission to quote from the *Standard Form of Cost Analysis* and *Rationalisation of Measurement*.

Crown copyright is reproduced with the permission of the Controller of HMSO.

Undertaking a work of this magnitude makes one aware of the numbers of people who have, directly or indirectly, contributed to it, and I am grateful to them all. To the following, however, I owe special thanks. Ray Thomas, for sharing in the excitement of research before and during the early days of personal computers; Norman Cross, for his encouragement and support; John Andrews, for demanding that I 'do something for builders'; Roy England, Joe Kilner and Geoff Morgan. My greatest thanks go to my wife, Jean, who made all this possible.

> For I am but a gatherer and disposer of other mens stuffe, at my best value. (Sir Henry Wotton, *The Elements of Architecture*, 1624)

Introduction

Buildings are some of the largest objects made by humankind. All of them are different in one way or another, and they are constructed on sites that are also all different. There, they will be exposed to the full rigours of the natural environment. Each building will have been individually designed to match its intended occupiers' likely requirements, and constructed to meet quality, time and cost targets.

THE PARTS AND THEIR CONTENTS

Each of the four parts contains an introduction and five numbered chapters. There is also a glossary (Appendix A) and two further appendixes. Every chapter has a number of sections, each with a number of subsections. Most chapters conclude with a commentary that supplements the information in the main text, a summary, sources of references, and some suggestions for further reading. The tables and diagrams of various kinds are all called figures and numbered sequentially within each chapter.

Part One **Our built environments** introduces the physical make-up of new and older buildings. The first two chapters discuss what can be seen from inside and outside our buildings. Another chapter is concerned with their structures. Readers learn to distinguish the various parts of buildings, to recognize their materials and appreciate the reasoning that lies behind their choice, and to use appropriate technical terms.

The prime purpose of many of our older buildings was to provide their occupants with shelter and security. Since the 1950s, research in the UK and other countries has shown how the qualities of our built environments can be improved. Hence, our interest in both new and existing buildings, and in the applied science, i.e. the **technology** of both their initial construction and their adaptation, renewal and upgrading. These improvements are now expected in new buildings, and looked for in existing ones, and are considered in Chapter 4. Some general ideas are used to simplify thinking about the large number of different kinds of construction materials (Chapter 5).

Part Two **Design** looks at the process whereby the requirements of an intending building owner are analysed, transformed into a design, and communicated to those who will be constructing it. The first chapter (Chapter 6) takes the viewpoint of the individual or the organization that needs to acquire a new building or refurbish an existing one. It considers the process whereby their requirements are investigated and re-expressed as instructions to the designers, and describes the work of those professionals who will undertake the design.

The main object of the next three chapters (7–9) is to identify many of the factors that are likely to influence the thoughts of those who participate in the design process. During the first stage in the design process, information is gathered, constraints on the freedom of the designers are identified, and the spatial requirements of the owner are incorporated into the scheme design. During the second stage, decisions are made on how this design will be achieved physically, including the materials to be used. The description of this procedure follows the order taken in Part One.

Design information will be produced by individual members of the design team and will be read and acted upon by others within the design team and by all those who work for construction organizations, local authorities and the owners. Thus, everyone must have some appreciation of drawings, specifications and other technical communications.

Part Three **Construction**. Of course, if the owner is prepared to employ construction managers and operatives directly and to accept the financial risks, the construction works can now be started. In most cases, though, the next stage in a project will actually be to decide, generally as the result of inviting competitive bids, which construction organization will undertake the work. Preparing a bid involves generating partial and short-cut predictions about the building process and its costs. To be able to do this, we must look first at the actual information to be generated when preparing to build. Then, when we consider how a bid is prepared, we can appreciate what is being cut short.

The part begins by considering the evolution and structure of the building industry (often called the construction industry) and the nature of employment within it. Construction processes and the skills of the operatives employed on them are discussed, along with the notion that construction on the site consists of a sequence of self-contained construction activities. The use of (preferably computer) graphics for planning and controlling these is introduced.

The remainder of Part Three is concerned with the management of these activities so that buildings are completed to time, quality and cost targets. From the relatively few dimensions given on design drawings, quantities of various kinds, e.g. materials, activity durations, have to be generated. These too are matters that will concern all members of the design and the construction team.

Part Four **The business environment** includes achieving value for money on a new project, the application of management accounting and risk management ideas to the affairs of businesses involved with buildings and the building process, and the rights and duties of the parties to the more usual types of contract. The law relating to buildings and building processes is excluded from this study.

The rate of interest influences many business decisions, including those of readers who have bought their home with a loan secured by a mortgage or who hope to do so one day. The three compound interest formulas that relate the values of financial transactions at different times are introduced, and tables based on them are included in Appendix C.

Most construction businesses obtain their work by the competitiveness of their bids (tenders). As they expend much effort in preparing these bids, the two main procedures are considered in some detail.

Many of the terms introduced in the book relate solely to the subject matter within that chapter. The definitions of terms that have wider application are repeated in the Glossary (Appendix A).

Appendixes B and C contain data that readers might wish to refer to directly.

CONVENTIONS

Where it has been necessary to be specific, the standpoint is practice in the UK. However, where UK terms are different from those used in North America, they are followed by their US equivalents in brackets (at least, until this gets tedious). 'British' or imperial equivalents are given alongside the SI units now used in the UK and some conversion factors from one to the other are included in Appendix B.

The procedures for generating fresh information usually involve some arithmetic, and all those who earn their livings from the built environment should be able to use an electronic calculator. Computer spreadsheets can be even more useful. Arithmetical formulas for generating fresh numerical information are expressed as if the sequences of data and operators were being keyed into a calculator.

Instead of £ or $, the neutral symbol # is used to denote the financial unit of a money value in a calculation. Readers are invited to adjust the values to suit their circumstances.

FURTHER READING

Texts for further reading are either generally available or are taken from the current catalogues of the following:

- the Building Bookshop at the Building Centre, 26 Store Street, London WC1E 7BT;
- The RIBA Bookshop, 66 Portland Place, London W1N 4AD. and at Bristol, Belfast, Birmingham, Cambridge, Manchester, Nottingham and Leeds;
- RICS Books, 12 Great George Street, Parliament Square, London SW1P 1AD, and at Coventry;
- the CIOB Bookshop, Englemere, Kings Ride, Ascot, Berks SL5 8BJ;
- the AIA Bookstore, The American Institute of Architects, 1735 New York Avenue NW, Washington, DC 20006, and at Atlanta, Indianapolis and Philadelphia.

PART ONE
Our Built Environments

INTRODUCTION

Part One is intended to expand the reader's appreciation of what buildings are and how they are constructed. It starts by considering the qualities of the room the reader is in, and goes on to consider how such qualities are achieved in both new and older buildings.

We spend most of our lives in buildings, and can learn much about them just by looking. Although we can only see the surfaces of many of their materials, an appreciation of their thicknesses and purposes will enable us to visualize their internal detail. Watching their demolition from the street can also be revealing. Construction text books will show details of the parts of buildings and how they go together, but they are only an expression of the real thing. You do not have to be able to draw such details yourself before you can understand them.

Construction details can also be learned by visiting building sites, but do this only with the permission of those in charge, and wear a hard hat when requested. Few direct references are made to the UK's current Building Regulations, but many matters in their Approved Documents are touched upon to create an awareness of their likely influence.

CONTENTS

1. Chapter 1 **The internal environment** Readers are invited to consider the qualities of the rooms and other spaces in the building they are in, and their reactions to them. Alternative ways in which the surfaces of the constructions that enclose these spaces can be given attractive and durable finishes are then described, and followed by descriptions of different kinds of doorways and windows. Readers are encouraged to look for examples. The internal circulation system, including staircases and lifts and means of escape in case of fire, and

its importance to both occupants and visitors, is discussed. The features of sanitary accommodation, their water supplies and their drainage are outlined, together with those of heating and electrical systems.

2. Chapter 2 **The external environment** is mainly concerned with those parts of a building that are visible from the outside, and with their materials and the technology involved. These parts include sloping and flat roof coverings of various kinds, and the surfaces of external walls and their applied finishes. External works, including gravity drainage systems, are outlined. Some of the characteristics of the natural and the built environment are introduced, including vernacular and other building styles.
3. Chapter 3 **The building structure** Pitched and flat roof structures, floors, walls and foundations are considered in some detail, open frameworks being distinguished from continuous masonry and concrete structures. Emphasis is placed on strength, stability, dimensional stability and continuity. The soils to be found on a building site and the use of bench mark levels are introduced.
4. Chapter 4 **Improving our built environments** Various ways in which buildings are becoming more 'hi-tech' are introduced. They include improving weathertightness, dealing with condensation, improving thermal insulation, reducing sound transmittance and improving fire resistance. We are also concerned with the health and safety of operatives and occupiers, with security, and with the buildability, durability, resistance to wear, energy efficiency and other economic aspects of buildings.
5. Chapter 5 **Construction products** The various types of building materials (or 'construction products' as we shall be calling them) are introduced. Each type of product that can provide a part of a building with its required qualities will have others associated with it, e.g. wood flooring and flooring nails. Such sets are associated in a classification system that is used throughout the book. Most of the chapter is taken up with examples of such sets. Readers should take every opportunity to look at the displays at builders' merchants and DIY stores and to collect and study manufacturers' catalogues.

The internal environment 1

A building is a space surrounded by bricks (Elizabeth Holes)

Buildings are built for people to use. Although we all enjoy being out of doors when the weather is kind, the quality of our lives depends on the artificial environments that past and present members of our communities have built and maintained over the years.

Almost all our buildings are unique, even if they differ from others only in their foundations or the colours of their paintwork. Fortunately for us, though, they are all likely to consist of the same kinds of parts. You will already be familiar with many of these, and will know their names and where they are to be found in a building.

The occupiers of a building are interested primarily in the qualities of the space they are in, and they are likely to take the construction of the building for granted. In contrast, those engaged in constructing a building may show little interest in the qualities of the interior spaces they are working to achieve.

Of course, we cannot have one without the other. Even so, in response to my question 'what is a building?', Elizabeth (see above) was right to put the spaces first as these are the purpose for which we build. When wishing to refer generally to these spaces, we shall call them either **rooms** etc. or **internal spaces**. The overall space within the external walls, i.e. the internal spaces and the floors and internal walls between them, we shall call the **building space**. The purposes of an internal space will include the following:

1. to provide sufficient space of a suitable shape for its designated activities to take place and for related objects to be accommodated;
2. to provide an environment that is safe and comfortable to the senses;
3. to complement the purposes of other internal spaces.

1.1 WHAT ENVIRONMENTAL QUALITIES SHOULD WE EXPECT?

We begin with some questions that could be asked about almost any building and any design. Initially, we shall assume you will be thinking about the building you are in at present. However, you should also take every opportunity to look for answers to these questions when in other buildings.

THE ROOM YOU ARE IN

Assuming that you are dressed normally for the time of the year, and that you have been sitting down for some time, are you **comfortable**? How satisfactory are the following:

- the temperature, humidity, purity and movement of the air;
- the visual quality of your surroundings including the ceiling and wall decoration;
- the quality of the natural and artificial lighting including the natural light reflected to the back of the space by the ceiling and wall decoration;
- the noise level – can you concentrate, or are you being distracted by, for example, the activities of other people, the operating of equipment in the room, the sounds of music, doors banging, raised voices, plumbing, or other noises elsewhere in the building?
- the furniture; and
- the amount of space around you? Normally, you should feel free to move without much constraint. Contrast this with how you feel when seated in a theatre.

Do you feel **safe**? Are you sure that:

- you can control your immediate environment, e.g. by using radiator valves, electric switches, curtains, door and window locks, and by opening or closing windows;
- the building will keep out unwanted visitors, animals, birds and insects (contrast this with how you might feel in a tent);
- harmful gases are not being emitted from any part of the building, nor from the ground below;
- the equipment in the building is functioning properly and is not polluting the internal environment in any way;
- the building is structurally sound, is weathertight, and will resist fires;
- you will be warned if there is a fire, you will be protected from smoke and you will be able to escape safely to the outside;
- the floor is level but not too smooth;
- you will not injure yourself by accidentally breaking an area of glass, perhaps not realizing it was there;

- you know where the sanitary accommodation, refreshment and other facilities are;
- you can move round the building with safety;
- you can find your way easily?

Now, consider the room itself.

- What is its purpose? Usually we can sum up the purposes of a room or other space by stating its generic name, i.e. the name we give to all those rooms that have those purposes, e.g. library, office, bedroom.
- What facilities does it offer, e.g. desk, chairs, washbasin, electric lighting and power points, windows that open? Are these adequate and suitably placed?
- Is the room big enough or too large for its purpose?
- Is it a lofty room, or is the ceiling rather too close to your head?
- Are its length and breadth in proportion?
- Which way do the windows face, and is the penetration of natural light sufficient?
- When the sun shines, is the contrast in light intensity sufficient to cause discomfort? Can this glare be regulated by curtains or blinds? Do some of the light fittings also cause glare?
- Is the doorway suitably placed and does the door open back against the wall?
- Would any of the contents burn and produce smoke?

THE REST OF THE BUILDING

What are your impressions of the rest of the building? That is:

- What do you think of the finishes to the ceilings, walls, and floors, the lighting level, temperature, noise level, smell, security, and standard of cleanliness?
- Do you feel the owner cares for the property and is looking after it properly?
- Do you feel the owner is concerned for your well-being? For instance, are there smoke alarms; are fire doors shut or wedged open; are fire exit routes clearly marked; are fire extinguishers provided; is there an emergency lighting system?
- How legible and helpful are the direction signs (if any), and is it easy to find your way to where you want to go?
- Can stairs be ascended and descended easily, or do they have tapered treads, i.e. 'winders'. How will a person in a wheelchair manage?
- Can you see any structural cracking?

1.2 THE INTERNAL FINISHES

From your chair you will see the surfaces of the ceiling above your head, the sides of the walls, and the top of the floor. Usually, these will be the surfaces of materials that have been applied to the structural parts of the building to provide **attractive, light-reflecting, easily cleaned and durable surfaces**. Collectively, they are called the internal finishes. The finished surface of a floor and the steps of a staircase should be **level and safe** to walk on. Some floor finishes are also required to be **warm** and **quiet in use** although in hot climates it may be more important for them to be cool to the touch. Only occasionally will the surface of the structure be visible, and then it may be painted or otherwise decorated.

We take for granted that the surfaces will be continuous, will exclude smoke and creepy-crawlies, and will give time for occupiers to leave in case of a fire. We should also be able to take for granted that the materials used for a finish will not:

1. contribute heat to a fire;
2. allow flame to spread across its exposed face; and
3. allow flame to spread across its inner face where it is exposed within the structure.

Most finishes are likely to consist of a surface finish and its base. We can regard each as a distinct part of the building. Each will have a different purpose and be constructed by different workpeople with different skills, e.g. by painters and plasterers. They will be the result of distinct and detailed design decisions on what and how much material to use and where it will go, which we call their **technical solutions.** The decision on the finish will depend to some extent on that for the base, and vice versa, that is, the decisions are **interdependent**, but we shall start with what the occupiers can see. You should have no difficulty in finding examples of the following types of surface finish.

TYPES OF SURFACE FINISH

Type A surface (paint type)

A thin decorative surface of hardened paint, stain or other decorative liquid. Two or more applications, or coats of liquid may be required to achieve the required quality of surface. The finished decoration must have a consistent colour and a smooth or otherwise suitable surface and must adhere to its base, which must also have a smooth surface. Redecorating the surfaces of ceilings and walls is an easy and economical way of renewing or changing their appearance to suit the wishes of the occupiers.

Type B surface (clear finish type)

A surface of hardened liquid which will reveal and protect the appearance of the base.

Type C surface (tiles type)

Small preformed rectangular units, e.g. tiles, of which there are many different kinds, are secured to their base with a layer of either a special adhesive or mortar. Hard floor surfaces mean noisy footsteps, and vice versa. Tiled floors often have skirtings to match. Wall tiles have many different **fittings** for use at corners, angles and edges.

Type D surface (flexible sheet type)

A layer of ceiling paper, wallpaper, carpet or other continuous sheet material secured to a smooth base.

1.3 THE BASES OF INTERNAL FINISHES AND THEIR BACKGROUNDS

Having decided what material to use for the visible part, and how it is to be finished, we can consider what materials will provide it with a suitable base. This will depend on the background to which the base will be secured, and which will provide it with stability and strength.

This background (see Figure 1.1) might be an open framework of spaced structural members, e.g. a floor structure of wooden members called joists spanning from wall to wall which will support both the base for the floor and that for the ceiling underneath, or a wall structure of vertical members. On the other hand, it might be a continuous construction, e.g. a concrete floor, or a wall made of bricks, blocks, or some other masonry. Note that both open frameworks and continuous constructions can be used for both horizontal and vertical structures. They are discussed in much more detail in Chapter 3.

TYPES OF BASES ON OPEN FRAMEWORKS

Open frameworks of structural members are mostly made of wood as this is strong, relatively light in weight, easily nailed, and readily available in most countries. Also, it comes from naturally sustainable sources. Open frameworks may also be constructed from metal sections. The spacing of the members must match the strength and rigidity of the material being used for the base. Types of bases suitable for fixing to open frameworks include the following:

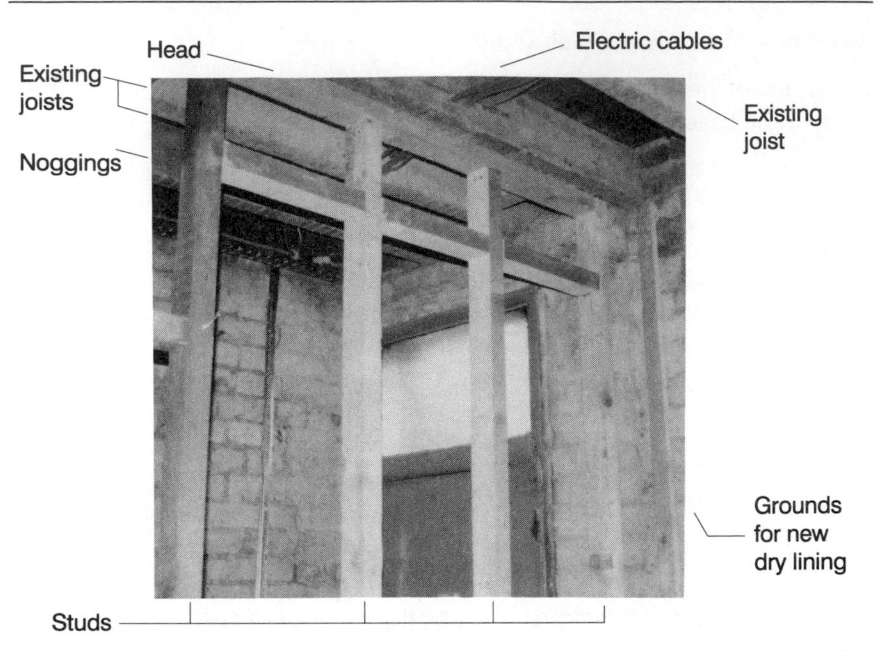

Fig. 1.1 Constructing an open framework for a new internal wall in an existing building and showing the existing continuous walls and open framework floor structure prepared for new finishes.

1. *Plasterboard type* Rigid preformed boards are nailed to open framework members. Crosswise members called **noggings** (see Figure 1.1) are cut and fixed between the framework members to provide support to otherwise unsecured board edges. Plasterboards consist of a hardened plaster layer about 9 mm thick with paper facings. The boards are finished with a thin coat of finishing plaster which will cover nail heads, gaps between boards and other imperfections. Historically, ceilings got their name from the plasterer's work of 'cieling', i.e. sealing, the otherwise exposed, or 'naked' structural members with wood strips called laths spaced about 5 mm apart, and followed by two or three coats of plaster. Irregular cracks in old ceilings may indicate a loose and possibly dangerous base, as one square metre might weigh 30 kg.
2. *Dry lining type* Large, broad, rigid preformed boards, e.g. plywood, plasterboards with tapered edges filled flush and with any surface imperfections made good with plaster or some other substance. In addition, plasterboard surfaces may be coated with 'Artex' and given a textured finish.
3. *Wood strip type* A continuous base is made by joining smoothed (i.e. 'planed' or 'wrought', or 'wrot') pieces of wood together in various ways and finishing their exposed surfaces to receive a Type A or B

thin coating. These types of internal finish can be used to line ceilings and walls. The simplest type consists of thin, narrow wood strips, i.e. **boards**, that have been machined to leave a tongue on one long edge, a groove in the other and with the corners of the exposed faces chamfered, i.e. bevelled, e.g. wooden 'matchboarding' (see Figure 1.2). Each board is fixed by pushing its grooved edge on to its neighbour and 'secret' nailing it through the tongued edge to create a continuous surface with 'V' joints. Being fixed at only one edge enables the boards to shrink crosswise without splitting, such shrinkage being disguised by the V-joints. A more sophisticated solution is the wooden **panelling** to be found mostly in important older buildings. A panel is any rectangular piece of material, and, in panelling, the panels are held in a rectangular framework of thicker wood. Most solid wooden doors are also made in this way, and are discussed later.

4. *Structural deck* A self-finished deck, or one to support, say, a carpet finish, can be constructed with sheets of moisture resistant flooring grade chipboard, or with planed, tongued and grooved (PTG) wood floor boards (see Figure 1.2). Hardwood flooring chosen for its appearance can be given a type B finish.

CONTINUOUS BACKGROUNDS

Two types of combinations of materials are available for constructing the continuous type of background, i.e. masonry and concrete. Any continuous construction made by bedding small preformed units in a matrix of mortar can be classed as masonry. It is convenient to think of these units as being either:

1. small rectangular units that can be held and lifted using one hand, e.g. clay bricks;
2. larger rectangular units requiring both hands, e.g. concrete blocks; and
3. rectangular and irregularly shaped units of natural stone which also need both hands.

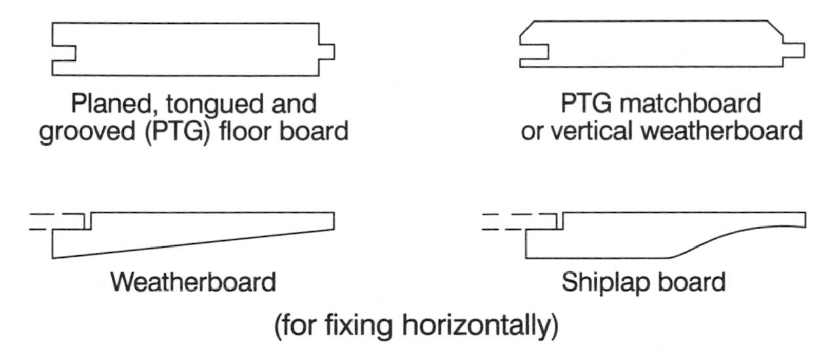

Fig. 1.2 Sections through various edge-jointed wood boards used to cover internal or external surfaces.

Concrete is a mouldable wet-mix of cement, sand, rather larger stones, and water, which can be given the required shape by being placed and kept in a suitable mould at least until it hardens. Specially made moulds are called formwork or shuttering. Concrete is used in the construction of all kinds of structures, i.e. floors and walls, foundations next to the ground, staircases and to make a framework of beams and columns. It is also used for the manufacture of preformed units, e.g. bricks, blocks, lintels, copings.

TYPES OF BASES ON CONTINUOUS BACKGROUNDS

5. *Plaster type* A thick (perhaps 13 mm) base formed from a hardened mixture of finely ground plaster and water or some other mouldable material. It will rely for its strength on adhering to the surface of a continuous background. As this can be somewhat irregular, at least two coats are required to achieve a smooth, reasonably planar surface. Different plasters will be used for undercoats on masonry surfaces, for undercoats that will bond to smooth surfaces (e.g. concrete), for finishing coats, and for producing specially smooth and hard surfaces. Non-rusting metal 'beads' with mesh 'wings' embedded in the plaster will give precision to angles and exposed edges. In older properties, a plaster mix of lime, water and sand will probably have been used. In higher quality work, angles (or arrises) with about 60 mm wide wings may have been formed with a harder (but then more expensive) hard-wall plaster. The connections between these and the general plastering are usually visible.
6. *Screed type* A rather thicker, harder and stronger base layer than (5) is called a **screed**. It is formed from a hardened mixture of cement, sand and water that has been applied to a continuous, stable, clean and rough background (see below). Shrinkage and the ensuing loss of bond can be minimized by using only enough water to hydrate the cement. The finish of its surface will depend on the requirements of the surface finish. A roughish surface is required to receive a mortar bed for, say, ceramic quarry tiles; a smooth surface is necessary under thin plastic tiles. The inert mineral particles used to bulk out a cement and water paste, e.g. sands, gravels, crushed stone, are called **aggregates**. While continuous structural floors will usually be of the same thickness throughout, floor finishes come in a variety of thicknesses. To **ensure a continuous level top**, the overall thickness of finish, bedding substance, and base is maintained by inversely varying the thickness of the base (see Figure 1.3).
7. *Granolithic type* A self-finished single thick (perhaps up to 40 mm) layer of a hardened mouldable mixture of cement, water and graded stone aggregate. Where granite is used, the grey-coloured result is called granolithic. Applying a hardening liquid to the surface will

reduce wear and the consequent production of dust. Where coloured cement and marble aggregates are used, and the surface is finished by grinding and polishing to reveal the colour and shape of the aggregate, the result is called terrazzo. All such thick layers of mouldable materials can be described as being **in situ** or **cast-in-place** finishes.

8. *Mastic asphalt type* A thick, probably two-layer coating of mastic asphalt to a continuous concrete floor next to the ground can provide a damp proof and smooth floor finish or a base for carpeting.
9. *Isolating membrane* Carpets, linoleum and other type D sheet floor finishes may be laid loose on a wooden or concrete floor but may need some protection against, say, irregularities in the surface, projecting nails, or moisture. Among the materials used are hardboard, damp proof membrane, reinforced building paper, carpet underlay.

INTERMEDIATE CONSTRUCTIONS

An intermediate construction is sometimes required between the base and its background. Examples include the following:

1. Metal frameworks suspended from the structural floor above to provide support for ceiling finishes. These may be used to lower the apparent ceiling height, or to hide pipes and the ceiling finish in a room in an existing building, or to enhance the acoustic or thermal

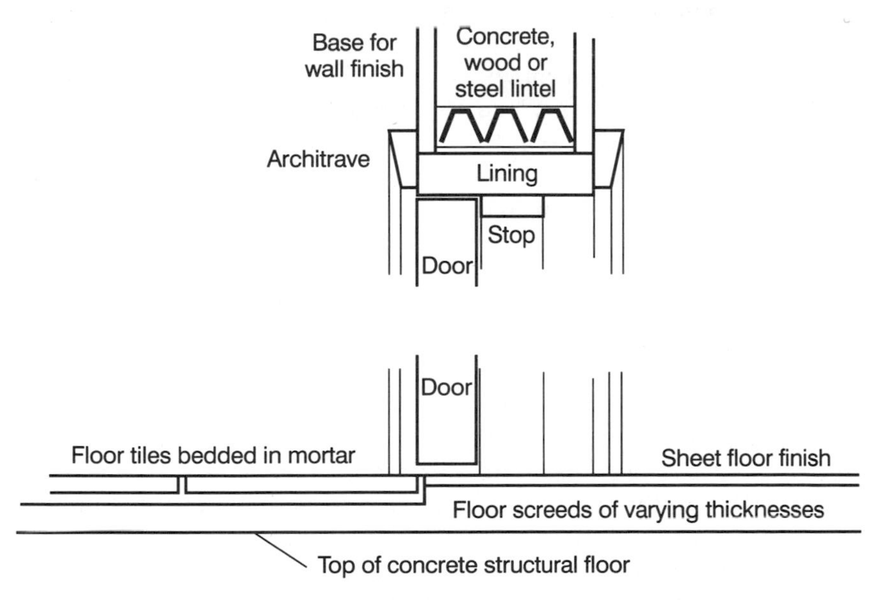

Fig. 1.3 Vertical cut through head and bottom of an internal doorway showing wall and floor finishes and bases.

insulation of a space, or to accommodate heating, lighting and/or ventilation units. They are most likely to be designed, supplied and installed by a specialist subcontractor although they can be constructed from traditional materials.

2. A resilient layer between floor boarding and a continuous structural floor to reduce the transmission of sound (see Chapter 4 and Figure 4.2). The boards will be secured to wood battens to form a 'ribbed floor' or 'platform' which is placed loose on a resilient layer of, say, mineral fibre. When supported by an open framework of joists, strips of resilient material are laid on the joists instead.
3. An intermediate open mini-framework of smaller section wood members, or grounds, secured to a continuous structural background to provide fixings for those bases that would normally need an open framework, e.g. wood panelling (see above).
4. A similar mini-framework encasing a pipeline or a steel column or beam to provide support for a type 1, 2 or 3 base (see page 8).
5. A sheet of expanded metal lathing (i.e. a metal mesh) secured to an open framework and with a coat of plaster or rendering to provide a continuous background for a type 5 plaster base.

JUNCTIONS BETWEEN FINISHES

Many junctions can be made at the same time as the bases to the finishes, e.g. a simple internal angle between ceiling and wall plaster, a tile skirting to a similarly tiled floor finish. Otherwise, a single rigid preformed strip can be used to cover the junction, e.g. a wood skirting (baseboard) which, not having to carry a load, need only be secured at intervals to a continuous or open background.

Skirtings add to the appearance of wall and floor finishes by covering the junction between them, which can then be left rough. Similarly, architraves (casings) are used to cover the junctions between wall finishes and the linings to internal doorways. Between some ceiling and wall surfaces there may be a cast in-situ or preformed plaster cornice which reinforces and hides the junction and enhances its appearance.

Wooden skirtings, whose top edges are usually shaped, i.e. 'moulded', may be fixed by nailing through the plaster base into the wall behind. High-quality skirtings may be fixed to grounds which are themselves fixed to the background. Grounds replace the base layer of the wall finish and will have the same thickness (see intermediate construction type 3, above).

The corners of skirtings, architraves, cornices, etc. are usually formed by diagonally cutting the ends of the intersecting lengths. Such cuts are called **mitres**. However, if the internal angles of wooden skirtings etc.

were mitred and the wood shrinks, the joint would gap. Instead, the end of one piece is usually shaped, i.e. 'scribed', to fit the outline of the other.

WHICH BACKGROUND IS IT?

To test whether a smooth, plain surface has an open or a continuous background, gently tap it with the knuckles. If the surface of a wall is generally unresponsive, its background is probably heavy masonry. A slightly resonant response will probably indicate a lighter masonry construction, e.g. lightweight concrete blocks. A slightly resonant sound occurring only here and there probably indicates where a finish has come away from its background. A distinctly hollow sound will indicate an open background, and some straight hair cracks will confirm the presence of rectangular baseboards. Lines of exposed nail heads, or small surface blemishes where the nail heads have been punched in and the holes filled, will locate the structural members. Tapping a ceiling or floor finish will give similar indications.

FURNISHINGS

Tables and chairs, curtains, blinds, room numbers, exit signs, pictures and loose floor coverings can all be classed as furnishings. They are removable, in contrast with the parts of the building proper which are all either directly or indirectly joined to each other. When the occupier of a building is not the owner, the tenancy agreement, i.e. the lease, should state clearly what furnishings, if any, belong to the building owner.

A letter plate is usually associated with the main entrance door, but this is only for the convenience of the postman and for ease of fixing by the joiner. It should be large enough to take an unfolded A4 envelope but positioned, perhaps in a side panel, so that no one can reach an arm through and open the door.

1.4 DOORWAYS

DOORS GENERALLY

From your chair you should be able to see a doorway and at least one window. The purpose of a doorway is to provide two-way access through a wall between two rooms or other spaces or between the inside and the outside that can be closed by a door to provide security, privacy, or simply to act as a barrier, say, against smoke. Thus, the choice or design of a door and its hinges, lock and other 'ironmongery' (or 'hardware'), its surround, and its fixings, will depend on its purpose. Notice that an internal door is fixed so that its hinged face is in line with the surface of the wall finish. Any architrave or other projection will limit its ability to be opened back into a room.

Before the advent of plywood, all doors were made of solid natural wood. Nowadays, these are just some of literally hundreds of different types, some made from natural wood, others from unplasticized polyvinyl chloride, a durable synthetic substance whose name used to be abbreviated to uPVC, but nowadays is more often seen as PVC-U.

Wooden doors are made with long, straight thin strips rip sawn (i.e. with the grain) from the trimmed trunks of trees, i.e. from **logs**, and then surfaced (i.e. planed) and cross-cut to length. Their tendency to shrink crosswise and to warp and split as their sap dries is minimized by careful drying, or 'seasoning'. Happily, the strips are dimensionally stable lengthways, i.e. parallel to the upwards growth of the tree. Doors and other **joinery** are made by joining narrow wooden members together in such a way that their lengths rather than their widths or thicknesses determine the size of the unit.

For instance, the height and width of a traditional wood-panelled door are established by having a frame consisting of vertical outer **stiles** (or **styles**) and horizontal top and bottom **rails**. Intermediate vertical members are called **muntins.** Mortice and tenon joints are used to connect pairs of members at right angles. A mortice is a narrow slot cut in the side of a member. A tenon to fit the mortice is made at the end of the other member by cutting away unwanted material. The joint is made with glue and the tenon is tightly wedged and possibly pinned with a wood or metal dowel. The ends of these tenons and the wedges can often be seen in the vertical edges of a panelled door.

The spaces between this framing are filled with wooden panels. While some panels are left plain, other, thicker ones will have their thickness reduced at the edges to fit grooves in the framing. In any case, to enable them to shrink crosswise, these panels will be only loosely secured to the framing. The junctions between the panels and the framing might then be hidden by flush or projecting wooden mouldings with mitred corners. Alternatively, the edges of the stiles and rails next to the panels might themselves be moulded 'on the solid'. Each variant has its own name, and these will be found in text books and catalogues. One or more spaces may be left open to receive glass panels which would be secured with small-section wooden **beads**. A single glazed panel for observing what is on the other side of a door is called a **vision panel**.

Another traditional type of door used mostly for outbuildings consists of vertical matchboarding (see Figure 1.2) secured to horizontal **ledges** and to diagonal **braces.** The door may also have a framed surround and should be hung so that the braces point downwards to the hinges.

The development of plywood has revolutionized the design and manufacturing of doors. A plywood is made from an odd number of thin wood veneers, each laid with its grain at right angles to that of its neighbours and glued. The veneers are obtained by 'peeling' a rotating tree trunk. Different types of plywood are made by varying the choice of wood species for the internal and the face veneers, their number and thickness, and the glue. Not only are plywoods dimensionally stable, they are also an economic way of using valuable decorative woods, which can be used just for the face veneers.

Plywood can be used for the panels of wooden doors and as facings to both sides of 'flush' doors. Such doors will have cores of wood or some other stable material, and may have their long edges 'lipped' with matching hardwood. Similar doors for internal use may be faced with hardboard. Simulated wood-panelled doors are manufactured using one-piece moulded facings. Fire doors must be able to resist fire for a stipulated time, usually half an hour or one hour. The substances of the doors and their frames, linings and trims will determine suitable kinds of surface treatment.

EXTERNAL DOORWAYS IN MASONRY WALLS

In an external doorway in a masonry wall, the door will most likely be secured within a frame. This will consist of a horizontal top, or 'head' and vertical 'jambs' (i.e. 'legs'), with a sill (sometimes called a threshold) which projects from the face of the wall. If made of wood, these members will be securely joined, or 'framed' together with mortice and tenon joints. The frame may be extended to provide sidelights on one or both sides.

Most likely, the frame would have been positioned in the wall as this was being built. It would have been secured with **frame cramps** screwed at about 600 mm intervals up each jamb and built in as the walls were raised. Alternatively, a temporary wooden outline or **profile** of the same overall size could have been positioned and loosely fixed to the wall as it was constructed. This would have been replaced by the door frame when the likelihood of damage by other construction activities had passed.

The door itself will fit flush into the rectangular sinking called a 'rebate' in the opening edge of the frame. Doors require a variety of ironmongery for various purposes. These include:

1. hinges for securing doors to their surrounds while enabling them to open and close;
2. knobs, lever handles, push plates, and other door furniture for opening or closing them;

3. door closers, or special hinges for making them close automatically;
4. latches, locks, bolts and other fastenings for keeping them shut and secure.

The glazing to the door and any sidelights may be subjected to accidental or intentional impact. Nowadays, depending on the risk, either safety or high impact resistant glass or plastic should be used.

The exposed edges of the wall (the reveals) and the equivalent lower face of the lintel over the opening (the soffit) are likely to be given the same finish as the walls. Their junctions with the door frame may be covered by small profiled wooden **moulds**. Although plastic units will be self-finished, the exposed surfaces of wood doors, frames etc. (collectively called 'joinery') will be finished either to improve or to preserve their appearance. Paints will obscure wood surfaces and blemishes, and provide renewable coloured finishes. Clear treatments will reveal and protect the surfaces of woods chosen for their appearance.

At home, the door you leave by (after securely locking and bolting the others) should be especially secure. If made of wood, it should be at least 44 mm thick and robustly constructed. Its opening edge should be held shut by two locks spaced well apart, a cylinder rim latch for normal use and a five-lever mortice deadlock manufactured to BS 3621 for added security. Alternatively, it may be fitted with a three-point lockable system of bolts operated by the door handle. It should be hung on three (i.e. a 'pair and a half' of) hinges to a robust frame which is, itself, securely fixed to a substantial wall.

INTERNAL DOORWAYS

A door frame is a structural unit, whereas the surround to an internal door is usually a rather thinner **lining** which relies for its strength on being fixed to the wall. Where a doorway is in a non-loadbearing partition, both it and the partition can be stabilized by using a storey-height lining (see Figure 1.4) secured to the floors below and above, and with a horizontal transom and a glazed or solid panel over the top of the door.

As should be the case with all internal joinery, a door lining will be installed after the building has been made weathertight. Its width will equal the combined thickness of the wall and the wall finishes on both sides, and its junctions with these finishes are likely to be hidden by architraves. By measuring this width and deducting what you think might be the thicknesses of the wall finishes, you can make a good guess at the thickness of the wall itself.

Instead of having linings that are thick enough to be rebated, it can be more economical to have ones that are just thick enough to be stable. A rebate to fit the thickness of the door is then formed by gluing and

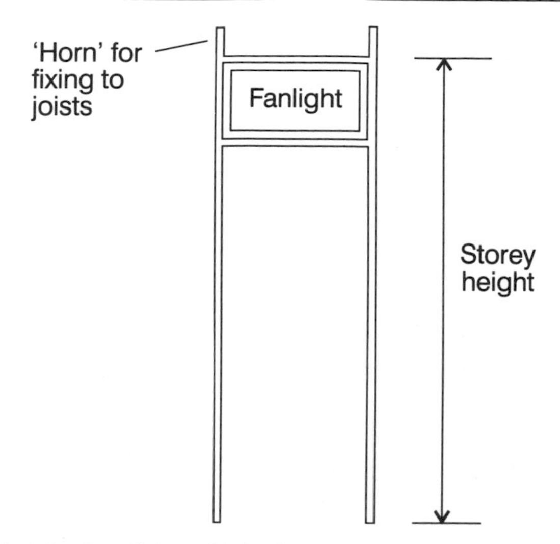

Fig. 1.4 Storey height door lining with fanlight.

screwing, i.e. 'planting on', lengths of small section wood called 'stops' to the inside faces of the lining (see Figure 1.3).

Some doors may help to create safe escape routes in case of a fire. The door manufacturer must warrant that the door will retain its integrity for the required number of minutes when exposed to fire. It must be hung to open in the direction of escape, and may have a vision panel so you can check where the fire is without opening the door. Doors should be self-closing, and kept closed except when in use. The passage of smoke through the gap between the edge of a door and its frame can be prevented by an 'intumescent' strip which expands to fill the gap when heated.

1.5 WINDOWS

The shapes and configurations of windows and glazed walls make a major contribution to the appearance of a building. Their choice or design depends on the need for natural light, visibility, privacy, security and ventilation. They must be weathertight, have an appropriate thermal and sound resistance, and their surfaces must be continuous with those of the structure and its finishes. The side frames should be securely fixed to the walls as described for door frames.

Most building substances are impervious to light, i.e. they are **opaque.** Those that can transmit light, e.g. glass, polycarbonate, are described as being **translucent.** However, any surface roughness scatters the light rays, and it is only where light passes between two planar and very smooth parallel surfaces that such substances become **transparent.**

Objects on the other side can then be clearly seen. Translucent or **obscured** glass with a textured surface that has purposely been made opaque will provide privacy. The term **glazing** can be either a noun or a verb, and can signify either the light transmitting sheet, or the process of securing it.

Windows regulate the passage of natural light into the building and provide visibility. Further regulation can be provided by curtains or blinds. Windows may also be used to regulate the passage of air into and out of the building. In many parts of the world, a mesh screen to exclude insects will also be fitted. Windows are normally in external walls. In internal walls, similar constructions are called borrowed lights, as they allow a room or corridor to receive light from a window in an adjoining space. Where a length of internal wall is partially or wholly glazed, it may be called a **glazed screen**. Neither is likely to have opening lights.

Although they vary considerably in their detail, windows and screens consist essentially of glazing either held in openable surrounds called 'sashes' that are secured to the window frame, or glazed directly to it. In addition to its head, jambs and sill, a window frame may have intermediate vertical members called mullions and intermediate horizontal ones called transoms. All will be profiled to increase their weathertightness and, possibly, provided with weather seals.

Glazing can consist of single or multiple sheets of glass or some other clear or translucent material possibly chosen for its safety in use, e.g. annealed glass. The edges of single panes of glass will be bedded in the glazing rebates with linseed oil putty or mastic and secured with either little headless nails called glazing sprigs or with wooden beads. A variety of methods are available for securing the sealed double-glazed units that are now the norm, depending partly on the type of finish to be applied to the sashes. These methods involve the use of supporting and spacer blocks, loadbearing butyl tape, non-setting mastic and beads (see Figure 1.5).

Sashes that are hinged at one edge are called casements, although small top-hung casements may be called fanlights. Others may be fixed to complex mechanisms that allow them to project outwards, and even to be reversed for cleaning (see Figure 1.5). All opening sashes should be fitted with key operated locking handles. Side hung casement windows of suitable size can act as escape windows from dwelling houses.

Present-day double-hung vertically sliding sash windows have solid frames and spiral balances rather than the hollow boxed frames and weights, cords and pulleys of their predecessors. They are delivered to the site as complete units possibly already glazed and ready for fixing.

Existing vertical sliding sash windows should always be viewed with suspicion (see Figure 1.6). When slightly open, they provide a most efficient way of achieving gentle ventilation while remaining weathertight.

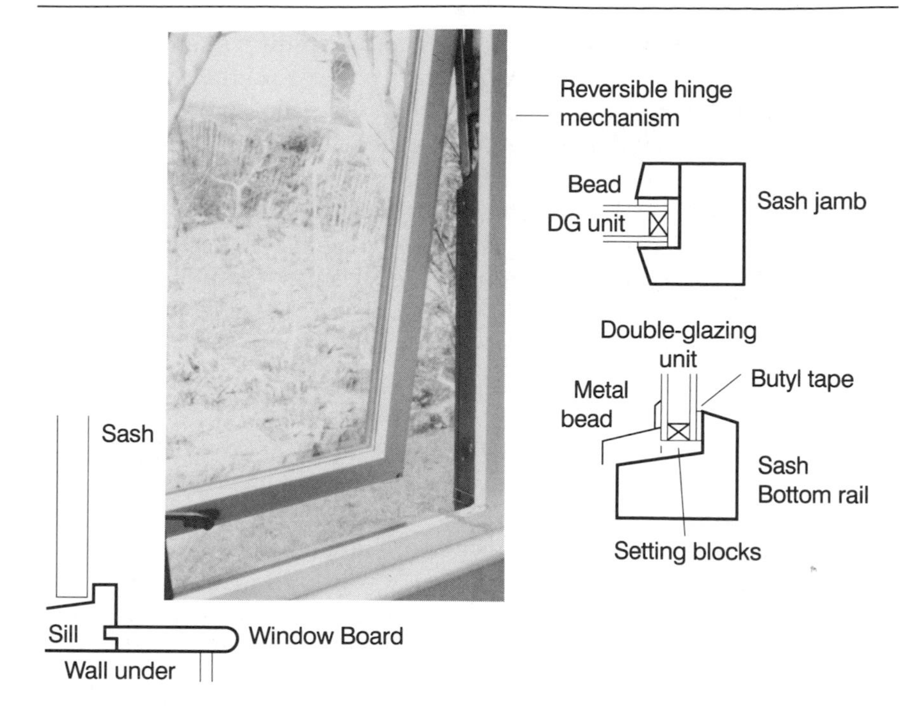

Fig. 1.5 A wood double-glazed reversing sash window.

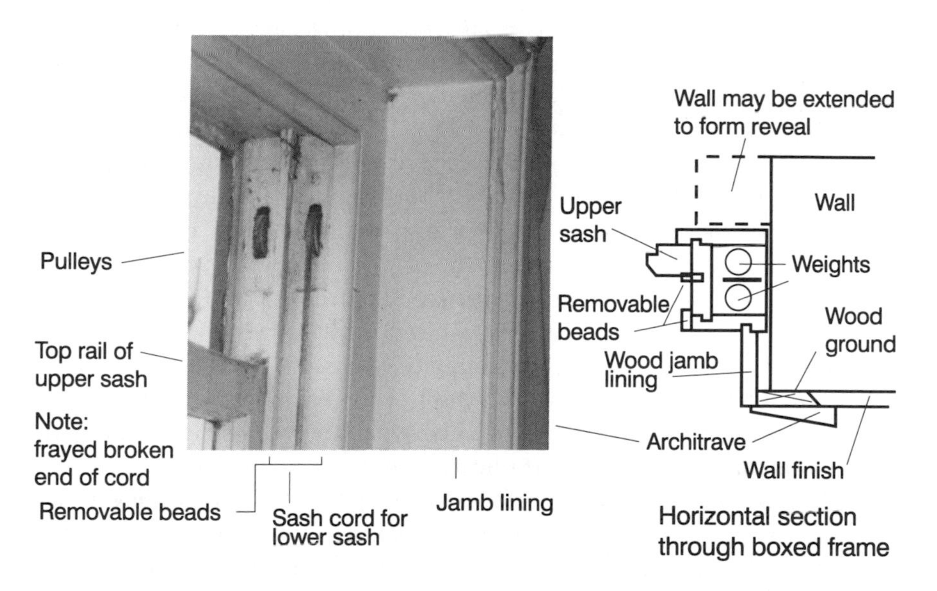

Fig. 1.6 The right-hand upper corner of a vertical sliding sash window showing a broken sash cord and a section through the side box frame.

Unfortunately, they rely on flexible sash cords running in small pulleys to connect the sides of the often very heavy sashes and their counterbalancing weights which move up and down in the hollow side frames. The original flax cords may have worn out or perished, and renewing them involves taking out some restraining members called beads, and removing the sashes and weights and reinstating. Because this is expensive and will damage the decoration, you are likely to find the sashes permanently fixed and sealed with paint to their frames. Do not attempt to move a sash unless you know it is in regular use. Even so, one day the cords will break unless they are renewed in time.

Window units may be made of galvanized steel, aluminium, wood or unplasticized PVC. Metal window units may be secured to wooden surrounds that are fixed to the walls. Although rustproof, their appearance can be enhanced by a factory applied polyester powder coating or a normal paint finish applied after fixing. Wood window units will require some protective treatment, e.g. a micro-porous stain.

Window units are usually positioned towards the front face of the wall, and, as in the case of doorways, will either be built in as the external wall is raised, or fixed into preformed openings made by building in and removing temporary profiles. Their projecting sills should discharge rainwater clear of the walls below. Look underneath them for grooves, or **throatings**, that stop the water from running back.

Where window units are fixed behind half-brick thick 'reveals', often the case in older buildings, their sills may not project at all. Instead, a one-piece projecting masonry sill, or the less durable sort made of small units, e.g. bricks, tiles, with vertical mortar joints, was often provided beneath the unit to keep the wall weathertight. However, over the years, the wooden sill would have shrunk, leaving a horizontal capillary path for moisture (see Chapter 2) that inevitably leads to its decay.

Internally, the appearance of the wall is maintained by returning the internal finish at the reveals and soffit as described for doorways or by wooden jamb linings and architraves (see Figure 1.6), and by providing an inner window sill. This can be a painted wooden window board with the lower part of one long edge rebated to leave a 'tongue' which will fit into a groove in the inner face of the sill of the window unit (see Figure 1.5). A window board will be notched between the wall reveals and fixed to the wall below, prior to the application of the wall finishes.

Both sides of all glazed surfaces must be accessible for cleaning. Opaque, i.e. **obscure** rather than clear, glazing will provide privacy, and polycarbonate or other high-impact resisting glazing will improve security and safety. The thickness of the glazing will depend on the pane size and the exposure to the wind, while double or treble glazing, the use of thermal or high-impact resistant glass and the substances used for fixing will affect security and the transmission of light, heat and sound.

1.6 INTERNAL CIRCULATION

VISITOR INFORMATION

First-time visitors to a building need to be assured that it is the correct one and that they are heading for the correct entrance. Inside, they should be presented with information on the internal arrangements, and the locations of the various activities. Rooms should be labelled with the functions of their occupiers.

Locational information should be offered throughout the building, to reassure visitors that they know where they are, that they are moving in the right direction, that they have reached their destination, and that they can find their way out to where they parked their cars. Fire safety signs will identify exit routes, smoke doors (which must shut automatically and must not be kept open), and exit doors.

SPATIAL SYSTEMS

An occupier or visitor needs to be able to move quickly and easily from one space to another in a building. The set of all the spaces within and around a building, together with stairways and lifts (elevators), forms a circulation system. One of the aims of the design process is to achieve efficient circulation subsystems.

Some subsystems will provide for the routine movements of people and objects, e.g. between entrances and workplaces; between classrooms and toilets. Others will provide escape routes in case of fire. Have you experienced any unsatisfactory spatial systems, e.g. bathrooms remote from bedrooms, a single staircase at one end of a long corridor, having to go through the general office to get to the manager's office?

ESCAPE IN CASE OF FIRE

It should be possible to evacuate a one- or two-storey house before a fire takes hold, provided everyone is awake and knows what is happening. Hence the importance of smoke alarms, there being no smoke without fire. New buildings must have electric mains operated alarms which, although independent, can be interconnected so that any alarm can set off all the others. Existing dwellings can at least have battery-operated ones installed, although these must be tested regularly and their batteries should be renewed at least annually. Alarms should be positioned in circulation spaces close to possible sources of fires, e.g. kitchen, living room and also outside bedroom doorways, with at least one on each floor. Bedrooms should open directly on to a circulation space.

If you are faced with a fire in a more complex building, it should be possible to turn round and find an alternative route to the outside. Readers should make a habit of checking the arrangements in any strange building, say, a hotel or college. Look first at the room you are in. Is the door to the corridor substantial and self-closing? In the corridor, are exit signs visible in both directions? Expect short dead-end lengths to have self-closing doors at the other end leading to alternative exits.

If the corridor is on an upper floor, do exit signs lead to at least two widely separated staircases, and are these isolated from the corridors by enclosed lobbies at each floor, with self-closing doors? Only if there is an automatic release mechanism for these doors in case of fire should they be allowed to remain open. If one staircase is affected by fire, would the remaining ones be wide enough to accommodate all the occupants?

STAIRWAYS

The purpose of a stairway is to enable people to move safely from one storey to another. A sufficient number of staircases should be so positioned that the building can be evacuated quickly and safely when one of them is unavailable due to fire and smoke.

The simplest wooden staircase is an open framework consisting of a **flight** of equally spaced horizontal **treads** with their ends connected to longitudinal **strings.** A wall string is secured to and supported by a wall. An outer string is open to a corridor or other space and must be strong enough to span the length of the flight. Their ends will be tenoned into vertical newels which also support the ends of the handrail. Further support will be provided by balusters to form a safe balustrade. (A ladder is also an open framework consisting of a series of **rungs** fixed to longitudinal **stiles.**)

However, most wooden staircases will have vertical **risers** inserted between the treads to provide strength and to enclose the space below. These will be secured to the back of the tread below and just behind the front edge, i.e. the 'nosing' of the one above, this possibly reinforced with a small mould. The ends of both treads and risers will also be housed and wedged to the strings (see Figure 1.7). Although most wooden staircases are finished underneath with a type 1 base, you should be able to find one that is open and shows its construction.

Sometimes, the outer string is cut to fit under the treads and risers. This enables the ends of the balusters to be dovetailed into the treads, these joints and the end grain of the treads being hidden by short nos-

Fig. 1.7 The (usually hidden) underside of a wooden staircase and one possible section.

ings. The end grain of the risers is hidden by thin ornamental **brackets** with mitred external angles.

Intermediate landings between straight flights enable users to pause for rest and allow the direction of the flights to change. An alternative is to have tapered treads, i.e. **winders.** While these may enable space to be used efficiently, they are awkward to use, can be dangerous, and may prevent the installation of electric stair-lifts. The parts of metal staircases tend to imitate those of wood.

Compare a number of different staircases for their ease of use. In each case, try to measure the vertical distance, i.e. the **rise**, between the tops of their treads and the horizontal distance, i.e. the **going (run)**, between their front edges. Use the handrail as you climb the stairs and watch what happens to your hand and arm as you change direction between straight flights or at the top where there might be a landing with a similar handrail. Then look for the individual pieces of wood that have been shaped from the solid and secretly bolted together to achieve a smooth and comfortable transition.

LIFTS (ELEVATORS)

The maximum rate at which persons will be expected to enter or leave the building or move between floors will indicate the locations, capacities and speeds of lifts. Lifts are normally chosen from those available from specialists who will both manufacture and install them, and will specify the electrical services required and their locations. Depending on the design, they will require shafts with suitably sized openings, and machine rooms. Although these must be protected against fire and smoke, when planning the strategy for escape routes in case of a fire, it must be assumed that lifts will not be used.

Landing doors, their frames, architraves and other finishes should be integrated with the rest of the internal finishes. Air must be allowed to move freely away from moving lift cars. Noise from the operation of the doors, and from motors, signalling devices and passengers, should be appropriately suppressed, depending on the use of the building, e.g. a lift in a multi-storey block of dwellings must be quieter than one in a department store.

1.7 SERVICES, INSTALLATIONS AND FITTINGS

SERVICES AND INSTALLATIONS GENERALLY

Those mechanisms that **deliver a direct benefit to the building user** we shall classify as services. They include providing cold and hot water, heating, or combined heating, ventilation and air conditioning (HVAC), electric lighting and power, telephone, computer networks, cable TV reception. Each service will have many parts, including:

1. input devices, e.g. the main switchgear and distribution board of an electrical installation, the open-topped storage cistern of a cold water service, a central heating boiler, a TV aerial;
2. conductors such as electric wiring and water pipelines, which are always 'live', and full of what they are there to conduct;
3. output devices, e.g. the power and lighting points of an electrical installation; the taps of a cold water service, heating radiators;
4. controls, e.g. switches, valves, pumps, fuses.

In contrast, other conductors are installed in a building to carry away undesirable substances such as the 'foul' water from sanitary fittings, rubbish (trash) from upstairs dwellings, rainwater from roofs, and the products of combustion from heating appliances. These conductors are usually empty, have no controls, and are open to the atmosphere. To distinguish them from services, we shall call them installations, although the terms 'service', 'installation', and 'system' are likely to be used in everyday speech without discrimination.

FITTINGS

A fitting is **something that enables a user to carry out a specific activity**, e.g. personal washing, clothes and dish-washing; seeing at night; watching TV; storing books. Some fittings will consist of single, isolated parts, e.g. a hat and coat hook, an electric light fitting. Others will have more than one part, e.g. a built in wooden cupboard or some other **joinery fitting.** Baths, water closets and other sanitary fittings will be connected to services and installations, e.g. a bath with bath panel, connected to hot and cold water services and the drains. **Water fittings** can include the pipework, valves, and cisterns, as well as the sanitary fittings. (The term **fitting** can also refer to a bend, elbow or similar object used to enable adjoining lengths of a particular type and size of pipe to be **fitted** together. Other examples are given later.)

WATER SERVICES TO FITTINGS

Many services have a tree-like, or **dendritic** form where, for example, a single, large-size distribution pipe is connected to a number of smaller distribution pipes (often called **branches**) which, in turn, supply the small pipes that are connected to washbasins and other sanitary fittings.

See if you can trace the service pipelines that supply cold water from the mains to the fittings in your home. Start with the stopvalve on the service pipe, or 'rising main', as it enters the building, usually in the kitchen. This should turn easily to cut off the fresh water supply. Above it there should be a drain tap for emptying the service pipe when, say, the house is being vacated for the winter. In a newer building, many of the sanitary fittings may be connected directly to the rising main, the supply pipes being fitted with devices that will prevent the contamination of water by backsiphonage or backflow. In older buildings, while fresh water for drinking will be supplied directly to, at least, the kitchen sink, many of the other sanitary fittings are likely to be supplied with cold water by distribution pipes from an open-topped cistern in the roof space. The water level in a cistern is maintained by the operation of a 'float-operated valve' or **ball valve**.

Hot water can be provided to a single fitting by an electric water heater connected to the cold supply, e.g. when installing a shower in an existing building. More extensive hot water services will consist of a dendritic system of pipelines from a cold water supply cistern together with some device for heating and storing the hot water.

Usually, the heat source will be enclosed in a hot water storage cylinder (see Figure 1.8). The cold supply will be connected to the bottom of the cylinder and the hot water will be taken from the top. A second pipe at the top will be a vent to the atmosphere to avoid air locks and maintain the cold and hot sides of the system in equilibrium. This can be visualized as a highly contorted 'U' with the top of the less dense hot water arm somewhat higher than the cold water level in the feed cistern. Thus, the vent must be taken higher and, in case it overflows, its top end may be bent over the cold water cistern.

The simplest source of heat is an electric immersion heater. Another source is a hot water boiler whose flow and return pipes are connected directly to the cylinder. Alternatively, and more usually, water from a boiler

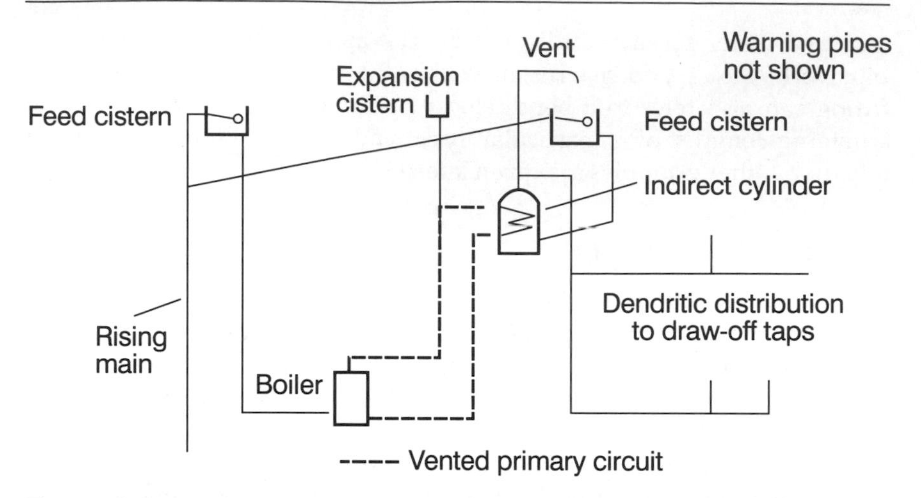

Fig. 1.8 Schema of a vented hot water service for a one- or two-storey dwelling.

can be circulated through a pipe coil within the cylinder (this type is called an indirect cylinder or a calorifier), thus heating its contents. In this case, the connecting flow and return pipelines are called the primary circulation.

Except for a vent to the atmosphere, the boiler, the primary circulation pipelines and the heating coil form a 'closed system'. If the boiler is positioned below the cylinder, the water in the return pipeline will be cooler and more dense than that in the flow pipeline and so will tend to fall, thus pushing up the freshly heated water and creating a circulation. This natural phenomenon is called a 'thermo-syphon'. The primary system will probably be supplied and kept topped up with cold water from a smaller cistern, also supplied from the rising main.

Unvented hot water storage system packages are now available. They must be demonstrably safe, have been approved by an accredited organization, and be installed by trained operatives.

As we turn on the hot draw-off tap to a washbasin and wait for the cooler water in the dead leg to be replaced with hot, we are aware of the energy that was wasted as that water had cooled. In larger buildings, dead legs are minimized by having a secondary circulation which brings constant hot water close to the various fittings.

Pipelines, hot water cylinders and supply cisterns should be insulated against heat loss or where their contents might freeze. Drain taps at low points will allow water conducting pipelines to be drained. To enable a tap to be rewashered easily, a servicing valve should be fitted to its branch distribution pipe. This will probably contain a slotted plug which will alternately open or shut a pipe with each quarter turn.

The horizontal slot at the back of a washbasin limits the amount of water in the bowl by allowing the excess to overflow into the waste outlet. Every other sanitary fitting or cistern supplied with water from a draw-off tap or ball valve should be fitted with an overflow warning pipe which will safely discharge the excess where it will be noticed.

WASTE WATER INSTALLATIONS

A foul water drainage installation that connects WCs, bidets, urinals and other sanitary fittings to the drains will convey an intermittent and variable volume of liquid. This will displace the equivalent volume of air, which must be allowed to move out and back through vents that are open to the atmosphere. All sanitary fittings, being open to the atmosphere within the building, must be fitted with 'U' shaped water seals called 'traps' which will isolate them from the air in the drains.

The depth of the seal is the vertical distance between where the two columns of water are separated and their surface level, which is determined by the outlet level (see Figure 1.9).

WCs have integral traps and are connected directly to a drain. While not desirable, other sanitary fittings on the floor next to the ground may discharge their waste water externally over the tops of receptacles at ground level called gullies. These will be connected to the drains, but their traps will isolate the air in the drain from the atmosphere. In older houses, upstairs baths and wash basins may discharge externally into a 'head' with a small diameter vertical stack and a 'shoe' at the bottom positioned over a trapped gulley.

Nowadays, sanitary fittings on upper floors will be connected by individual branch discharge pipes to a (probably 100 mm diameter) vertical discharge stack. While this will be mainly inside the building, its top must be in the open air well away from windows and fitted with an open cage to

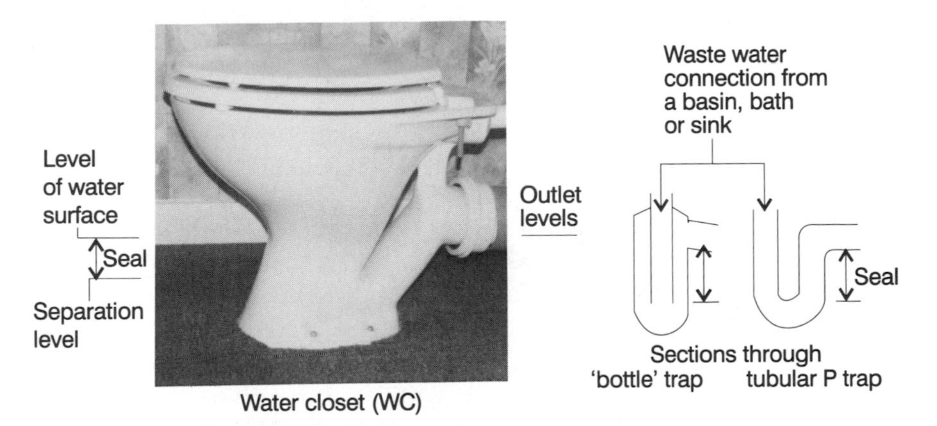

Fig. 1.9 Water seals for sanitary fittings.

discourage birds from nesting. Where an existing building has been provided with new sanitary accommodation, their waste pipes and discharge pipes may have been taken through to the outside face of the building (see Figure 1.10). The bottom of the discharge stack will be connected to a branch pipeline of a gravity drainage installation. This is one where the pipelines slope downwards sufficiently for gravity to cause their contents to flow down to the sewer. External drainage installations are discussed in the next chapter. Readers should take every opportunity to identify the services and waste discharge pipes and stacks in a building. The latter are also dendritic systems, although the flow is from smaller to larger pipes.

If such branch discharge pipes can be made fairly short, and given a gentle slope, a discharge from one fitting will cause little fluctuation in the air volume and will not disturb the water seal in other traps. However, suppose two WCs are connected to one branch discharge pipe and both are flushed at the same time. If the foul water fills the discharge pipe, the air in front will be pushed violently through the pipework and the pressure of the air behind will be diminished. Such surges in pressure may draw the water seals from other fittings.

Thus, in some circumstances, vent pipes must be introduced to normalize these variations in pressure, and the design of sanitary pipework, particularly in multi-storey buildings, needs great care. Lightweight pipework will convey sounds from the use of sanitary fittings to other parts of the building, and this can be rather upsetting for residents of blocks of flats, particularly at night.

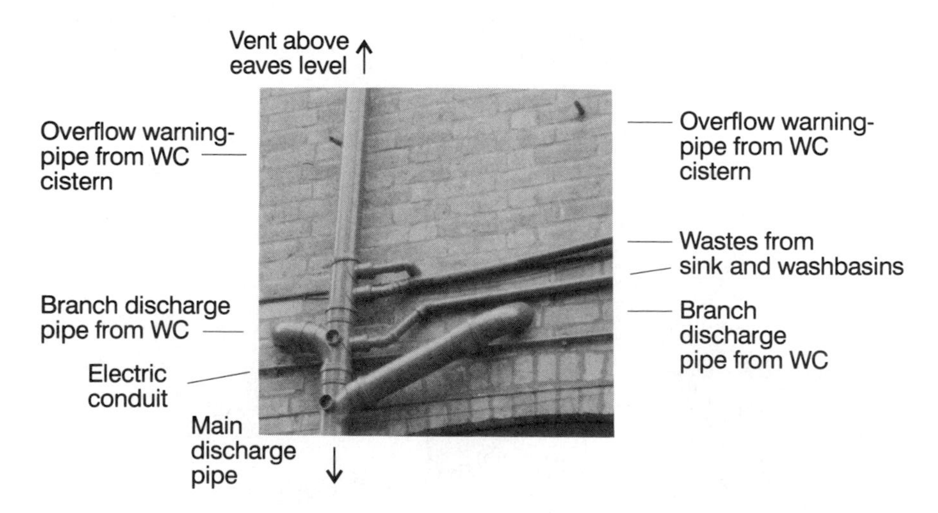

Fig. 1.10 Pipework connecting outlets from sanitary fittings to a vertical discharge pipe.

HEATING SERVICES

Low-pressure hot water heating services are larger versions of the primary circulations mentioned earlier, with radiators installed where heating is required. Some older heating circuits with large-bore pipelines will depend on the thermo-syphon effect for their circulation. Nowadays, small-bore pipelines rely on 'circulating pumps' which simply move the water along and keep radiators supplied with hot water.

Although radiators do radiate heat energy, they function mainly by conduction. That is, they conduct heat from the water within to the air at their surfaces which, as it becomes hotter and lighter, is displaced upwards by the heavier air elsewhere. The heating service may share a boiler with the hot water service. There should be some control of the room temperatures, e.g. thermostatically controlled radiator valves, room thermostats connected to motorized valves.

ELECTRICAL SERVICES

The following will require a supply of electricity:

1. internal lights, cookers, washing machines, dishwashers, immersion heaters, unit heaters, extract fans;
2. emergency lighting, internal telephones, fax machines, smoke, carbon monoxide and fire alarms;
3. lifts, industrial plant;
4. boiler controls, fans and pumps;
5. entryphone systems, external security and other lighting, parking control barriers.

In addition, power points for occupiers' appliances should be available at likely locations. TV aerials and cabling and computer terminal cabling, although installations rather than services, will probably be regarded as electricians' work. At home, take a look at the meters and control panels. These may be configured so that certain equipment can use cheaper off-peak electricity.

VENTILATION SYSTEMS

Where mechanical extractors are used to change the air in toilets, kitchens, rest rooms etc., inlets will be needed to supply an equivalent volume of fresh air. Inlets are also required to those internal spaces that will be ventilated passively through chimneys or other vertical extract ducts with high-level outlets. These will take advantage of the 'stack effect' of the air in the building being warmer and thus lighter than that outside and the reduction in air pressure over the tops of the ducts as the wind speed increases.

1.8 COMMENTARY

When speaking generally of those physical objects from which our buildings are made, we have been following the common practice of calling them **materials**. Unfortunately, the word can also mean what the material itself is made of. For instance, bricks are materials, but the material of a brick might be fired clay or concrete. To avoid this difficulty, we shall refer to the hardened paints, hardened mortars and concretes, fired clay, wood, steel, lead, glass etc. that we find in buildings as **substances**. Each sort will have distinct and different properties that can be tested, e.g. their mass density, thermal resistance, permeability.

Even so, when referring to the objects from which our buildings are built, we shall, in the main, be calling them 'construction products' or just 'products'. This is the term used in the Construction Products Directive (CPD) adopted by the European Commission in 1988. A product is a specially designed object or service that is offered for sale and, nowadays, a building is always constructed with products purchased from others.

Construction products, their processing, and the 'finished work', or 'in-place work' made with them are closely related. They will be studied in groups, or classes. Concretes, plasters and other thick coatings, and paints and other thin coatings, all harden, or 'set', and change their substance as a result of chemical action after they had been placed in their final position in the building. Before that, they would have consisted of **formless** mixtures of small particles or liquids, e.g. cements, aggregates, bagged plaster, water, paints, stains.

In contrast, the form and substance of the other materials will have been decided by their manufacturers before delivery to the site, i.e. they are already **formed**. Examples include clay tiles, plywoods and other rigid boards having one minor dimension, timber and other solid sections with two minor dimensions, plastic pipes, doors, sanitary fittings, boilers, The substances of such products will be unchanged by their construction processing.

Masonry is a composite construction made by bedding bricks, blocks or stones, i.e. formed products, in a matrix of formless mortar. More classes of products will be identified in the Commentary to the next chapter.

Building construction textbooks tend to follow the order in which buildings are constructed. In contrast, by starting with the qualities of spaces and their surfaces, we are more able to understand the purposes of the underlying parts. This is closer to the design process.

1.9 SUMMARY

It was suggested that buildings are constructed to provide suitable internal spaces. Some of the purposes of these spaces were listed. Readers were invited to answer questions on the qualities of the room they were in and the other rooms in the building and how they felt about them. Internal circulation systems including those for escape in case of fire were introduced.

This was followed by an introduction to the desirable qualities and the construction of those parts of a building that are visible from the inside. These included the internal finishes, doorways, windows, stairways, hot and cold water services, internal drainage installations, heating and electrical services, and visitor information.

FURTHER READING

Hall, F. (1988) *Essential Building Services and Equipment*, Newnes, Oxford.

Illston, J. M. (ed.) (1994) *Construction Materials, Their Nature and Behaviour*, E & FN Spon, London.

Reid, E. (1984) *Understanding Buildings*, Longman Scientific and Technical, Harlow.

Specification, Emap Business Publications, London.

Builders' merchants' catalogues and manufacturers' literature.

The external environment 2

In the first chapter we considered the qualities we can expect to find in a room or some other space within a building, and we looked at its visible parts and those that provide occupiers with various facilities. In this chapter we shall be looking at buildings from the outside.

A building is best understood when looked at from its top downwards. Then, the paths of the building loads and the rainwater, and escape routes in case of fire, can be traced down to the ground. The climatic conditions in which the building must function and survive are also important to our understanding of it. General climatic conditions will include the wind speed, the rainfall, the snowfall, the temperature, and the exposure to chemicals, e.g. sea salt and industrial pollution.

However, every building will have its own micro-climate, as the general climatic conditions will be modified by local factors. These will include the surrounding buildings and trees, the altitude, the exposure and configuration of the site, and the shape and size of the building.

We shall start as if we were looking from a window at the outsides of the roofs and the external walls of neighbouring buildings, although a closer inspection will be necessary later. Obviously, it is the outsides of buildings that ensure that they are **weathertight**, i.e. they exclude wind, rain, hail, snow and airborne particles, and this will depend to a large extent on the properties of their substances.

Glass, lead and other metals, plastics, bitumens and some painted surfaces are impervious to water, whereas fired clays, mortars and concretes are porous, water having been used in their manufacture. However, their pores may not always be interconnected and of a size to make the substance permeable through **capillary attraction**. This is the natural force which also causes a liquid to rise in a wick or spread through blotting paper. Wood is permeable to some extent. However, a substance that is impervious to water may still be permeable by water vapour, and this is discussed in Chapter 4. Generally, rainwater, snow, condensation and water vapour are all referred to as moisture.

2.1 ROOF COVERINGS

SLOPING ROOF COVERINGS GENERALLY

Of course, the roof covering should present an acceptable appearance, but its chief purpose will be to exclude rain, snow, airborne particles, and wind from the top storey, i.e. to be continuously weathertight. It must also exclude birds, insects, and reptiles. All coverings must be secured to the structure so that they will resist the effect of the strongest winds. Higher buildings nearby can channel winds, thus increasing their speed. When a strong wind blows across roofing units, the air pressure directly above them is reduced, causing the pressure below the units to become positive and to tend to push them off.

As you will see, most pitched roof coverings consist essentially of two opposing surfaces sloping down from a central **ridge.** This simplifies both the covering and its supporting structure. However, you will find many variations on this central theme. The configuration of any one roof is determined by the plan shape of the building being roofed, changes in direction being effected by diagonal external junctions called **hips** (see Figure 2.1) and internal junctions called **valleys**. The space created by a sloping roof we shall call the **roof space** or **loft**.

The otherwise exposed end of a roof space may be closed by a triangular sloping covering with hips at the two junctions with the main covering. This is called a **hipped end**. Alternatively, the roof space can be

Fig. 2.1 The eaves and the lower part of a hip in a double-lapped plain tile roof covering.

closed by extending the end external wall upwards in the form of a triangular **gable** wall (see Figure 2.2).

UNIT PRODUCTS FOR ROOF COVERINGS

Slates are made from naturally occurring sedimentary rocks which, as a result of pressure and heat, have become highly durable and capable of being split into uniformly thin sheets. Their properties, and the thicknesses and sizes which can be produced, depend on their geological source. While most are 'slate' grey in colour, purplish ones come from parts of North Wales and greenish ones from the Lake District. Many quarries have now closed. **Stone slates** are made from other kinds of stratified rocks. Note that all slates are more or less flat. On roofs with a low angle of slope, or **pitch,** capillary attraction may cause water to be drawn up between them.

Most slated coverings consist of overlapping rows, i.e. **courses** of the same size of unit. These are laid to the same sloping spacing, i.e. **gauge** with the joints between the sloping edges of the units in one row coinciding with the centres of units in adjoining rows, i.e. staggered. However, some quarries were only able to produce a mixture of differently sized slates, and these will have been sorted and laid with the largest at the bottom and 'to diminishing courses' with the smallest at the top. Slates have to be perforated either near their upper edges or near the middle, for fixing with durable nails. Stone slates are traditionally hung on wood pegs pushed through holes near their tops.

Most tiles are made of either fired clay or concrete. Clay tiles are naturally red to brown in colour. The upper faces of concrete tiles are given a coloured finish which is liable to fade. Their shape and colour will determine the appearance of the roof covering. Tiles are manufactured with nail holes and supporting nibs and, depending on the exposure, whole courses may be left unnailed.

Plain tiles are about 165 mm × 265 mm in size. Like slates, they are laid in overlapping rows with staggered joints and are described as being **double lapped** as their nail holes will have two rows of tiles above them (see Figure 2.1). Larger **single-lap** tiles are specially shaped to exclude moisture and are lapped only at their edges (see Figures 2.2 and 2.5). The actual angle of slope of all units will always be somewhat less than the pitch of the roof because their lower edges are lifted by the units in the course below. A number of different types of technical solution are available for covering roofs and excluding the weather.

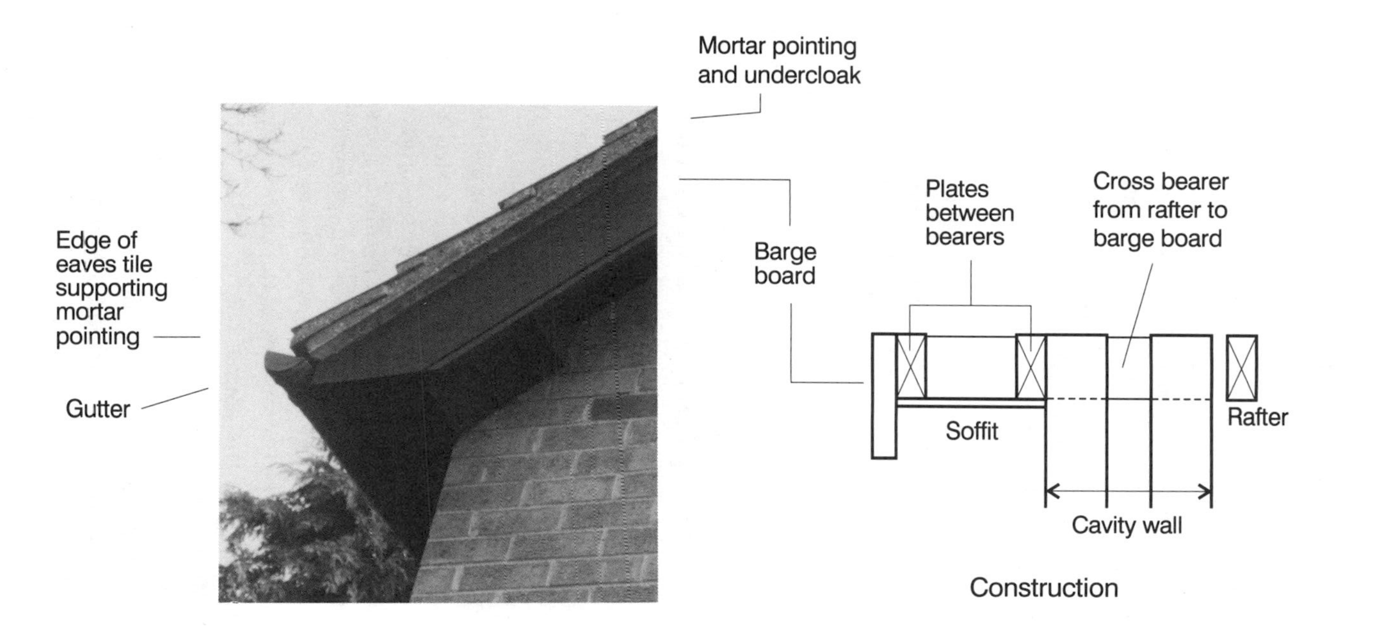

Fig. 2.2 The junction between the eaves and the verge to a gable end wall, showing the edges of the single lap roof tiling overhanging the bargeboard.

UNIT COVERINGS TO SLOPING OPEN FRAMEWORKS

Type A1 (tile type) unit covering to a sloping open framework will consist of:

1. overlapping horizontal rows, i.e. **courses** of loosely fitting small units, e.g. slates or tiles. They will be supported by and secured to
2. wood strips called **battens**. These must have been treated with preservative and be nailable without splitting. They must also be strong enough to sustain the mass of the roofing units and the wind loading across the span between the sloping roof members (see Chapter 3). As the roofing units will not totally exclude wind, snow, dust and small insects, etc, they should be backed by a
3. continuous, flexible breather membrane, i.e. one which excludes water but allows the passage of water vapour. By allowing the membrane to droop between the structural members, a space is provided for moisture from melting snow or driven rain to run down into the gutters.

EAVES, VERGES, HIPS, AND VALLEYS

The additional construction that provides continuity between the bottom of the roof covering and the top of the external walls is called the **eaves.** The equivalent construction at the edge of a sloping roof covering overhanging a gable wall is called the **verge**. In both cases, their purpose is to provide continuity at the junctions between roofs and external walls by excluding insects and birds and keeping the building weathertight including protecting the upper portions of the walls below.

The vertical facing to the ends of sloping (or flat) roof structural members is called a **fascia**, and the equivalent at a verge is called a **bargeboard** (see Figures 2.1, 2.2 and 2.5). Where eaves or verges project, the lining between a fascia or bargeboard and the wall is called the **soffit**. Insect proof openings in the soffits of projecting eaves will promote cross ventilation in the roof voids and minimize condensation (see Chapter 4).

In a type A1 covering, the bottom edges of the units in any one course are supported by the units in the course below. At the eaves, however, there is no next lowest course. On a roof covered with slates or double-lap tiles, shorter, eaves slates or tiles (see Figure 2.1) are used instead. If a roof is covered with single lap tiles, eaves tiles may still be required to support the mortar pointing to their exposed lower edges (see Figure 2.2). The fascia can be fixed so that its top provides any further support, it being stiffened by a wood triangular section **tilting fillet** (see Figure 4.1). Where not done, the bottom course of units will appear to droop.

Notice the different materials employed with each of the different types of unit covering to keep them weathertight at various kinds of junctions. Look for:

- alternate wider slates and plain tiles at the sloping edges of **verges**;
- a bottom course of shorter slates or tiles at **eaves** (see Figures 2.1 and 2.2);
- ridges and hips covered with angular or half round section tiles made of fired clay or concrete and bedded in mortar (see Figures 2.1 and 2.5); in slate districts, they may have a two-piece slate covering;
- variously shaped tiles at external and internal right angled junctions between roof slopes, i.e. at hips and valleys;
- lead gutters to valleys and coverings to hips;
- lead soakers covered by lead stepped cover flashings where slates or plain tiles butt against a higher masonry wall or chimney (see Figure 2.3). Soakers are small rectangular pieces of lead or an equivalent substance bent at right angles. Their sloping portions are laid on the sloping unit below, with their vertical portions against the masonry. Each is secured by having its upper edge bent over that of the unit below, and will be fully covered by the unit above. The portions against the wall will be covered by a sloping lead cover flashing whose own top will probably be cut to form a series of steps to fit the masonry courses. The upper edges of these steps are bent, turned into the raked out horizontal masonry joints and wedged with small pieces of lead. The masonry joints are then pointed.
- Lead 'slates' with vertical sleeves which maintain weathertight junctions where discharge stacks and other pipes project through the covering.

PARAPETS

An alternative technical solution for the eaves is to finish the covering

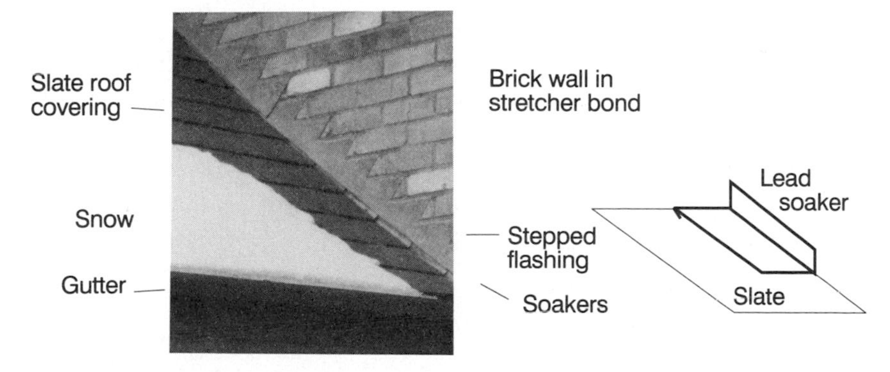

Fig. 2.3 A weathertight junction between a sloping slated roof covering and a higher wall showing lead soakers and a stepped cover flashing.

perhaps 300 mm short of the inner face of the external wall, and extend the (usually masonry) external wall upwards to form a parapet. This will have an impervious coping on the top and damp proof courses to keep the wall weathertight. The gap between the parapet and the roof eaves is made weathertight by a lead **parapet gutter** with outlets at intervals (see Figure 2.8). The building then presents a clean and attractive wall surface to an observer. Examples of such buildings, many constructed in the last century or earlier and with gracious sliding sash windows, can be found in most towns in the UK. However, all such internal rainwater installations are likely to fail in time, and are expensive to maintain.

OTHER COVERINGS TO SLOPING OPEN FRAMEWORKS

Type A2 (profiled sheeting type) unit covering to a sloping open framework.

1. overlapping rows of long, close fitting, profiled units spanning between horizontal supports, e.g. galvanized corrugated sheets, corrugated PVC sheets, double-skinned profiled coated steel sheeting with thermal insulation in between.

Type A3 (asphalt shingles type) unit covering to a sloping open framework.

1. continuously supported bitumen strip slates (asphalt shingles) such as those used to cover house roofs in North America, bonded to each other and nailed to
2. a continuous base or **sheathing**, which is highly resistant to the effects of the weather, and strong enough to resist the various loads and transfer them to the structure, e.g. external quality plywood made with a weather-proof and boil-proof (WBP) adhesive. An almost horizontal base may be referred to as a **deck**, or a **substrate**.

In exposed locations in the UK, many buildings are given a double covering consisting of a type A1 unit covering but with the membrane supported on a type A3 base. A drainage space is created between the battens and the membrane by small section wood **counter-battens** fixed in the direction of the fall.

Type B1 (sheet lead type) continuous covering to a sloping or nearly horizontal open framework (see Figure 2.4).

1. Malleable non-ferrous, e.g. lead, copper, zinc, sheets joined together in suitable ways to form a continuous and impervious surface and fixed to
2. an isolating membrane to disconnect the covering from
3. a type A3 base.

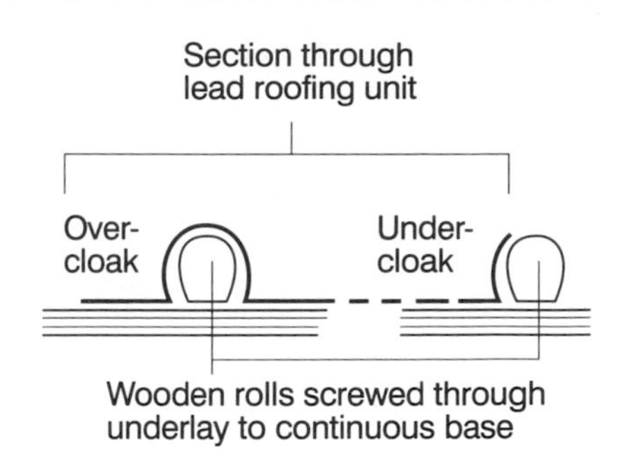

Fig. 2.4 A sheet lead roof covering showing how the sloping long edges of the units are connected and weatherproofed by being lapped over wooden rolls.

Type B2 (polycarbonate sheeting type) continuous covering to a sloping open framework.

1. Sheets cut to shape and spanning crosswise between sloping glazing bars, e.g. translucent polycarbonate multiwall sheets covering the roofs and walls of conservatories. Rooflights etc. in older buildings may have been constructed with wire reinforced glass supported by one of the many specialist glazing bar systems that incorporate drainage channels. These may be referred to as 'patent glazing'.

Type B3 (thatch) continuous covering to a sloping open framework.

1. A thick layer of reeds, straw, or heather, secured using traditional methods and supported by
2. either wood battens or a reed mat.

FLAT ROOF COVERINGS

Type C1 (asphalt on concrete type) continuous covering to any continuous background.

1. Mastic asphalt applied in two layers to a felt isolating membrane on a
2. concrete or mortar screed laid on a continuous background with a sloping surface shaped to direct moisture to gutters or outlets.

Type C2 (waterproof membrane type) continuous covering to a sloping or flat open framework or a continuous background.

1. One, two or three layers of flexible impervious sheeting with water-tight bonding, e.g. a built-up bituminous covering with the first layer either
2. nailed to a type A3 continuous base or partially or totally bonded to a type C1 screed.

A 'flat' roof covering on an open framework is likely to have a type C2 covering. Its base, or **deck**, will be secured to wood joists which, if not themselves strong enough to span from wall to wall will need intermediate supports such as steel beams. (The strength of constructional, i.e. **carcassing** timbers depends mainly on their width, i.e. their depth when fixed. This is limited by the size and straightness of the trees from which they are cut.)

The top of the deck should slope slightly to direct moisture to the gutters or other outlets. This may be achieved either by fixing the joists to slope or, more likely, by securing tapered pieces of wood called **firrings** to the tops of the horizontal roof joists.

Junctions between a flat roof covering and higher walls can be kept weathertight by turning up the covering to form vertical skirtings. These should be supported by the sides of wooden curbs fixed to the base and have their tops covered by lead flashings fixed to the walls. If, instead, the skirting is bonded solidly to a higher wall, the covering is liable to be torn from the skirting as the joists shrink transversely. This can be as much as 10 mm. Rainwater can then enter the building.

Although most continuous (usually concrete) backgrounds to a type C1 or a type C2 covering will be horizontal, some will be sloping or vertical, e.g. at changes in level of flat roofs. Vertical skirtings will be required at junctions with curbs for rooflights and where abutting higher walls. Junctions between Type C1 (asphalt type) coverings and upstands and skirtings will be reinforced with a triangular section of additional asphalt. Type C2 flexible membranes will be bent up in two stages over triangular section mortar or wooden 'fillets'. Thermal movement can be minimized by a white reflective top layer of e.g. stone chippings, paint, paving tiles.

At all the eaves, either the covering can be turned down to discharge over a gutter, or the rainwater can be held back by a curb possibly with an aluminium trim to provide a durable edge to the covering.

'WINDOWS' IN ROOF COVERINGS

A sloping roof covering may be penetrated by various kinds of 'windows' which will provides natural light and ventilation to the useful spaces within a roof space or below it. A **dormer-window** is just a normal window set vertically in a sloping roof (see Figure 2.5). It will be connected to the main roof by either a pitched roof, possibly with a little gable wall at the front, or a flat roof, and have triangular shaped, i.e. **spandrel** side walls.

An alternative solution is to have a sloping **roof light** (which may be called a **roof window** or **skylight**). **Lantern lights** have glazed tops and upstands and may be found at the ridges of pitched roofs and on flat roofs. **Domelights** will be found in flat roofs. All will be lifted above the

Fig. 2.5 A dormer window in a roof with a covering of single-lap tiles.

plane of the roof covering by a curb which will provide vertical support for whatever material is being used to keep the junction weathertight, e.g. a lead flashing, an asphalt skirting.

THE RAINWATER INSTALLATION

The rainwater-drainage installation should conduct rain and melted snow away from the building. Its purpose is to keep the external walls weathertight and durable by collecting the rainwater and conducting it to the drainage installation. It is likely to consist of (see Figure 2.1) gutters (eaves troughs) which intercept the water from off the roof, and rainwater pipes, i.e. 'downpipes' that connect the gutters to the drainage installation in the ground.

The bottom edge of the covering to a pitched roof should be centrally above the gutter where this is fixed to the eaves fascia. Older buildings without fascias may have their gutters supported on metal brackets secured to the feet of rafters or fixed to the wall. Alternatively, cast-iron gutters may have been placed on the outer portion of the top of the external walls, particularly when these are thick. Such gutters will almost certainly leak into the wall as a result of corrosion, particularly at the joints. The exceptionally high thermal expansion and contraction of plastic gutters may cause joint failures in the future.

All except the largest gutters will be fixed to give a slight 'fall' to the nearest outlet, where they will discharge into downpipes. A pair of offset bends or a one-piece swanneck fitting may be required to connect gutter outlets to the downpipes as these will usually be fixed back against the walls. Their lower ends will either be connected directly to the drainage system, or end with a **shoe** which directs the water into a receptacle. See the discussion on drainage, later.

The damage caused by a broken or blocked installation is usually obvious. If water enters the walls, it will affect the internal finishes and may cause wood structures to rot and fail. The external surfaces can become stained and damaged by frost, and vegetation may develop. In very cold climates, ice dams can form at roof overhangs, causing water from melting snow to back up and enter the building. In some parts of Africa, traditional houses have walls built of sun-dried 'mud' blocks made from a mixture of clay and water. Such walls may fail at ground level where they have been eroded by water dripping directly from the roofs or splashing back from the ground.

2.2 EXTERNAL WALLS

MORTARS AND THEIR CONSTITUENTS

Any substance that will stick two separate objects together can be described as being cementitious. A mortar is a wet-mix of a cementitious substance, e.g. cement, lime, or a mixture of the two, plus a suitable sand and water. Its properties in use may be enhanced by an **admixture**, e.g. to improve its plasticity. Mortar is used mainly to stick small structural units together to make masonry. A mortar-mix must be stiff enough to be picked up on a trowel, plastic enough for the units to be slid into place on their mortar beds, yet capable of remaining in place as other courses are laid on top, i.e. it must be **workable**. The mix must also be consistent, adhere strongly to the units, and be of comparable strength to them.

Until the patenting of ordinary Portland cement in 1824, the only cementitious materials in general use had been limes. These are made by burning limestones or chalk to reduce their calcium carbonate to calcium oxide. When mixed with water to form calcium hydroxide, this readily absorbs carbon dioxide from the atmosphere, including that dissolved in rain, and forms crystals of calcium carbonate. Nowadays, limes for mortars are supplied as an already hydrated powder as this is more reliable and easier to use. Limes and cements for special purposes, e.g. to withstand the effects of sulphates in the ground, are also available.

To make lime mortar, the hydrated lime is mixed with clean 'building sand' of suitable particle size before adding water. The sand increases its bulk, adds to its strength, counteracts its tendency to shrink, and provides airways for the carbon dioxide. When repointing or repairing existing walls, the mortar should be similar in strength to the original. (The wall and ceiling plasterwork of many existing buildings will consist of coats of a lime and sand mix, possibly strengthened by the addition of hair and gypsum based plaster. Plastering sand is different from building sand.)

What we refer to in everyday speech as 'cement' is more properly called ordinary Portland cement (OPC) as there are others. It is so called because it looks rather like the stone quarried at Portland, on the coast of Dorset. OPC is made by burning a mixture of clay and limestone or some other form of calcium carbonate, and grinding the resulting clinker to a fine powder. When mixed with water to form a paste, the two substances begin to combine chemically. After two hours or so, the wet-mix stiffens and then begins to harden and develop its strength. Provided it is kept moist and not disturbed, the mortar may go on getting stronger for months to come.

OPC can replace some, or all of the lime in mortar wet-mixes. However, a mortar should not be stronger than it need be, and one containing some lime will have better adhesion, workability, and toughness than a simple mix of cement and sand.

The sand for mortar and the sands, gravels, crushed stone or other larger particles used in making concretes are all called **aggregates**. Almost all mortar wet-mixes are specified by giving the proportions of the volumes of the cementitious products and those of the sand, e.g. 1:3. The amount of water is seldom mentioned as usually this is freely available and is added after the other constituents have been mixed dry. Even so, it is the cement and water paste that sticks the particles of the aggregates together and the strength of the resulting mortar or concrete will also depend on their properties.

Cements and limes (and plasters) only become active when mixed with water. Cement requires a minimum of 28% of its weight of water for hydration. However, to make a workable mortar (or concrete) mix, the aggregates that will bulk out the cement paste must be wetted as well. This increases the water:cement ratio to about 55%, and when this excess evaporates, it leaves a permeable substance. Using even more water will leave the substance even more permeable, less strong, and more susceptible to attack by chemicals in solution.

BRICKS AND BLOCKS

The variety of materials available locally will be obvious when walking through any built-up area. In the UK, the outermost part of many of these walls will be made of **facing bricks** that have been manufactured for their appearance and durability. They will be strong enough for their purpose, and most types will absorb any surface moisture into their pores and later allow it to evaporate, thus ensuring a weathertight wall. **Common bricks** are manufactured solely for their strength and durability although when walls are built entirely of these, their exposed surfaces should be finished neatly, i.e. with a **fair face**.

The manufactured, or **work size** of a metric brick is about 215 mm long × 103 mm wide × 65mm high. Adding a 10 mm thick mortar joint to each gives a **co-ordinating size** of about 225 × 113 × 75 mm. Note the proportions between these dimensions of 6:3:2. Bricks used in some older buildings in the Midlands and northern parts of the UK may be some 8 mm higher.

While most bricks are made from clay, they can also be made from a mixture of sand and lime and from concrete. Most are made by pressing the raw materials in a mould, and will have a depression, called a **frog,** in one bedding surface. Others, particularly those from the Midlands, will be 'wire cut' from an extruded ribbon of prepared clay. Clay and sand-lime bricks harden by being heated to a high temperature. Concrete bricks harden through chemical changes.

Walls with right-angled junctions and rectangular openings can be built entirely of standard-sized bricks. However, a range of differently shaped standard special bricks will usually be available to suit special circumstances, e.g. to form a 45° junction. Special shapes can also be fabricated by cutting pieces from standard bricks and bonding them with a durable adhesive. If, instead, standard facing bricks are cut to shape and these cut faces are exposed, the result is seldom acceptable.

In contrast, the co-ordinating length and height of most blocks is 450 mm × 225 mm, i.e. twice the length of a brick and three times its height. Blocks are available in thicknesses to match the thicknesses of walls. Solid or cellular concrete blocks are made using aggregates to suit the attributes of the wall, i.e. with dense aggregates for strength, lightweight aggregates for thermal resistance. These may have smooth or rough surfaces. Aerated blocks are made from a mix of cement, water, a fine aggregate, and a special bubble forming admixture. They are wire cut to size (their faces usually have a torn appearance) and steam cured at high pressure. Because of their unacceptable appearance, most blockwork walls will have an applied finish (see later).

BRICK AND BLOCK WALLS

All masonry walls depend for their stability on their being vertical and on the overlapping of the units in successive horizontal layers, i.e. **courses**. Then, the force of gravity will ensure that the constituent units are held together, the selfweight of the vertical wall will act within the width of its base, and there will be no tendency for the wall to be pulled over. Even so, while we might speak of a 'brick wall', the mortar is just as important as it:

1. ensures that the load from each unit in each course is evenly spread over the units in the next lower course;

2. creates a single continuous mass by sticking separate units together; and
3. can provide as much as a quarter of the volume of the wall.

The exposed edges of the mortar beds should be so finished that there are no ledges to intercept the moisture which would otherwise run down the face of the wall. This is called **jointing**. For appearance, the outer 20 mm or so in width of the mortar beds can be replaced by a special, possibly coloured mortar, the process and the result being called **pointing**.

To ensure the mortar is solid, all exposed edges should be formed by pressing with a tool of some kind. The horizontal slightly sloping **struck** or **weathered** joint finished flush with the wall face at the bottom is made with a trowel. A slightly hollowed but otherwise flush joint is called a **bucket handle** joint, although other tools may be used.

THE FACES OF BRICK EXTERNAL WALLS

If only the long sides of the bricks show on the face of a brick wall, and they overlap those in the course below by half the length of a brick and joint, i.e. about 113 mm, the wall will consist solely of the bricks you can see (see Figure 2.6). This arrangement is called **stretching** or **stretcher** bond. As the thickness of the wall will be the same as the width of the bricks and this is about half their length, it is likely to be described as being a **half-brick** thick wall. Differently coloured units in rows or some other arrangements, perhaps set on end to form **soldier courses**, can provide decorative features.

On the other hand, if the surface of a brick wall shows a pattern of long and short sides, i.e. of **stretcher** and **header** faces, it is more likely to

Fig. 2.6 A corner of a wall showing a half-brick thick wall on a one-brick thick plinth and details of stretching (or stretcher) and English bonds.

be a thicker wall (see Figures 2.6 and 2.7). However, there is no way of telling how thick, except by measuring at an opening.

To create a stable thicker wall, the bricks must be laid overlapping the ones below with frequent headers which will tie together what would otherwise be vertical half-brick thick layers. The different arrangements of the overlapping units are called **bonds**.

English bond is where the bricks in alternate courses are either all stretchers or all headers. **Flemish bond** shows stretchers alternating with headers in every course. Walls exposed on one side may be built with two or three stretchers to each header, as this economizes on the more expensive facing bricks and can be easier to build.

Their overlap is created by inserting a half width **closer** next to the long side of each corner brick (see Figure 2.6). All bricks are approximately the same size and can be bonded in a variety of ways to make walls of any multiple of a half brick in thickness. (In contrast, blocks are almost invariably laid in stretching bond.)

A triple thickness horizontal joint at about the level of the lowest floor finish will probably indicate the edge of a two-layer slate damp-proof course (dpc). An intermittent line of a black substance will reveal the edge of a sheet bitumen dpc (see Figure 2.6). Try tracing this line to a doorway and compare its level with that of the internal floor finish. Any dark blue bricks visible near the ground will probably be **engineering bricks**, chosen for their durability and low water absorption. Two or three courses of these are often used as a dpc. The purpose of a dpc is to stop moisture rising up into the wall from the foundations and the ground.

Water will rise naturally against gravity through hair's breadth spaces such as the pores in bricks, or where an external rendering has become detached from its base. This phenomenon of 'rising damp' is another example of capillary attraction. The limit reached by such rising damp can sometimes be seen as an irregular line of whitish salts on the outsides of old brick buildings. This probably means that the walls are without dpcs, or the dpcs have failed or are bridged by soil. A horizontal row of small diameter filled holes at the dpc level will indicate that the wall has been chemically treated to resist rising damp (see Figure 2.7).

Air bricks or vents at regular intervals just below the dpc level will indicate that the floor next to the ground is of ventilated hollow construction (see Figure 2.7). Higher up, you might also see the projecting ends of overflow warning pipes from cisterns, baths etc. (see Chapter 1 and Figure 1.10).

Water used in construction processes or from rain can provide a medium for dissolving and transferring chemicals from the air and from the ground into the building structure, and from one building substance to another. Adverse reactions may then take place. Examples include:

1. calcium or magnesium sulphates from the ground, or from bricks or filling materials, or from aggregates (but this is not likely in the UK)

Fig. 2.7 Visible evidence of an injected chemical dpc in an external wall. Also, a ventilation inlet to a hollow floor, and a soil-bridged blue brick dpc.

can react with the hydrated cement in concretes and mortars; as a result, there is an increase in volume which has a bursting effect on the construction;

2. chemicals left on the surface of a brick wall after water has evaporated may become visible as efflorescence (see Figure 2.8). Although unsightly, this is usually a short-lived feature of new buildings. Persistent efflorescence near the ground may indicate a more serious chemical attack. Where crystals are formed just behind the surface (crypto-florescence), they can damage it.

THE FACES OF STONE EXTERNAL WALLS

Local building stone that is available in roughly squared blocks, albeit of various sizes, can be laid with horizontal joints although not always in

Fig. 2.8 Efflorescence on the face of a newly built parapet wall.

continuous courses. Other, more irregular **random rubble** or **polygonal** units will be laid as nearly horizontal as possible, possibly with continuous roughly horizontal beds formed at intervals for stability. Usually, their mortar joints will be carefully finished. 'Throughstones' from face to face may be incorporated to reinforce the integrity of the wall. The Lake District in the UK, which has a notoriously high rainfall, is an exception as here the thick external masonry walls are built with large pieces of slate set sloping outwards with their outer edges close together and with the mortar bedding kept well back and hardly visible.

Stone used on important private, public and commercial buildings may have been transported long distances, e.g. buildings in London faced with Cornish granite and Portland stone. Large, rectangular stone units can be produced, their exposed faces being finished with either a smooth or a textured, i.e. **rusticated** surface. Decorative joints can be formed by recessing some of their edges, and the units are laid overlapping on thin mortar beds, which makes for great stability. Such stones and the walls made with them may be called **ashlar**. The appearance of external walls can be enhanced by the addition of features that imitate those of classical buildings, e.g. statues, columns, pilasters (which look like columns, but are actually piers attached to the wall), pediments.

Window sills, copings, horizontal **string courses** (see Figure 3.5), and other projecting parts that are intended to throw rainwater clear of the walls often fail to do so. (Look underneath them for grooves, or **throatings**, that stop the water from running back.) Indeed, under the direction of the wind, these projections may even concentrate the water on to particular areas of the wall surface where small particles previously held in suspension may be left as unsightly staining.

Walls that are otherwise built with smaller or less regular units of e.g. stone, brick, flint can be given greater apparent and actual strength and stability by having their corners and the jambs of doorways and other openings constructed with larger and more rectangular units with thin mortar joints. In fact, it may be impracticable to make suitable rectangular corner, or **quoin** stones from irregular units such as the polygonal granite shown in Figure 2.9.

FINISHES TO EXTERNAL WALLS

If the appearance of the products used for constructing continuous external walls is not satisfactory, or the wall is not weathertight, they must be given a separate finish. Open frameworks will always need a finish.

Type A (cement rendering type) external finish on a continuous background.

1. A thin hardened layer of some decorative material that may also have weather excluding properties. This will have been applied to

Fig. 2.9 Squared and dressed corner stones to a 1.8 m high polygonal granite wall.

2. a base consisting of a thick hardened layer of a cement, sand and water mix (sometimes called a rendering) with a smooth or textured surface. This mix might be made with a coloured cement. Alternatively, particles of some mineral aggregate chosen for its appearance might be applied to the surface before it hardens. A strong bond to the background is vital (see Figure 2.10) as it must be able to withstand the pressure exerted by water as it freezes.

This type of finish can also be applied to an existing external wall to improve its weathertightness. Perhaps because of movement in its background or shrinkage as the wet mix hardens, this type of base can crack or become loose and allow rainwater to penetrate behind the rendering. It may then be easier for the moisture to move on through the wall and wet the inside than to move back through the rendering and evaporate from the outside. Identify loose areas of rendering by tapping with your knuckles as before.

Type B (tile hung type) external finish on a continuous or an open background. This is similar to roof covering type A1.

1. Horizontal rows of overlapping small units, e.g. plain roofing tiles,

Brick external wall of old industrial building rendered and painted to improve weathertightness

Fig. 2.10 A rendered and decorated wall finish to a brick wall showing where loose rendering has been removed for safety.

possibly with shaped lower edges to add interest, are hung on and nailed to

2. wood battens secured to a continuous background, e.g. lightweight block wall, or to a wood open framework, and backed by
3. a continuous, flexible breather membrane.

Type C (boarding type) external finish on a continuous or an open background.

1. A thin hardened layer of some decorative and weather excluding liquid, e.g. microporous stain or paint. This would have been applied in one or more coats to
2. durable boards ('siding') of wood, wood products, or plastic with moisture-excluding edge joints, fixed with their lengths either horizontally or vertically (see Figure 1.2). Plastic boards do not need to be decorated.

CURTAIN WALLS, CLADDING AND RAIN SCREENS

Whole elevations of some buildings are faced with large panes of glass in a regular open framework. Such coverings may be referred to as curtain walls or rain screens. Elevations of other buildings may have an arrangement of glazed and opaque cladding, perhaps incorporating thermal insulation and a vapour control layer. All must have durable weathertight joints capable of excluding wind-driven rain. Cladding can be added to the external walls of an existing building to improve its weathertightness, insulation and appearance.

Now that steel sheets can be given durable surface treatments, the roof and the walls of single storey buildings, e.g. warehouses, can both be clad with the same kind of double-skinned profiled steel sheet cladding (colloquially called 'crinkly tin').

2.3 EXTERNAL WORKS

A **Work** is something made or done, e.g. brickwork is something made of bricks. However, in our context, **works** is a convenient way of referring to a variety of different constructions. External works are those constructions in the vicinity of a building, including walls, fences, gates, pavings, electrical and water mains, security lighting, building services connections, and drainage installations. Trees and shrubs, turfing, grass seeding and other planting may also be included. Outdoor activities may call for cycle racks, seats, sandpits and play and sports equipment. They are all likely to be included in the design proposals for a new building and in the contract for its construction.

Clearing the site and preparing it for the new building would usually include demolishing all existing small buildings. This is discussed in Chapter 7. Any major demolition will usually be carried out by a specialist organization well before the start of the new work.

EXTERNAL DRAINAGE INSTALLATIONS

The design of a drainage installation will depend on the locations of public sewers (if any) and on the policy of the local water services undertaking regarding their use. In built-up areas, each building will have one drainage installation for rainwater, i.e. for **storm water**, and another for **foul water** and sewage. These will almost certainly be connected to separate public sewers below an adjoining street, their presence being indicated by the covers to their manholes. The actual connections to the sewers and the work in the public highway, including its reinstatement, will normally be done by the constructor.

In open and rural areas, the only public sewer may be for foul water only. Then, the rainwater from roofs and pavings must be taken to sumps on the site called **soakaways**, and allowed to drain away into the subsoil, assuming this is permeable. Alternatively, it may be possible to get permission to have a combined drainage installation, and connect that to the foul sewer. Where there is no foul sewer either, the foul water must be stored temporarily on the site in a **cesspool** to await collection by vehicle and disposal elsewhere. Alternatively, the site must be provided with its own sewage treatment works.

Where branch pipelines join, or where a pipeline changes its direction, provision is normally made for open connections between the individual (straight) pipelines to enable blockages to be cleared. Depending on the pipeline diameters and depths, this will involve installing or constructing either:

1. access pipes or junctions,
2. inspection chambers (which you inspect or work in from the top), or
3. manholes (which are large enough to get into).

Pipelines under roads will require more cover than those in fields and gardens, and those near to or below building foundations must be cased in concrete.

A drainage pipeline must not be distorted or fractured by any ground movement. Either rigid vitrified clay pipes with flexible joints, or PVC-U pipes, which are themselves somewhat flexible, will be used in most new below-ground gravity drainage installations. In older buildings, expect to find short vitrified clay pipes with integral socketed joints. Each type and size of pipe will have its own range of bends, junctions, and other fittings. A blockage in an older, rigid pipeline might be due to the fracturing of a pipe and the relative displacement of the broken ends. This might be due to differential settlement where, e.g. a pipe enters or leaves an inspection chamber.

Pipes can be laid directly on the bottom of the trench if the soil is non-cohesive, e.g. gravel, and can be loosened and trimmed. Otherwise, the trench must be deepened and the pipes placed on a graded bed of granular filling at least 100 mm deep and the full width of the trench. The trench must be wide enough to allow for granular filling at least 150 mm wide to each side of the pipe and finishing level with its top. This will provide support for its sides. The pipeline is completed with a 300 mm top layer of either granular fill or suitable 'as dug' backfilling free from large stones. Pipelines that cannot be given their minimum cover must be given additional protection, e.g. by being cased in concrete. Where a pipeline is taken through a foundation wall and under a building, special bedding and jointing arrangements will allow for differential movement.

Manhole covers are invariably heavy, particularly where they must be strong enough to sustain road vehicles. The inspection covers used in domestic drainage installations will probably be relatively light in weight and not difficult to lift. Take every opportunity to trace pipelines from discharge stacks, gullies etc. to inspection chambers and thence to the sewer, and observe their flow when in use. Air will move out of and back into a drainage installation as the flow of water fluctuates just as it does in the sanitary pipework discussed in Chapter 1. The vents provided there will probably be adequate for the drainage installation as well.

SERVICE CONNECTIONS

Most buildings will be provided with water, gas and electrical services from connections to mains in an adjoining public road. Each service undertaker will have their own regulations regarding the depth of underground services, protecting the services from damage, entry into the building, meter box and controls inside the building. The water supply pipe can usually be isolated from the undertaker's communication pipe by a stopvalve below a small iron cover in the pavement just outside the property. Another stopvalve on the supply pipe will be found just inside the building.

REFUSE (TRASH) DISPOSAL

The occupants of every building must be able to store their solid waste for collection at intervals by those responsible for its disposal. For single dwellings, the standing for a 'wheelie bin' should be convenient for the residents and have paved access from the road. In larger residential buildings, occupiers may share a larger wheeled container, and in high buildings, they may share a vertical chute that discharges into their container. Chutes can be noisy in operation, and should be isolated from living areas.

2.4 THE BUILDING IN ITS ENVIRONMENT

OVERALL IMPRESSIONS

Suppose you were new to the area, and were visiting the building you are in for the first time. Consider the following questions:

- Would you have been able to find the building easily?
- To what extent would you be disappointed by what you saw?
- Has the building a name or number, or some other identification?
- Given its location, is the building an appropriate use for that site?
- Are there any open spaces around the building? If so, are they landscaped appropriately and can they be kept under observation by the occupants of the building?
- Is there a convenient and adequate car park?
- Are there shops, restaurants and other services close by?
- What public transport is there to the site?

Our first impression will be based on qualities such as:

- the size, shape and proportions of the building, i.e. its form;
- the colours, textures and condition of the materials used externally;
- the window arrangements and their detailing;

- the number of storeys, any ornamentation and, possibly, the shape of the roof;
- the cleanliness of the building and any open spaces;
- graffiti and other evidence of vandalism.

But to appreciate a building from ground level, there must be sufficient space round it for us to stand back and take it all in. Thus, the space surrounding a building is also important. Then, we shall be able to see how it harmonizes with its neighbours and we can get an impression of the wider built environment with its streets, squares and other small spaces, parklands and tree planting. These are interdependent, e.g. the buildings shade each other, obstruct extensive views, cause wind currents, and create both small, unsafe passageways and open relaxing public spaces. Streets encourage vehicles and promote atmospheric pollution. Trees, waterways and grass cool and cleanse the air and relax observers.

A building will always 'say' something to us. For instance, houses with facing brick front elevations and common bricks elsewhere will have been built at minimum cost. Weeds growing in eaves gutters, and decaying woodwork will indicate the owner's lack of awareness of the importance of maintenance or lack of funds. On the other hand, classical columns and pilasters, ornate cornices and mouldings, and hardwood joinery will indicate opulence. Presumably, the owner accepted the extra cost of such visual features to achieve non-financial goals such as impressing visitors or customers, or staying ahead of the neighbours. You will find examples in any town by walking round with your eyes lifted above the street level and by visiting banks.

BUILDING STYLES

In any particular locality, domestic and other buildings may have been built in the same way by successive generations, using the same kinds of local materials. This is sometimes called **traditional** (or **folk) building**, or **building in the vernacular style**. Examples in the UK include:

- cottages in Norfolk with roofs covered with clay pantiles, and flint walls reinforced with brick corners;
- cottages in Devon with thatched roofs and cob walls with rounded corners. Cob is a mixture of clay and straw.

Conscious building styles, or fashions, tend to be international, although they may be modified to suit the materials and skills available locally. Each will offer fresh solutions to design problems, and will be recognized as a style when a sufficient number of practitioners are following it. It can be spread by personal example, or by giving information in books or other media. For instance:

- the ancient Romans sometimes built in the style of the even more ancient Greeks, who had, themselves, also built in Southern Italy, Sicily and Asia Minor;
- in medieval times, the Gothic style was developed by successive generations of masons and spread by those who went abroad to work;
- the designs of the Renaissance architect Palladio were studied by visitors to Italy, who returned home and applied his solutions to their own work. His rigorous approach to design became more widely appreciated after the publication of English versions of his book in 1715 and 1738, and these influenced the design of country houses in both the UK and the USA. Some of the ones that remain are open to the public.

Nowadays, through publicity in the technical press, a newly completed building can have an almost immediate influence on the thoughts of architects world-wide.

2.5 INVESTMENT IN THE BUILT ENVIRONMENT

As we travel through and look at our built environments, we should remind ourselves that it all has to be paid for. As well as houses and blocks of flats, we are likely to see factories, farms, offices, warehouses, hospitals, schools and colleges, garages, shops, churches, libraries, theatres, sports grounds with spectator stands, pubs and restaurants, railway and bus stations, airports, lawcourts, and prisons.

In addition, large-scale constructions will have been built for the general benefit of our communities. These can include electricity generating stations, oil refineries, dams, water supply and treatment plants, pumping stations and reservoirs, pipelines and cable networks, roads, railways, and other **civil engineering works**.

Some idea of the magnitude of the initial investment in construction can be gained by considering the number of homes in your area and their price range. There are also the continuing costs of owning and/or occupying a building. For example, consider the total cost of providing and maintaining your family home, usually the most significant item in any family budget. In the following, we shall use the symbol # to indicate the financial unit.

If a house is bought with a loan of #100 000 secured by mortgaging the property, and if the annual repayment of the loan, including interest, will be 10% of the sum borrowed, then its annual cost over the repayment period will be #10 000. What is the current repayment rate for mortgages in your area? The annual cost of providing your family with a home can be estimated by totalling the following:

1. the annual repayment of a 100% mortgage, based on an appropriate purchase price;

2. the annual costs of local taxes and water supply and treatment;
3. the annual costs of gas or other heating fuel, electricity, insurances; and
4. a contribution to the costs of future maintenance and renewals. The *English House Condition Survey 1991* (1993) reports that within five years an average of £1050 would need to be spent on general repairs to each owner-occupied dwelling.

What proportion of your family income is this annual total? The topic is considered in Chapter 20.

2.6 COMMENTARY

Many of the kinds of construction products mentioned in this chapter belonged to one or other of those classes mentioned in the Commentary at the end of Chapter 1. In this chapter we have added fillings (e.g. under and around drainage pipelines), large units (e.g. at corners of masonry walls), profiled sheets, flexible sheets and malleable sheets. In the next chapter we shall meet reinforced concrete, soils and rocks. This products analysis is used frequently throughout the book and is illustrated in Figure 5.1. The actual type and size of product which is chosen during the design process to provide a part with its main qualities, we shall call the 'main product'. However, they are seldom adequate on their own.

In looking for the various types of internal finishes or water supply pipes mentioned in Chapter 1, you may have noticed that other products were sometimes present as well. For instance, the crisp, straight external angles of a plaster wall finish might be due to the use of metal angle beads; the corners of a tile skirting and the external angles of a tiled wall finish may be formed with specially shaped tiles. Similarly, when looking at a water supply pipe, you will have noticed the fittings at angles and intersections, called elbows and tees. Each type and size of pipe will have its matching set of fittings which will not fit any other type or size, and this will also apply to drain pipe fittings as well.

Those fittings and other products that are used only in association with a particular main product we shall be calling 'dependent products'. They are considered in more detail in Chapter 5 and the analysis is fundamental to our study of the duration of construction activities.

Chapters 1 and 2 also introduced some of the qualities of the parts of buildings, which we shall be calling their **attributes**. These included attractive appearance, smooth surface, safe and level (floor surface), durable, light reflective, continuous, energy efficient, strong, stable. However, these are only some of the qualities we might expect in a building, and in the next two chapters we shall be looking at some more.

The values of these attributes are seldom if ever tested. Instead, they are predicted from a knowledge of some of the properties of their substances, and of how these substances have behaved in similar buildings in similar circumstances. For instance, fired clay facing bricks of a particular type may have proved durable over the years when used above the damp-proof course in the external walls of two-storey houses. However, such bricks may have failed to withstand freezing and thawing when used in constantly damp foundation walls.

2.7 SUMMARY

After considering the influence its environment has on a building, the chapter was mainly concerned with the alternative technical solutions of those of its parts that are visible from the outside. These included pitched and flat roof coverings and their eaves and verges, wall faces and finishes, rainwater installations, and external drainage.

The impression given by the outside of a building was considered and the idea of vernacular and other building styles was introduced. It concluded by inviting the reader to consider the investment represented by the built environment, and the financial implications of being a building owner/occupier.

REFERENCE

Department of the Environment (1993) *English House Condition Survey 1991*, HMSO, London.

FURTHER READING

Cook, Geoffrey K. and Hinks, A. J. (1992) *Appraising Building Defects*, Longman, Harlow.

Illston, J. M. (ed.) (1994) *Construction Materials, Their Nature and Behaviour*, E & FN Spon, London.

Reid, E. (1984) *Understanding Buildings*, Longman Scientific and Technical, Harlow.

Specification, Emap Business Publications, London.

Builders' merchants' catalogues and manufacturers' literature.

The building structure 3

In the first chapter we considered those qualities we can expect to find in a room or other space within a building, and we looked at its visible parts and those that provide occupiers with various facilities. In the second chapter we looked at the parts of a building that can be seen from outside, and at associated works on the site.

This chapter considers those parts of a building that give it strength and stability and which we call the structure. (In the UK's Building Regulations, some of these loadbearing parts have stipulated minimum standards of fire resistance and are called **elements of structure** – see Chapter 4.) The toughness of a structure can be demonstrated by using some empty cardboard boxes. A shed-type building is rather like a single empty box. When the lid of the box is secured, or the box is inverted, a surprising amount of force is required to squash it. If the box is filled tightly with smaller boxes, to simulate a more normal building, it becomes even more difficult to destroy. Even stronger boxes are made from corrugated cardboard consisting of two thin sheets held apart by a wrinkled sheet.

Yet a single flat piece of cardboard will not even stand up by itself. Fold it to form two vertical planes at right-angles, though, and it will be remarkably stable as one will prop up, i.e. **buttress** the other. In the main, buildings are composed of thin, flat, planar, vertical and horizontal structures of adequate stiffness, firmly joined to each other at all vertical and horizontal junctions to ensure the maximum mutual support.

Also, when the piece of cardboard is horizontal and supported at two opposite edges, it will bend under the slightest load. But when two opposite edges are pushed inwards so that the cardboard arches upwards, it becomes much stronger.

As we saw in Chapters 1 and 2, some structural parts are of continuous construction, e.g. masonry, while others consist of an open framework of (usually wooden) spaced structural members. Most buildings consist of a mixture of the two.

3.1 STRENGTH AND STABILITY

BUILDING LOADS

In everyday speech, the force, or pressure, which a construction itself exerts on its supporting parts is called its **weight,** and we shall be continuing this practice elsewhere in this book. However, when using SI units of measurement, we must describe the quantity of matter in a body as its **mass**. The force it exerts is called its **dead load**.

This can be calculated by multiplying the volume of the construction by the **mass density** (weight density) of its substance. Mass density is expressed in kg per m^3. In the USA, the equivalent weight density is given in lb per cubic foot (for water, this is 62.4 lb/cu ft, or 1000 kg/m^3).

Another value in use for this purpose is the **relative density**, or specific gravity (SG or G), which is the ratio between its mass density and that of water. This is the same, whatever the units. Some approximate SGs are: structural timbers 0.5; brickwork 2.0; reinforced concrete 2.4.

In addition to the dead load of its materials, a structure will be designed to sustain a maximum **imposed load**. On the roof, this will be caused by snow and people, perhaps when carrying out repairs.

The imposed loads on floor surfaces will arise from their having to support people, furniture, stored objects etc. and these must be allowed for when designing the foundations. However, as their locations are not predictable, the floor design must allow for the maximum concentration of this load occuring anywhere. For example, every square metre of the upper floor of a two-storey house should be able to support three persons. The design must also allow for extreme **wind loads** due to the force of the wind from any direction acting on roof coverings and external wall surfaces.

STRENGTH

Strength refers to the ability of a substance to withstand the forces applied to it. Provided the substance itself does not move, any force which is applied to it generates an equal and opposite resistance within the substance. In the SI international metric system, the unit of force is the newton. This is related to the acceleration of an object under gravity of 9.8 m (or 32.2 ft) per second/per second. Every kg of mass (or weight) exerts a force of 9.8 (say 10) newtons, and this will be resisted by its supports.

All the different substances which we might use to make a strong and stable building can be described as being **strong**. Most of these substances will be chosen because of their ability to resist being squeezed, or pushed, and this **compressive strength** is also expressed in units of force. For example, the compressive strength of a concrete block can be expressed either in mega-newtons (mega = millions) per square metre, i.e. MN/m^2, or in N/mm^2, which are numerically the same. However, the strength of any masonry wall will be limited by the strength of its mortar joints.

To some extent, a substance will also be capable of resisting being pulled or stretched, and this **tensile strength** may be more or less than its compressive strength. For example, the tensile strength of a wood might be four times as great as its compressive strength although, to make use of its tensile strength, it has to be connected to whatever is doing the pulling. In contrast, that of concrete might be only one-tenth of its strength in compression. Concrete blocks, clay bricks and walling stones will normally be used only where they will be in compression, e.g. by being placed on top of one another in masonry walls.

In any construction which is built, like Stonehenge, with horizontal **lintels** or **beams** on top of vertical **posts** or **columns** (or the equivalent portion of wall directly underneath), the self-weight of the horizontal members acts all along their length, but their mass is resisted only at the ends. Thus, they tend to bend, or **deflect**, squeezing their upper portions and stretching their lower portions as they generate their resistance to the load (see Figure 3.1). However, the greater the span, the more the self-weight, and with any particular section and substance, there will be a limit to the safe span. Of course, the thicker the sections are, the stronger they are, but the heavier their self-weight becomes. In any case, most lintels, beams, joists and other horizontal members are expected to carry applied loads as well as their self-weight.

The forces from these loads and the opposing resistance of the substance of, say, a lintel and its supports also give rise to **shear** forces where one portion tries to slide past its neighbour. Where a column is rather slender, unevenness in its loading may cause it to buckle.

A rather different structural member with only one end connected to a supporting structure and with its other end free to act as a kind of lever is called a **cantilever.** In high winds, tall buildings behave like vertical

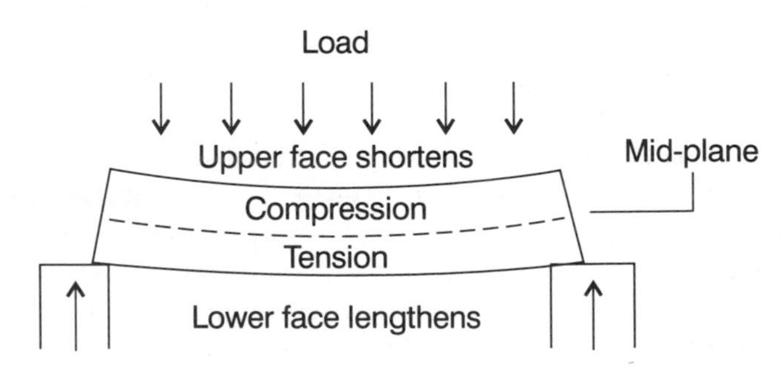

Fig. 3.1 Schema of the forces in a freely supported beam when bending to sustain a load.

cantilevers, their tendency to be uprooted being resisted by their mass, their structure and by their foundations.

Stress, strain and stiffness Woods and steels are some of the substances that are, for practical purposes, reliably strong in both compression and tension and in resisting shear. As, within limits, they will resume their shape after being compressed and/or stretched, they are also **elastic**. Knowing by how much the substance will be shortened or extended when a load is applied enables the deflection of a structural member to be predicted.

The load per unit of the sectional area of a member is called the **stress** (**s**), and the resulting change in one unit of length (i.e. the rate of 'extension') of the substance in the direction of the stress is called the **strain** (**e**). Provided that the substance behaves elastically under load, this strain will be in proportion to the stress. Energy is necessary to displace the atoms and produce the strain, and when the stress is removed, the atoms snap back and the energy is released.

An object which behaves in this way, e.g. a steel spring, or an archer's bow, is described as being **resilient**. The result of dividing a stress by the resulting strain is a constant for that substance called 'Young's modulus of elasticity', or E. Obviously, the less the proportionate strain, the higher the value of E. Thus, E is a measure of the **stiffness** of a substance.

Fractures A substance which is being stressed in tension will be prevented from reaching its full theoretical strength by the development of fractures. These begin at flaws in the surface and other imperfections, and their extent will depend on the effectiveness of the chemical bonding of the substance.

Substances that fracture easily, including glass and the many types of clay bricks that can be 'cut' by hitting them smartly with the edge of a trowel, are described as being **brittle**. In contrast, under increasing tension, a mild steel member will pass its **elastic limit.** While it may continue to stretch rather than fracture, it will not now be able to return to its original dimensions and will break eventually. Such substances are described as being **tough** and **ductile**. Ductile substances (usually metals) can be permanently deformed without fracturing by stretching them beyond their elastic limits, e.g. deformed steel reinforcing bars for concrete, sheet lead.

Tough buildings Of course, the less the stress in a structure, the less the resulting strain and consequential bending. As we like to have **rigid** buildings without any obvious deformations, we tend to use much more substance in our buildings than is strictly necessary, and this **redundancy** improves their **toughness,** safety and stability, and keeps them well within their elastic limits.

Most buildings are remarkably tough, their reserves of strength enabling forces to be transferred between them as foundations settle, parts deflect under loads etc. Under continuous stress, a substance may give up resisting to some extent. This phenomenon is called **creep**.

STABILITY, OR FIRMNESS

In Europe, national building codes and regulations must conform to the Essential Requirements of the Construction Products Directive (1989), known as the CPD. For example, under the heading of 'Mechanical resistance and stability', the CPD states

> The construction works must be designed and built in such a way that the loadings that are liable to act on it during its constructions and use will not lead to any of the following:
>
> - collapse of the whole or part of the work;
> - major deformations to an inadmissible degree;
> - damage to other parts of the works or to fittings or installed equipment as a result of major deformation of the load-bearing construction;
> - damage by an event to an extent disproportionate to the original cause.

These requirements are echoed in the UK's Building Regulations. The ability of a horizontal or vertical structural part of a building to remain firmly fixed in position will depend on:

1. its dead and imposed loads and their direction, e.g. vertical, sloping outwards;
2. any wind effects;
3. its substance and its strength;
4. its thickness;
5. its size between constraints provided by other parts, e.g. between buttressing masonry walls;
6. its ability to maintain its internal integrity, particularly when pierced by openings; open frameworks can often be made more rigid by triangulation, that is, by including strongly jointed diagonal crossing members (a triangle with its three straight sides and rigid joints cannot be deformed);
7. its being sustained by parts lower down and, ultimately, by the ground.

DIMENSIONAL STABILITY

Variations in the amount of heat from the sun, heat from fires, and changes in temperature due to internal heating or cooling systems all cause building substances to expand and contract, although each will respond at a different rate. Changes in moisture content affect porous substances similarly. What is important is whether this **dimensional instability** is significant or not.

Thermal movement The proportionate change in the length of a substance for each 1° C change in its temperature is known as its **coefficient of linear expansion**. Even in a temperate climate such as that of the UK, a 10 m length of steel will extend and contract by 3 mm during the year. The movement of fired clay and stone will be less and that of non-ferrous metals will be up to three times more. Polycarbonate sheets, plastic gutters and other synthetic polymers are the most unstable, and expand and contract by more than five times that of steel. Thermal movement, particularly the differential thermal movement between steel framing and masonry, can cause damage in any long or high construction unless expansion joints are incorporated.

Moisture movement All porous substances will swell or shrink with the absorption or release of moisture. Moisture movement can be expected in many recently manufactured substances as their moisture content adjusts to that of the environment, e.g. freshly fired clay bricks take up moisture and expand; new precast concrete products, including concrete blocks, dry out and shrink. Such dimensional changes tend to be irreversible. They are most significant in the first few months, and the products should not be built in until they have almost stabilized.

Woods readily give up and absorb moisture to stay in equilibrium with the moisture content of the surrounding air. They shrink as both they and the building dry out, although only in directions at right angles to their length. (Timber buildings should be so designed that the combined vertical shrinkage of horizontal members is much the same all round.) Suppliers should be asked to supply timbers within a specified range of moisture content.

In an old building with wood floor joists built into masonry walls there will usually be a gap between the bottom of the skirting and the boarded floor. Try to find an opportunity to measure this gap even though it may be covered by carpeting, and estimate the percentage the joists have shrunk. It can be as much as 5%. Perhaps you can find other examples.

Chemical changes may also cause dimensional instability, and moisture is usually involved, e.g. the rusting of carbon steels; the formation of sulphate (sulfate) crystals in mortars and concretes.

ACHIEVING CONTINUITY

One of the characteristics of a building is that, to some extent, all its parts depend on each other. That is, the parts are **interdependent**, and when any one part is missing, incomplete or not connected to its neighbours, the building as a whole and its occupiers suffer. Consider, for instance, the effect of having no front door, or a length of water pipe missing, or not having the staircase connected to the upper floor.

Substances that are created as they harden in position (see the Commentary at the end of Chapter 1), e.g. concretes, mortars, paints, plasters, are **naturally continuous** within themselves. When adjoining parts with the same or some other of these substances are made at the same time, continuity between them is automatically achieved. Examples include the continuity between concrete beams and columns, between concrete roofs or floors and masonry walls, between plastered and painted walls and ceilings, and between the mortar beds and joints and the bricks, blocks or stones in masonry walls.

To achieve continuity elsewhere, the substances of adjoining products must be **connectable** in one way or another. Fixings are either **continuous** or **discrete**. Most continuous fixings perform by **adhesion** and will be made by the hardening of substances such as mortars, adhesives and mastics. These may also be used for spot fixings. Metals may be joined by the addition of molten metal, as in welding and leadburning, or by drilling and either riveting or bolting.

Pipes with sockets at one end can be jointed by inserting the plain end of the next pipe and filling up the annular space with an appropriate jointing substance. Cast-iron rainwater pipes on older properties are a good example, and are identifiable by being fixed through the 'ears' cast on to their sockets, and by their resonance when tapped. Other pipes may be in one piece as they are available in long lengths, e.g. plastic rainwater pipes. Otherwise, pipes with plain ends, including those with small bores that are often called tubes, will require special connector fittings. Continuity between steel reinforcing bars and concrete is achieved mechanically as the surfaces of bars are naturally rough. The bond can be increased by hooking their ends and by using twisted or indented bars.

Direct mechanical fixings include various kinds of metal nails, screws, bolts and nuts, and wall ties. Nails and screws can only be used with substances that are **penetrable**, have adequate **holding power** and are able to **resist splitting**.

Most mechanical fixings are intended to hold two adjacent substances closely together and prevent them from sliding past each other. They must be **strong** enough to resist the forces that might act upon them and be at least as **durable** as the substances being fixed together. (Loose roofing slates provide a common example of the premature failure of fixings.) The strength of such fixings is in their resistance to being bent or sheared, i.e. in their sectional areas. Only a few will be subjected to a direct pull.

Special **mechanical fixing units** with their own fixings can be used for special purposes. For example, metal straps, struts and hangers can be either nailed or screwed to wooden members of a roof structure; pipes are secured to walls with clips, brackets and screws; doors are hung from their frames with hinges and screws.

Substances to be connected that are not directly penetrable will need to be bored to receive bolts etc. Bolts act primarily by providing shear resistance. However, when the nuts have been tightened, their tensile strength will pull the objects together, the friction between their surfaces then increasing their resistance to sliding.

A building may also contain some **discontinuities**, e.g.

1. thermal expansion joints in long and/or high external walls;
2. a space over the top of a non-loadbearing internal partition to accommodate the deflection of a structural floor or beam above;
3. around sound-resisting timber separating walls.

MAKING THE MOST OF WHAT YOU'VE GOT

Until the last century, woods were the only commonly available substances that were strong in both compression and tension, and roof spans were limited by the sizes and shapes of available trees. This applied to King Solomon, (see 1 Kings, chapter 5: verses 6–11) who, some 3000 years ago, had to import trees from Lebanon in order to roof his temple in Jerusalem. It continues to be true for traditional styles of buildings in countries with limited access to world markets.

Without trees, buildings have had to be constructed with commonly available substances that were strong in compression only, such as fired clay and stone. Figure 3.2 shows a street scene in Alberobello in southern Italy. The buildings, called **trulli**, are square on plan, with double-skinned conical roofs, and are constructed entirely of thin pieces of limestone bedded in mortar. Nowadays, we are fortunate in having a wide choice of home produced and imported products, although countries with limited foreign currency may restrict their imports.

3.2 OPEN FRAMEWORKS

OPEN FRAMEWORKS IN GENERAL

A planar structure can consist of a set of spanning members of constant cross-section spaced at suitable intervals to form a frame (see Figure 3.3). In a roof, these spanning members are called **rafters** or **common rafters**. In a floor or ceiling they are called **joists**, and in a wall or partition, they are called **studs**. Although made mainly of wood, metal sections may be used. The members are connected to form a frame by either:

1. having their ends secured to cross members that are, themselves, secured to an adjoining structure, e.g. the ends of vertical studs fixed

Fig. 3.2 A street in Alberobello, Italy, showing *trulli* constructed of small pieces of limestone bedded in mortar.

to a horizontal sill that is secured to the floor below, and to a head secured to the floor or roof structure above (see Figure 1.1), or
2. having their ends secured directly to an adjoining structure, e.g. the ends of floor joists either fixed to metal joist hangers themselves built into a masonry wall, or secured to the head of a stud wall.

Note the mutual support that is provided when a frame is connected to an adjoining structure, e.g. an upper floor is secured to the external walls, a pitched roof structure is stabilized by being connected to the ceiling joists and the tops of walls. Further internal rigidity is provided by:

1. continuous bases for finishes, e.g. boarded floor and roof decks, plasterboard wall and ceiling linings;
2. continuous linings to the outsides of wall and roof structures, e.g. plywood sheathing; and
3. battens supporting slates or tiles.

Openings wider than the space between adjacent members are formed by 'trimming' the unwanted portions of members and supporting their otherwise free ends by similar-sized 'trimmers'. A frame can be given further rigidity by diagonal bracing. Wood members are supplied in a limited range of widths and thicknesses and in lengths ranging from

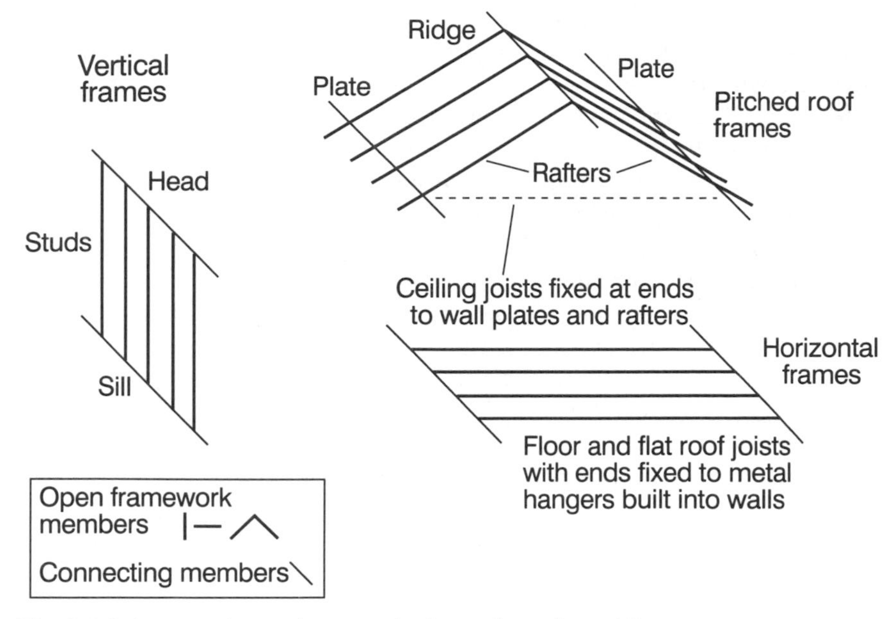

Fig. 3.3 Schemas of open frameworks for roofs, walls and floors.

1.20 m to 4.80 m or beyond in 300 mm intervals. They may be called **scantlings** or **carcassing** or, in North America, **dimension lumber**.

SIMPLE PITCHED ROOF OPEN FRAMEWORKS

The limited spanning ability of horizontal battens or continuous sheathing supporting a roof covering will restrict the maximum spacing of the supporting roof rafters to perhaps 450 mm. The dead load of the materials used in the roof covering, the imposed load caused by snow and people on the roof, and the force of the wind on its surfaces must be sustained by a strong and stable roof structure usually of open construction. Its purpose will be to transfer these loads safely to a supporting structure of walls or to a structural frame of beams and columns.

The simplest roof structure is a **lean-to** consisting of a series of rafters sloping down from a higher wall to a lower one and with their upper and lower ends fixed to wall plates to form a frame. It gets its stability from the walls.

Most roof structures consist essentially of one or more pairs of open framed structures leaning on each other at the top, and sharing the upper connecting member called a **ridge** or a **ridgeboard.** Alternatively, they can be visualized as a **coupled roof**, i.e. one constructed with a series of couples, or pairs, of opposing rafters with the same pitch.

Where the building space being spanned is no more than about 3 m wide, e.g. a garage, the pairs of rafters may be strong enough on their own to span from the ridgeboard to horizontal **wall plates** secured to the external walls at the eaves. For larger spans, either the sectional size of the rafters can be increased, or they will need intermediate support from one or more horizontal **purlins** (see Figure 3.4). These purlins can be supported by the principal rafters of roof trusses, by **props** off internal walls, or by having their ends built into walls. Where two roof slopes come together at right angles to form a hip or valley, the shortened **jack rafters** in the adjacent supporting frames are stabilized by being fixed to a shared diagonal hip or valley rafter.

Roof loads tend to flatten a pitched roof, but the tendency of the sloping rafters to move outwards can be resisted if the ceiling joists are connected to the bottom ends of opposite rafters as well as to the wall plates. The ceiling joists can also be tied together at intervals in their span by horizontal **binders** which are supported by vertical **hangers** secured to the purlins. The members of a pitched-roof framework can

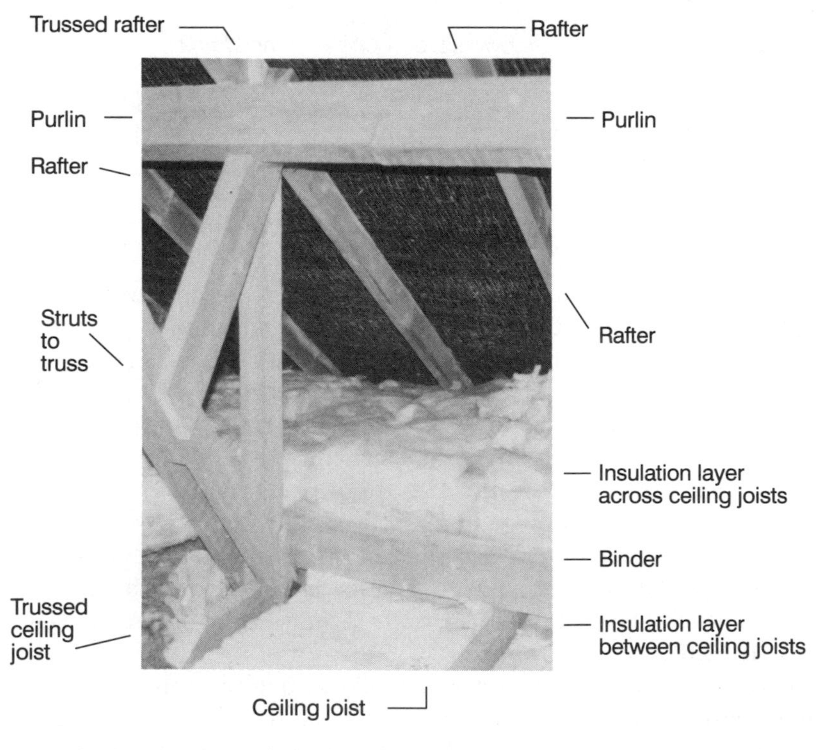

Fig. 3.4 A view inside a pitched roof structure supporting a tiled covering and showing portions of rafters and ceiling joists, an early form of trussed rafter, a purlin and a ceiling binder.

often be seen from below, perhaps through a trap-door in the ceiling finish (see Figure 3.4).

In this traditional type of construction, the members have to be thick enough to withstand being secured by nailing, although this does not enable the full strength of the wood to be utilized. Developments in metal plate connectors have led to the use of thinner and lighter wood sections and the evolution of prefabricated **trussed rafter** units consisting of a pair of rafters, a ceiling joist and internal triangulating members.

Figure 3.4 shows an early form of trussed rafter where every fourth pair of rafters were connected to a ceiling joist and provided with struts and hangers. These supported purlins and ceiling binders, which supported the other rafters and ceiling joists in the traditional way.

Nowadays, the whole roof structure will most likely consist of trussed rafters. They will span between parallel external walls, thus allowing flexibility in the layout of the rooms below, and are secured to wall plates with metal truss clips. The wall plates themselves will be anchored down to the walls by vertical restraint straps. The units must be stabilized against the 'domino effect' by diagonal bracing nailed under the upper portions of the rafters and by being anchored to the walls at gable ends (if any). Thus, continuity will be created all round between the roof and wall structures. Increasingly, metal clips, restraint straps, hangers and ties are being made of stainless steel.

Even so, where the top floor of a building is partially or wholly within the roof space, the bottom portions of the rafters will be free to splay out when loaded, and they will rely for their stability on the resistance of the walls to that outwards thrust or the support of trusses.

PITCHED ROOF WITH TRUSSES

Where a roof spans over a single space that is open to the roof framing, e.g. an assembly hall, a railway station, a sports stadium, there is no opportunity for propping, and, instead, the purlins will be supported at suitable intervals on even stronger **principal rafters**. These members are incorporated into structures called **trusses** that span the whole space.

The many different roof truss designs that have evolved by trial and error over the centuries may incorporate wood or metal tie beams in tension and internal vertical and sloping compression and tension members. The ends of the principal rafters might rest on **padstones** or on wood plates bedded on the tops of thickenings to the walls called **piers**. In addition, some of the loads on the trusses may be taken down to a lower point in the walls, thus increasing the continuity between roof and walls and allowing the mass of the upper portions of the walls to resist any outwards thrust. It may be possible to discern an arch shape which springs from these lower points and follows the line of the lower truss members.

To sum up, the traditional roof structures we have been describing consist of a double criss-cross of horizontal and sloping members, with the upper ones supported by the ones below, i.e. battens on rafters on purlins (and plates) on principal rafters. However, Type A2 profiled sheet coverings (see Chapter 2) will only require support from purlins on roof trusses. The roof structure is usually the most difficult structural design problem in buildings and a large variety of different solutions have been evolved, each with their own name.

FLAT ROOF AND FLOOR OPEN FRAMEWORKS

The purpose of flat roof and upper floor structures is to sustain and transfer their dead and imposed loads safely to supporting walls or a structural frame of beams and columns. Floors are horizontal so that the force of gravity will have a neutral effect on the contents of the room and on the occupants. The roof or floor deck, which supports the applied loads of e.g. furniture and people, must have the strength to span between the individual structural members and to transfer their loads safely to them.

Domestic floors of up to about 4.5 m span can be achieved with wooden joists, but beyond that, some intermediate support will be required, e.g. one or more steel beams. These will receive and sustain the loads from the joist ends and transfer them through their own ends to some stable structure. When the beam ends are built into masonry walls or piers, their loads are often spread over a wider area of wall by stone or concrete padstones. To minimize the reduction in the clear height of the room below, such beams may be partially or wholly within the depth of the floor.

Transverse lines of pairs of crossing wood or metal struts called **herringbone strutting** will be used to stiffen floor and roof joists and to help spread their applied loads. The stairwell through an upper floor will probably offer the only opportunity for measuring its overall thickness and deducing the thickness of the floor construction.

OPEN FRAMEWORK WALLS

Platform framing is where one-storey planar wall frames each consisting of a horizontal sill, vertical studs, diagonal bracing and horizontal head are prefabricated, perhaps on the already constructed ground floor. The sill and head may both be called **plates. Transverse** solid strutting called **noggins** may be used to stiffen wall studs and provide fixings for the edges of board finishes. Openings will be formed and trimmed for doorways and windows, and the framework will be sheathed externally, possibly with plywood. Insulation and vapour control layers may also be added (see Chapter 4).

The frames are constructed in turn, raised, braced temporarily and later joined to their neighbours. Small sections of wall, particularly non-loadbearing internal partitions may be constructed in place. A two-storey open framework wall with full height studs connected to both ground and first floor joists and the roof framing is called **balloon framing**.

Open frameworks can be used to support a variety of claddings. For example, bands of glazing and other impervious vertical cladding provide the **curtain walls** of many tall buildings. These will be secured to an open steel or aluminium framework of vertical mullions and horizontal transoms that will transfer their dead and wind loads to the structural floors or to a structural frame. Curtain walls of low-rise buildings may have wooden open frameworks.

STRUCTURAL FRAMES

The strength and stability of the entire building may be provided by an open structure of trusses, horizontal beams and vertical columns which supports lightweight roof and wall coverings as well as any upper floors. Sheet cladding to roofs and walls will also require purlin supports. On the other hand, the walls might be self-supporting and a frame might be used just to support the roof and the upper floors.

The purpose of a frame will be to intercept, sustain and concentrate the loads being carried by individual structures and to transfer them safely through vertical columns to the column bases and thence to the ground. Floors and walls must be firmly connected to frame members at their junctions. A frame may be made of steel, aluminium, concrete reinforced with steel or wood. Its successful design will depend on using substances of known and reliably consistent strength. Depending on the use of the building and their substance, structural members may need to be protected against fire.

A variety of different hot rolled steel and aluminium sections is available to the designer who seeks to match a section to the forces in a member. For instance, in a simply supported beam (see Figure 3.1), the most squeezing or stretching strain and the maximum demand for strength occurs at the upper and lower surfaces. In between, these effects lessen to zero at the mid-plane. Thus, an economic section for a steel beam (or column) will be an exaggerated 'I' where most of the substance will be in the upper and lower **flanges**, with a somewhat thinner **web** in between. Such sections can be connected to each other either by welding or by having short lengths of steel angle riveted or bolted to their webs (i.e. **cleats**) and/or their flanges (i.e. **brackets**).

The economic design and construction of wood structural frames depends on:

1. being able to measure the strength of timber sections by non-destructive stress grading; and
2. having a range of timber connectors for reliably transferring forces at joints.

A finish may be applied to the exposed surfaces of structural members or to their casings to improve their **appearance.** Their foundations are likely to be deeper than those for walls, and may be cast before the strip foundations.

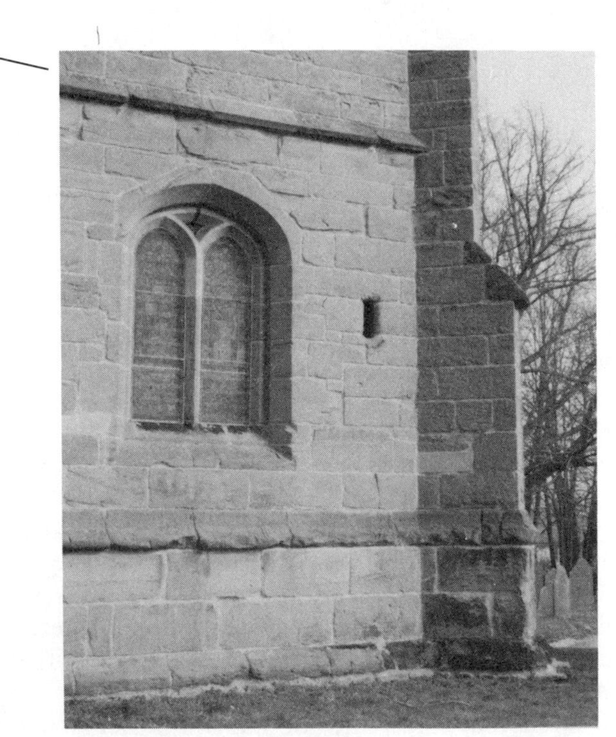

Fig. 3.5 The buttressed corner of a sixteenth-century church tower.

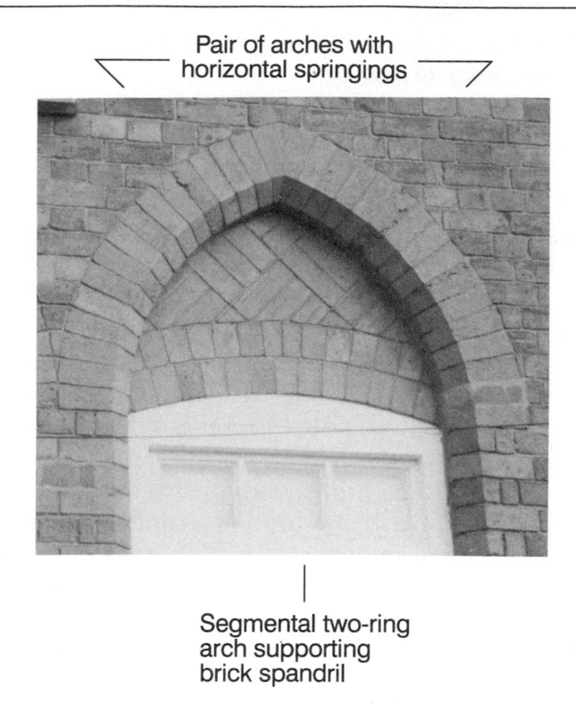

Fig. 3.6 A Gothic style brick arch over a window in a gable end to a village school built in 1913.

3.3 CONTINUOUS STRUCTURES

THE STABILITY OF MASONRY CONSTRUCTIONS

This continues the discussion on masonry construction in Chapter 2. Historically, rectangular blocks of stone or fired clay were the obvious shapes to use for vertical walls, where gravity would pull their constituent units together rather than pull the walls over, i.e. they would be in compression. Also, the design could ensure that their loads, including any sideways ones from roofs and floors, would act well within their thickness, so as to avoid overturning. Where overturning was a possibility, the effective thickness of the wall could be increased by frequent attached piers or by short lengths of projecting walls, called buttresses.

Figure 3.5 shows a corner buttress providing extra stability to a church tower. Notice also how the walling stones are corbelled over the window to relieve the load on its two curved arch stones. The string course is there to protect the wall below by throwing off rainwater. The base of the wall is thicker and made of larger stones to increase stability.

The history of the style of religious building we now call Gothic is linked with the development of vaulting (which spanned the spaces between walls), and pointed arches. The latter are really pairs of arches that lean on each other at the top rather like the two arch stones in Figure 3.5. As their lowest units were laid horizontally (see Figure 3.6), or almost so, they transmitted most of their loads vertically, and minimized sideways thrusts. With experience, pointed arches allowed larger and larger windows to pierce loadbearing walls. By adjusting their slope, the tops of openings of differing widths could all finish at the same level.

Buttresses were also pierced, the sideways roof thrusts being transferred to outer walls by half-arches called 'flying buttresses'. Thus, every part was balanced against other parts and the whole was intended to continue in a state of equilibrium although, alas, many failed to do so.

Italian architects who built during the fifteenth and sixteenth centuries (the Renaissance period) seem to have despised this style, and were responsible for nicknaming it Gothic. Instead, they found inspiration in the buildings of ancient Rome, with their semicircular arches.

Nowadays, lintels and arches are used to maintain stability over window and doorway openings by transferring the loads of the walls above them to their sides. Such loads are mainly from the triangular portion over the opening, as the remainder of these wall loads will be automatically transferred sideways by the overlapping masonry units acting as corbels. Cavity walls and other developments are considered in Chapter 4.

CONCRETE STRUCTURES GENERALLY

Concrete is essentially a cement and water paste extended by being mixed with some hard mineral aggregates. The aggregates are usually a mixture of graded sand particles (the **fine aggregate**) and larger graded pieces of crushed rock or gravel (the **coarse aggregate**). Nowadays, ready mixed concrete of the desired specification is readily available from specialist suppliers.

Plain or **mass** concrete is concrete that is used primarily for its ability to sustain compressive forces. Where a construction will be subject to both tension and compression, e.g. when bending, carefully positioned steel bars of suitable diameters will be incorporated to provide the tensile strength.

The principal use of a combination of concrete and steel bars is to form continuous structural parts with unique shapes. These vary from the mundane shapes of flat roofs, floors, walls, beams and columns, to thin curved roof shells and spiral staircases that depend on their form for their strength. Most concrete staircase are, in effect, sloping slabs with the surfaces of treads and risers formed in their upper faces, and with sloping soffits.

The surfaces of all concrete structures will have been defined by temporary moulds referred to as **formwork**, **falsework** or **shuttering**. These will consist of the **forms** (sometimes called **shutters)** that will encase the wet-mix, together with strong and stable **supports** that will resist the mass of the concrete and reinforcement yet still be easily and safely dismantled.

The forms must be capable of withstanding the hydrostatic pressures of the wet concrete without deflecting. The quality of their inner faces will determine what the surfaces of the hardened concrete will look like and, to assist their **release** from the hardening concrete, these faces can be treated with a **mould oil**. Where a part is to be constructed from a broader horizontal base, e.g. columns or walls rising from a floor, stub **kickers** will be cast first to locate the forms. These, too, will need their own formwork.

The reinforcing bars may be delivered already bent to shape, or they may be bent on the site. **Cages** of bars may be assembled before being placed within the formwork. Otherwise, the bars are wired together in position. Their distance from the faces of shutters, from other constructions, e.g. from concrete blinding, or from the surfaces of excavated voids, must provide them with a sufficient protective cover of concrete.

After a final cleaning of the shutters, the concrete is poured into them and compacted around the reinforcement. When the concrete is strong enough, the shutters are stripped away and the surfaces of the exposed concrete repaired or finished as required. Some kind of skeleton propping may be retained, particularly where floors are to be used as workplaces, or to support the construction of the next floor. Where formwork is being reused, it must first be cleaned, repaired where necessary, and re-oiled.

Having to wait while the concrete hardens complicates the construction process. An alternative is to construct the work from precast units small enough to be hoisted and placed in position, e.g. floor and roof beams, chimney caps. A staircase can be precast as a single unit spanning from floor to floor. Where between masonry walls, it can be constructed by building in individual precast concrete steps and landings.

3.4 SUBSTRUCTURES

The roof, the upper floor and the wall structures considered so far have all helped to enclose or subdivide the building space. Those parts of a building that are below the plane of the top of the floor next to the ground are often considered jointly as the **substructure**. Although the need for a horizontal damp-proof course below the external walls has already been mentioned, this should be seen as part of a continuous moisture-excluding layer below the building space.

FLOOR STRUCTURES NEXT TO THE GROUND

In the UK, the floor nearest the natural ground level is called the **ground floor**. In the USA, this is the **first floor**. The purposes of the floor structure itself will be to provide a level support for floor finishes, people and things and to transfer the dead and imposed loads safely to the ground. Other purposes of the combined structure and finish will be to exclude plants and other vegetation, moisture, water vapour, soil contaminants, and living creatures. Special precautions must be taken where the natural level of radioactive radon gas in the subsoil may be unacceptable, e.g. in parts of the West Country, Northamptonshire and Derbyshire.

The technical solutions chosen for these floors will be influenced by the natural ground level, the finished floor level, and the slope of the surface of the ground after the vegetable soil has been removed.

One approach is to have a ground-supported, i.e. **solid** floor (see Figure 3.7). Here, the top of the ground slab is lifted to the level of the floor finish by filling the space below with some strong, hard, and chemically inert **hardcore**. This will also interrupt the upward movement of moisture by capillary action. Ducts may be required below the lowest floor to accommodate heating mains and other services. These will require access for maintenance and renewals.

An alternative approach is to have a suspended or **hollow floor** consisting of the following.

1. An upper structural floor spanning either from wall to wall or, to reduce the span, between intermediate ventilating 'honeycomb' sleeper walls (see Figure 3.7). The floor may be constructed of either cast-in-place or precast reinforced concrete. Alternatively, it can consist of an wooden open framework of floor joists with their ends supported by joist hangers built into the walls, and secured to wall plates bedded in mortar on the sleeper walls and with a dpc in between.
2. An airspace that contains the sleeper walls and is cross ventilated by airbricks set in the external foundation walls.

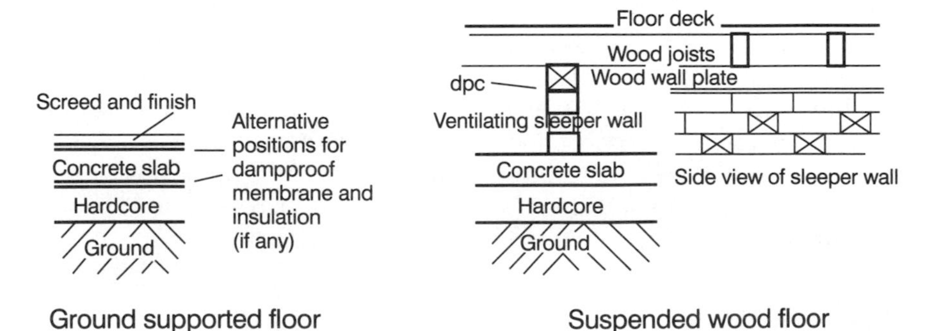

Fig. 3.7 Vertical cuts through ground supported and suspended timber floors with an elevation of a ventilating sleeper wall.

3. A stepped or sloping concrete ground slab on a bed of hardcore to seal the surface of the ground and support any sleeper walls.

In both approaches, moisture and water vapour can be excluded by a dampproof membrane below or above the ground slab and continuous with the external walls' damp-proof course.

WALL FOUNDATIONS

The loads at the bottom of a loadbearing enclosure wall structure will be sustained by the foundation walls (sometimes called **rising walls**). These must be at least as strong and stable as the ones they support, and possibly more durable. Their appearance will only matter where their faces are exposed. A damp-proof course at ground floor level will disconnect the enclosure walls from their foundation walls, which will be kept moist by the ground and must be capable of withstanding being frozen.

Below these foundation walls there will be some (usually) continuous concrete foundations. These must be let into the ground to a sufficient depth and have sufficient spread that the ground there is able to sustain the loads placed upon it, whatever the earth movement and climatic conditions, e.g. extended freezing, drought. Where the ground is sloping, these foundations are likely to follow the slope in a series of shallow overlapping horizontal steps. Where the subsoil is variable or subject to ground movement, foundations may be reinforced with steel bars and, in extreme circumstances, may be supported by driven or cast-in-place piles.

Many older buildings in the UK have brick walls built directly off the trimmed and levelled bottoms of trenches, usually with the lowest courses consisting of brick projecting **footings**. Nowadays, the cast-in-place concrete projecting strip and trench-fill foundations spread the wall loads so that their force is safely within the bearing capacity of the ground. They also absorb any irregularities in the bottom surfaces as well as any variations in trench depths and levels.

Each storey of a masonry external wall to a house will have a mass of about 1 tonne (1000 kg) per linear metre. Thus, it will exert a force of about 10 000 newtons, or 10 kilo-newtons (kN) on each linear metre of any supporting wall as well as on its foundation. Note that this does not include the dead and imposed loads from the roof and floors, nor the wind load. The width of its strip foundation can be calculated by dividing this force by the bearing capacity of one square metre of the subsoil.

Footprints in the sand at the seaside or boot marks on a muddy track remind us that all soils will compress to some extent when resisting a load. Thus, the sequence followed when constructing a new building should be such that the loads on its foundations and their consequential settlement are increased evenly all round. For the same reason, an extension to an existing building should remain independent and be allowed to settle fully before being connected.

3.5 BUILDING SITES

Every site as well as every building is unique, each having a different location, size, shape and set of natural surface levels. These levels will always have to be modified to some extent in order to receive the new construction, and conditions below the surface are never certain. Preparing the site involves a quite different set of processes from those of construction, and often results in the removal of some of it. In addition to providing direct support for a building, the ground on which it stands must also resist the leverage of wind on the structures and the effects of earth movement.

VARIETIES OF SOILS

The hard, natural masses of consolidated minerals which we call **rock** are usually distinguished from the **soils** that have been derived from them by weathering and abrasion, and which may have been transported to their present locations by glaciers, water or wind. Soils are usually classified by their particle size, which will range from coarse through fine to organic. We describe them as being either non-cohesive, e.g. gravels, sands, or cohesive, e.g. clays, silts. Gravels and sands allow the passage of water. In contrast, clay which has been mixed with water, or 'puddled', can be used to construct water-retaining structures such as the cores of dams and the bottoms and banks of canals. The study of the ability of soils to support buildings, roads, and other constructions is called **soil mechanics**. Peat offers no support for buildings.

Changes in the moisture content of soils, particularly clays, will cause them to shrink or swell. These changes may be caused by the growth or cutting down of trees, or the lowering or raising of the natural level of water within the ground. When silty, porous soils become frozen, they may be holding sufficient water for the expanding ice to cause 'frost heave'. Repeated freezing and thawing cycles may weaken the foundations of roads and other pavings.

TOPSOIL

Where land is still in its natural state, or has only been used for agriculture, the top layer will be rich in herbage, organic matter, seeds and roots. The thickness of this layer will vary. Normally, it can be expected to be between 100 mm and 200 mm thick. It can be as much as 1500 mm thick in fenlands, and earlier on, it had been this deep on some of the prairies of North America.

As topsoil will decay and compress in time, it is always removed from the location of new construction. Where the natural levels are to be modified by cutting or filling, this covering of topsoil should first be carefully stripped off and stored temporarily out of the way. Any turf and the top 40 mm or so of soil may be cut and stored separately. When the excavation and filling has been completed, the topsoil and turf can be restored.

SUBSOIL

The nature of the subsoil can be rather unpredictable due to the complex geological history of some areas of the Earth's surface. It can, of course, be investigated by making borings or by digging 'trial holes', but these will only reveal what is at those specific positions. Quite unexpected formations may be revealed when the site is excavated, including the discovery of highly non-cohesive or **running** sand, or water, or rock. The **water-table** is the level below which voids within the subsoil are saturated with water. There may also be wells and other old constructions.

The term **rock** is applied loosely to any material that is so hard it has to be broken up using compressed air equipment, wedges and crowbars, or even explosives before it can be excavated. Sometimes, isolated pieces of rock are found in otherwise soft material. These can be particularly troublesome when they are only partially exposed at the side or the bottom of a void, and have to be cut through, or dug out, and the additional space backfilled with concrete.

CHANGE IN BULK DENSITY

When soil or rock is disturbed, it breaks up into irregular shapes, and takes up more space than before. This process is known as **bulking**, or **swell**. Its extent is expressed as a percentage of the solid measure, called the **coefficient of bulking**. Sands and gravels have the lowest coefficients (5–15%) as they already have voids. Solid rocks have the highest (20–60%).

SITE CONTAMINANTS

The UK is thought to have some 100 000 contaminated sites with a total area equivalent to that of Northamptonshire. See Building Regulations Approved document C2.

BENCH MARKS

The work of the UK Ordnance Survey has included establishing permanent bench marks (OSBMs) and publishing lists of their locations and other physical details and their heights above the national datum. This datum is the mean sea level at Newlyn, in Cornwall, as calculated from hourly recordings taken between 1915 and 1921. There, the tidal flow is least affected by the narrowing English Channel.

The height of any point above this datum will be the height of the nearest OSBM plus or minus the vertical distance between the OSBM and the point. OSBMs are mostly cut into vertical brick or stone walls (see Figure 3.8), the published level being at the centre of the horizontal bar. Others consist of brass bolts or rivets let into horizontal surfaces. An OSBM can usually be found within 450 m in urban areas and 1000 m in open rural areas.

In any case, a **temporary bench mark (TBM)** must be established on the site for a new building. It must be given a notional positive level that is high enough for all the site, drainage and building levels to be positive also. Its location and form will be chosen by the design team, and it must be preserved for at least the period of the Works. If desired, its height above mean sea level can be calculated by levelling back to an OSBM.

Fig. 3.8 An Ordnance Survey Bench Mark (OSBM) cut into the wall of a derelict shieling to the west of Ben Lawers in the Grampian Mountains in Scotland. Its altitude is 505.24 m above the mean sea level at Newlyn, some 500 miles away.

3.6 COMMENTARY

Although the roof, floors, walls etc. may appear to be merely static assemblies of materials, they are, in fact, mechanisms actively carrying out their functions. For instance, the building responds as you walk across a room – although not as dramatically as if you were in a small boat. A masonry external wall will be interacting all the time with adjoining floors, the roof, other walls, ground movement, and conditions inside and outside the building. Various forms of energy, water vapour and moisture will be being rejected, absorbed or allowed to pass through, and static and dynamic loads will be transmitted. Its size and shape will vary as the temperature and atmospheric conditions change.

Thus, we can say that, in combination, these constructions constitute a **system**. Ideas associated with systems have developed since the 1950s following the work of Bertalanffy (1967) who writes 'A system is defined as a complex of components in interaction, or by some similar proposition.'

3.7 SUMMARY

This qualitative study of the strength and stability of buildings began with the mutual support provided by their roofs, floors and walls and the need for continuity between these structures. Dead, imposed and wind loads were distinguished, and aspects of structural behaviour were introduced. The requirements of the Construction Products Directive (1989) were listed. The thermal and moisture movement of important substances were considered in some detail, as were the various ways in which continuity can be achieved between the structural parts.

Open-framed structures were distinguished from continuous ones and masonry and concrete constructions were discussed. Some technical solutions for pitched and flat roofs, roof trusses, upper floors, structural frames, floors next to the ground, and walls and their foundations were described in some detail.

REFERENCES

European Union (1989) Council Directive 89/106/EEC of 21 December 1988 on the approximation of the laws, regulations and administrative provisions of the Member States relating to construction products, *OJ L* 40, 11.2.89, p. 12.

von Bertalanffy, Ludwig (1967) *Robots, Men and Minds*, George Braziller, Inc., New York.

FURTHER READING

BS 6031. *Earthworks*, British Standards Institution, London.

Everitt, A. (1994) *Materials*, Longman Scientific and Technical, Harlow.

Gordon, J. E. (1991a) *The New Science of Strong Materials, or Why You Don't Fall through the Floor*, Penguin Books, London.

Gordon, J. E. (1991b) *Structures, or Why Things Don't Fall Down*, Penguin Books, London.

Illston, J. M. (ed.) (1994) *Construction Materials, Their Nature and Behaviour*, E & FN Spon, London.

Levy, Matthys and Salvadori, Mario (1992) *Why Buildings Fall Down*, W. W. Norton & Co. Ltd, London and New York.

Reid, E. (1984) *Understanding Buildings*. Longman Scientific and Technical, Harlow.

Salvadori, Mario (1990) *Why Buildings Stand Up*, W. W. Norton & Co. Ltd, London and New York.

Improving our built environments 4

In earlier chapters, we have considered the construction of what we might call basic buildings with their essential attributes of weathertightness, opacity or transparency, strength and stability, dimensional stability, continuity and appearance.

During the last 150 years or so, our buildings have become more and more complex in response to the public demand for improvements in environmental and safety standards and our ability to afford them. In this chapter we shall consider those extra parts which a modern society requires in their new and refurbished buildings and the extra care required in their construction. We begin by considering further how to deal with three different forms of moisture, i.e. as rain and snow, as moisture in the ground, and as a vapour.

4.1 KEEPING DRY

RESISTANCE TO RAIN AND SNOW

Some of the technical solutions available for covering roofs and excluding moisture were mentioned in Chapter 2. In both Chapters 2 and 3, we assumed that external walls, possibly with applied finishes, would also keep out moisture. However, where exposed to rain driven by very strong winds, even one and a half brick thick (328 mm) solid walls with 20 mm rendering (i.e. type A external finish) may still not be adequate. In any case, that is a costly solution.

In brick-producing areas of the UK, older dwellings would have been constructed with one-brick thick (215 mm) solid external walls sometimes rendered externally. The external walls of larger buildings would have been thicker except, perhaps, at the top storey. Only in the last 50 years or so have hollow or **cavity** external walls become the country-wide norm for new buildings.

These have a masonry external leaf and a masonry or a wood open framework inner leaf, with a continuous air space, i.e. a cavity, in between and extending to below ground level to prevent moisture from being carried across. An open framework will probably be sheathed with plywood or similar durable sheet material on the side facing the cavity. The leaves will be connected with metal ties and the inner leaf will bear most if not all of the loads from roof and floor structures. To maintain the isolation of the external leaf, dpcs must slope outwards over openings or where an external cavity wall becomes internal lower down. Of course, all bridging of the cavity round openings etc. must be impervious.

RESISTANCE TO MOISTURE FROM THE GROUND

As mentioned in Chapter 3, a continuous barrier to moisture from the ground is essential. This must extend through the floor next to the ground and through the external walls although it will not always be made of the same substance nor will it necessarily be in the same plane. The dpcs in the walls must be able to transmit the wall loads without deforming, and should be positioned at least 150 mm above the finished ground level.

The floor membrane to a ground-supported solid floor usually consists of substantial polyethylene sheets with lapped and sealed joints. First, the top of the hardcore filling is covered by a level bed of sand **blinding.** The polyethylene sheets are then laid on top where they will be held down by the concrete floor slab (see Figure 3.7).

One alternative is to lay the sheeting or apply a multi-layer thick coating of some impervious substance on top of the slab, where it will be held down by the screed for the floor finish. Another is to make the top membrane of two coats of mastic asphalt which can be finished level and smooth to receive a sheet covering, e.g. carpet. This solution is particularly useful when replacing a floor next to the ground in an existing building, perhaps after discovering wood rot.

Basements and other spaces within the ground will usually be kept dry by being constructed within a moisture-excluding 'tank', i.e. a continuous membrane. The technical solutions for the floor, walls, foundations for structural columns etc., including the choice of substance for this **tanking** membrane, and its location within the construction, will depend on the extent to which water will be present in the soil and the resulting hydrostatic pressure. The safest solution is to apply an impervious substance, e.g. three coats of asphalt, where it will be protected and sustained by the structural floor and the external walls.

CONDENSATION AND THERMAL BRIDGES

Water can become a vapour, i.e. a gas, at ordinary temperatures. Water vapour is always present in the air, and we add to it as we breathe and do the washing, and as we burn natural gas or other hydrocarbon fuels. Every day, each of us produces water vapour equivalent to a litre of water or more. Thus, in an occupied enclosed space, the quantity of water in the air in the form of vapour will increase with time. Also, as the temperature of the air is increased, the amount of moisture that can exist in it as a vapour also increases.

In the UK, we are mainly concerned with what happens when it is cold outside. The presence of water droplets on the inside of a window pane after a cold night is evidence that the air next to it had been cooled below its **dewpoint** temperature, and could no longer hold this moisture as a vapour. Such **condensation** may also be seen on other impervious surfaces, e.g. on ceramic tiles or gloss paint in a kitchen or bathroom, particularly on a north-facing external wall.

Problems with condensation are likely to arise where any dense substance such as glass, steel or reinforced concrete, i.e. one with a high thermal conductivity, is continuous from the outside to the inside of a building. The temperature throughout that substance will tend to be the same as that at its colder face. Thus, during prolonged periods of cold weather, the exposed internal surfaces will also be kept cold. The air next to this surface may then be cooled to below its dewpoint, the excess moisture being deposited on its surface. This can lead to mould growth and the breakdown of decorations.

A glass window pane is an extreme example of such a **thermal bridge**, often called a **cold bridge**. A concrete floor which is extended to form an external balcony is another. Thermal bridges are likely wherever the thermal resistance of a construction is lessened, perhaps at gaps in a layer of insulation. Also, the internal environment becomes more sensitive to thermal bridges as the general level of thermal resistance is increased.

Where condensation occurs on the surfaces of a porous substance, the moisture is likely to be absorbed. Water vapour will permeate most building substances and as it does so, it may be cooled below its dewpoint and condense within them. This 'interstitial' condensation can cause some substances to deteriorate. The traditional solid brick walls mentioned above seem to be able to absorb moisture from rain or from interstitial condensation which evaporates when conditions improved.

DEALING WITH CONDENSATION

Up to 50 years ago, most buildings were draughty. Where chimney stacks projected from roofs, the 'stack effect' of the reduced pressure where the wind blew over their open tops encouraged the natural upwards movement of the warmer internal air and the products of combustion from open fires.

To enable these stacks to function, they needed an equivalent volume of air from the outside which would have been drawn in through gaps round doors and windows, between skirtings and hollow ground floors, and between the units covering pitched roofs, as well as through open windows. Also, a positive wind pressure on one side of the building and the consequential reduction of pressure on the other side caused similar air movements inside the building, and air leaked in and out. This often excessive replenishment with fresh and relatively dry air kept internal conditions reasonably healthy and acceptable, except for the draughts, although these helped to evaporate any interstitial condensation.

However, in many existing buildings, central heating has been installed, chimneys have been sealed, and old windows and external doors have been replaced by highly efficient units. Many of these do not have trickle vents or small opening lights which would provide background ventilation.

Thus, unless occupiers can and do open some larger windows (which means losing some warm air), the relative humidity will increase and internal conditions will become uncomfortable and possibly polluted. In extreme cases, such a polluted environment can have an adverse affect on the health of occupiers, the phenomenon being sometimes referred to as 'sick building syndrome'. In addition, condensation is likely to occur in the roof space and within the external walls.

In a new building, or where thermal insulation is being added to an existing one, continuous and sealed **vapour control layers** should be introduced to prevent water vapour from diffusing into the constructions. These can be made of polyethylene sheet strong enough to withstand accidental perforations, and have been known as 'vapour barriers'. Perforations for pipes, cables etc. must also be sealed. As a rule, most if not all the added thermal insulation should be placed outside these layers.

Putting this another way, vapour control layers should be on the warm side of a construction. Vapour control layers will also stop the air inside being forced into any cavities in the construction by e.g. the wind outside, the action of mechanical ventilation systems and the stack effect of flues.

CONDENSATION WITHIN THE ROOF

Gaps in a traditional slate or tile covering to a pitched roof usually provided the roof space with sufficient ventilation. To restrict this, the spaces between the battens were often partially or wholly 'torched' with mortar, much of which will by now have fallen on to the ceiling below.

The impervious underfelt used in more recent constructions not only keeps out any windblown rain and snow but also acts as a vapour control layer. Being on the cold side of any loft insulation, though, in winter the ability of moisture in the roof space to remain as water vapour is reduced and condensation is liable to occur. Nowadays, a breather membrane, i.e. an underlay that is permeable by water vapour, is used under a slate or tile covering instead.

In new buildings, the humidity and the risk of condensation within pitched and flat roofs is reduced by providing continuous or intermittent insect-proof eaves vents with ducts into the roof space (see Figure 4.1). Pitched roofs may also be ventilated at their ridges.

4.2 MAINTAINING A COMFORTABLE ENVIRONMENT

PROPERTIES AFFECTING THERMAL TRANSMISSION

Under the heading of 'Energy economy and heat retention', the following is required by the CPD:

> The construction works and its heating, cooling and ventilation installations must be designed and built in such a way that the amount of energy required in use shall be low, having regard to the climatic conditions of the location and the occupants.

When considering thermal transmission, the significant properties of the substances involved are as follows.

- Thermal conductivity and resistance Heat flows through a substance when there is a difference in temperature between its opposite faces. The thicker the substance, the less the heat flow. This property is measured by its thermal conductivity (k), which is the wattage (W) transmitted between two opposite faces of a 1 metre (m) cube of the substance when their temperatures differ by one degree Celsius (or Kelvin). A substance can also be thought of as resisting the transmittance of heat. This thermal resistance (r) is the reciprocal of k and has the merit of varying directly rather than inversely with the thickness (L). It is calculated by dividing the thickness in metres by k.
- Surface resistance The surface of a substance also affects its heat transmission. Its nature, and the difference between its temperature and that of the surrounding air will determine the amount of heat absorbed or dissipated. Aluminium foil, polished silver tea pots, and other bright surfaces gain or lose the minimum of heat, and dull black ones the most, e.g. solar panels. This surface coefficient and its reciprocal, the surface resistance, is also affected by air movement, e.g. the wind chill factor.
- The total thermal resistance (R) of a complex construction such as a cavity wall is the sum of the individual thermal resistances and surface

resistances. Its reciprocal is the 'steady state' thermal transmittance, or 'U-value' used for calculating energy gains or losses in the UK. The resistance value is more likely to be specified in North America. In 1972, the maximum U-value for the roofs of new houses in the UK was 1.42 W/m^2K. This is now almost down to the Canadian standard, which is equivalent to 0.2 W/m^2K or less, depending on the locality.

Air is a very poor conductor of heat. However, where there is a difference in temperature within a body of air, the cooler, denser air will tend to fall, thus displacing the warmer, lighter air upwards. As discussed in Chapter 1, heat energy will be transferred from a so-called 'radiator' to a building space partly by such convection. The insulating property of a substance tends to be based on trapping air or some other gas in closed cells that are so small that any convection is negligible.

IMPROVING THERMAL INSULATION

The essential parts of a building considered earlier will provide some measure of thermal resistance. What has to be decided is:

1. what total thermal resistance is to be achieved; the official minimum requirements for roofs, external walls and doors and windows will each be different;
2. what thermal resistance will be contributed by the essential parts; lists of these resistances or their equivalent reciprocal U-values are available from manufacturers and can be found in official publications; and
3. how the extra resistance will be provided.

Insulation products include:

1. glass and mineral fibre quilts,
2. loose fibre fills,
3. urea-formaldehyde foams,
4. expanded polystyrene and polyurethane rigid boards and batts,
5. plasterboards with insulating or reflective backings,
6. insulating blocks.

Pitched roofs

Unless there are rooms in the roof space, insulation quilts can be laid between and over the ceiling joists (see Figures 3.4 and 4.1). Although this is the most efficient position for the insulation, unless a vapour control layer can be placed underneath, it creates a cold roof space. Walkways should be constructed over the top to provide access. Alternatively, a rigid board insulation can be secured between the rafters to create a warm roof, and this can also act as a vapour control layer.

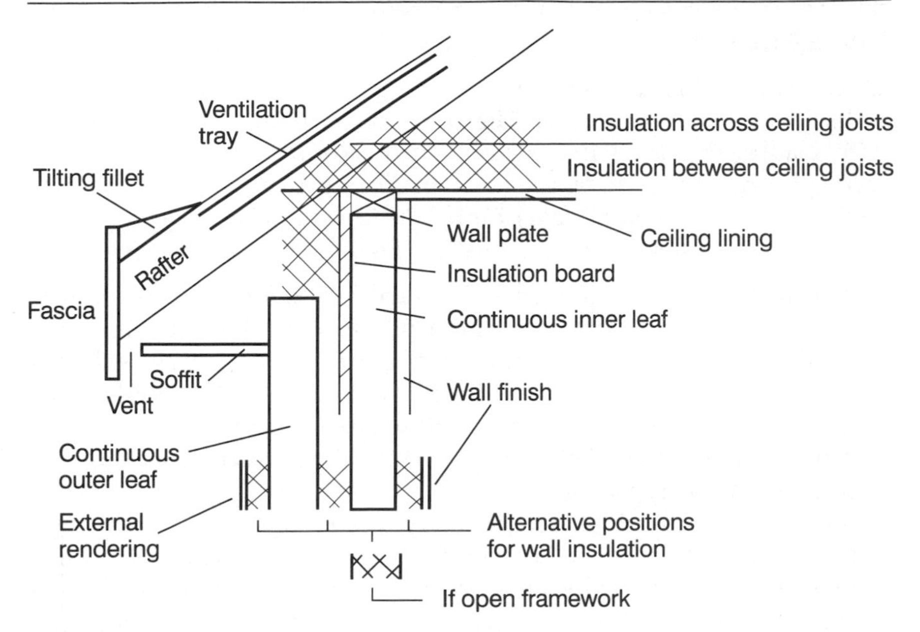

Fig. 4.1 Vertical cut through the eaves of a pitched roof supported by a cavity wall and showing alternative positions for thermal insulation.

Another type of roof covering consists of a pair of profiled metal sheets with an insulation layer in between, preferably all supported above the structural purlins. If the insulation is close fitting and impermeable, the lower sheet acts as the vapour control layer, and the arrangement will provide a 'warm' roof. If not, the upper profiled sheet becomes the vapour control layer and condensation is liable to occur on its underside.

The tops and sides of cold water feed and expansion cisterns must also be insulated, the material being linked to any ceiling insulation below. Where possible, a cistern should be exposed to heat from an uninsulated portion of the ceiling below or from a heating stack. Service, expansion and overflow warning pipes should also be insulated.

Flat roofs

Examples include continuous concrete constructions with screeded tops, wood open frameworks with boarded decks, and metal structures and decking. To create a warm deck, a vapour control layer is placed on top of the structural deck followed by the insulation layer and the weather-tight membrane. Alternatively, the insulation layer can be placed above the membrane and weighted down, i.e. ballasted by e.g. a layer of gravel or paving slabs.

External walls

Of course, the first improvement to the thermal resistance of external walls was to use cavity walls. The next was to build their inner leaves with loadbearing insulating blocks. Further improvements have been made by the inclusion of layers of insulating material, preferably all between a vapour control layer on the inside and a weatherproof layer on the outside. Alternative positions (see Figure 4.1) are as follows.

1. Incorporate with the internal wall finish. For example, a layer of insulation, a vapour control layer and a plasterboard finish can either be fixed separately or as a composite board to the structural wall. The choice may depend on whether the background is continuous masonry or a wood open framework where some or all the insulation can be fixed within the thickness of the framework. This technical solution can be applied to both solid and cavity walls.
2. (a) Partially or (b) wholly fill the cavity of a masonry cavity wall. In (a), a suitable thickness of rigid insulating board is held against the outer face of the inner leaf by retainer disks clipped to the wall ties. The space between the leaves must be wide enough to leave a normal (50 mm) thickness of cavity unobstructed. In (b), the cavity is filled with foam or mineral fibres after the wall has been constructed.
3. Secure a layer of insulation to the outside of the wall. This must be protected from the weather and can be incorporated in any one of the external finishes described in Chapter 2. Alternatively, it can be incorporated into a self-contained external cladding system or rainscreen.

Where the roofs of a building are at different levels, some walls that were internal ones at lower levels become external ones at higher levels and their attributes will change. Thermal insulation will be required, and moisture must be excluded both generally and at the junctions where the changes occur.

Floors next to the ground

Rigid board insulation strong enough to sustain the floor loads can be introduced either below or above the structural floor (see Figure 3.7). If not on the room side of the damp-proof membrane, it may need to be protected from ground contaminants. Vapour control layers may be introduced to isolate the insulation.

The thermal resistance at the edges of the floor can be improved by vertical rigid board insulation on the inside face of the foundation walls or within a cavity. Alternatively, insulating blocks can be used to build the foundation walls or a layer of insulation can be placed below the perimeter portion of the ground slab.

Windows

Nowadays, in new buildings, only windows with two layers of glazing will provide an acceptable thermal resistance. While pairs of single-glazed casements can be used, double-glazing units in single specially profiled sashes are the norm. To preserve their edge seals, double-glazed units should be secured so that drained and ventilated spaces are maintained round the edges of these seals (see Figure 1.5). To avoid thermal bridges at the head, jambs or sill of an opening for a window or door frame, the wall insulation must be maintained up to the frame.

Thermal bridging generally

Increasing environmental standards demand an increasing attention to detail in the design, supervision and construction of our buildings. In particular, insulation layers and vapour control layers must be continuous, any gaps being potential thermal bridges and entry points for moist air. Mortar joints between insulating blocks, wood open structural members with insulation in between, and other intermittent thermal bridges will reduce the effectiveness of the insulation and must be allowed for when calculating U-values. The bridging effect can be minimized by providing an additional and continuous layer of insulation, e.g. on the cavity side of an inner leaf, and over the tops of ceiling joists.

4.3 HEALTH AND SAFETY

THE HEALTH AND SAFETY OF OPERATIVES AND OCCUPIERS

Under the European Union's Construction Products Directive, 1988, designers must ensure that buildings can be constructed without endangering the construction operatives and the general public. In the UK, the Construction (Design and Management) Regulations 1994 (CDM) requires most new construction work to be notified to the Health and Safety Executive (HSE). The HSE administers many Regulations affecting the responsibilities of building owners, occupiers and construction organizations.

A health and safety plan for a project has to be created by the **planning supervisor** and developed and implemented by the **principal contractor**. On completion, a file of information that might affect the health and safety of occupants and workpeople in the future is passed to the owner, i.e. the **client**. More detail is given in Chapters 6, 7 and 11.

Every year, one in every 50 or so construction operatives will be injured and off work for more than three days. Ozone-destroying chlorofluorocarbons (CFCs), paint solvents, formaldehyde, chemicals for poisoning wood-destroying fungi and insects, and asbestos and other mineral fibres are all used in the manufacture of construction products. These may also endanger the health of operatives, and can have a continuing adverse effect on occupiers.

Occupiers may face additional risks. For example, the general level of atmospheric pollution inside a building can sometimes be higher than that outside; concentrations of tobacco smoke can make it worse. The staircases may include tapered treads. Violent contact with easily broken non-safety glazing to a door or screen can result in injury. A large uninterrupted area of transparent glazing may not be immediately obvious to someone walking past unless made visible, i.e. 'manifested' by some physical marking, e.g. etched images. Safety glass and safety plastics, security, blast resistant and bullet resistant glass, anti-bandit glass and fire-resisting glass are all available nowadays.

FIRE

Fire is a chemical reaction between oxygen in the air and vapours that are given off from certain substances when they are heated above their ignition temperature. Although this can occur spontaneously, a fire is usually started by a spark or pilot flame.

Under the heading of 'Safety in case of fire', the CPD states:

> The construction works must be designed and built in such a way that in the event of an outbreak of fire:
>
> - the load-bearing capacity of the construction can be assumed for a specific period of time,
> - the generation and spread of fire and smoke within the works are limited,
> - the spread of fire to neighbouring construction works is limited,
> - occupants can leave the works or be rescued by other means,
> - the safety of rescue teams is taken into consideration.

Fire resistance is the ability of a part of a building to continue to carry out its purpose for a specified length of time during a fire, i.e. to bear a load, to exclude a fire, to insulate against high temperatures. Substances are classified as being **combustible** if they do not withstand the test for being **non-combustible** (BS 476).

Fired clayware, concretes, plasters, metals and other inorganic substances are non-combustible as they will neither burn nor contribute heat to a fire. Most non-combustible substances are poor conductors of heat, the exceptions being steel and other metals. However, a fire may so raise the temperature of a non-combustible substance that it softens and loses strength, or even disintegrates.

Organic or partially organic combustible substances will be either **easily ignitable**, or **not easily ignitable**, and will contribute to a greater or lesser extent to the growth of a fire. This contribution will be initiated when an ignitable substance is heated to its **ignition temperature**, and will depend on the amount of heat produced, i.e. on its **mass** and its **calorific value**. Fire can be propagated by the **spread of flame** across a hot surface, the rate of spread depending on the substance. Wood may char where the oxygen present is insufficient for complete combustion, thus partially inhibiting further burning.

In addition to producing heat and flame which destroys buildings, burning is likely to be accompanied by smoke and other vapours, all of which can kill people. The chief culprits are polymers of various kinds, e.g. woods and plastics. People die in house-fires mainly because smoke from burning furnishings and furniture prevents them from breathing.

ACHIEVING APPROPRIATE FIRE SAFETY

In Chapter 1, we considered what escape routes an occupier might expect to find in case of a fire. Various strategies are available to contain and minimize the effect of a fire and to preserve those escape routes, including the following.

1. Subdividing the building into compartments separated by fire-resisting compartment floors and walls. These must provide complete barriers to fire except for openings for fire doors, pipes and ducts. The maximum permitted floor area of each storey in a compartment will depend on the purpose of the building, the height of the topmost floor, and whether there is an automatic water sprinkler system. A wall that is shared by two buildings must also be treated as a compartment wall.
2. Constructing protected shafts between compartments to enclose staircases, lifts, pipe ducts etc. and, possibly, sanitary accommodation. These shafts will have the same construction as compartment floors and walls except they may include glazed screens.
3. Sealing concealed spaces, e.g. between a suspended ceiling and the structural floor above, and subdividing them with fire resisting cavity barriers.
4. Ensuring that all parts will resist fire for a suitable period of time including using non-combustible products wherever possible.
5. Limiting the risk that fire might spread from one building to another.

MAINTAINING A SECURE ENVIRONMENT

While prisons have to be designed to keep some of the occupiers in, all buildings need to be designed and equipped to discourage unwanted visitors and to make it difficult to gain unauthorized entry. Decisions should be based on an appraisal of the crime risk, including the possible effects of a loss. Nowadays, losing computers and data can be more disastrous than losing money and goods.

Security aspects of doorways and windows were mentioned in Chapter 1. While more secure or additional locks and bolts can usually be fitted, many existing external doors will offer little resistance to physical attack. Their glazing, or that of side panels, may be easily broken and in large enough panes to enable an intruder to gain access to the door locks.

Many existing buildings will have been designed with little consideration for security. While existing windows can be provided with locks and reglazed with laminated safety and security glass, they may be out of sight from the ground or easily accessible from a lower roof. Intruder alarms can be fitted, and movement actuated security lighting installed to reveal the presence of intruders. A closed circuit television (CCTV) system may be helpful. Fire and smoke detectors and alarms will provide some defence against arson. Telephone lines to the police or a security organization should enter the building below ground level.

The boundaries of the site can be strengthened, and lockable gates provided. Unnecessary pathways and access points can be closed and the remainder and their lighting improved. Surveillance from the building should be encouraged by improving visibility and minimizing hiding places for intruders.

MAINTAINING A QUIET ENVIRONMENT

A **separating floor or wall** is intended almost to eliminate the transmission of airborne and impact sound from one space to another. Although the UK's Building Regulations are concerned only with having separating floors and walls between dwellings, desirable performance standards should be set for all constructions as they will all transmit sound to some extent.

The absorbing of airborne sound within a room is a different matter from reducing its transmittance to other rooms. The sound level within a room can be reduced by using **porous, sound-absorbent** substances for ceiling, wall and floor finishes and for furnishings.

The vibrations resulting from the 'impact' of footsteps and from repositioning furniture are best absorbed at source. It may be possible to have a soft floor covering such as carpet. Alternatively, a floating floor deck (see Figure 4.2) suited to the floor finish can be supported on a resilient layer of, e.g. mineral fibre or expanded polystyrene board (see composite finish type 2 in Chapter 1).

The noise of flushing a WC, running a shower, or emptying a bath is

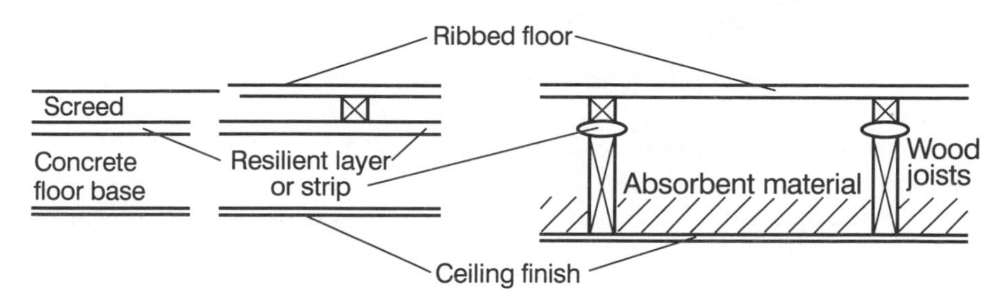

Fig. 4.2 Some of the many ways of increasing the resistance of a floor to the passage of sound.

also a kind of impact sound. It is easily transmitted by thin, lightweight, rigid plastic pipework and the effects are minimized by using thick, heavy cast-iron pipework.

The vibrations of speech, music, machinery, vehicles and other airborne sounds cause equivalent vibrations on the surrounding constructions which, in turn, set up new airborne sounds in adjoining rooms etc. Usually, all the walls and floors surrounding an internal space are continuous with one another and will extend round adjoining spaces as well. Thus, they will all carry some of the airborne sound from a noisy space to adjoining spaces.

This is called 'flanking' transmittance. It will not be lessened simply by reducing the direct sound transmittance of the walls between the noisy space and its neighbours. Air will also convey noise through gaps in construction, along the air space in a cavity wall, and in roof spaces over the tops of ceilings, as well as along direct pathways such as through open windows and doorways.

The principal agency for reducing the transmittance of unwanted airborne sound from one internal space to another in a building is the mass of the surrounding constructions, e.g. masonry, concrete. The heavier the construction, the greater its resistance to vibrating in harmony with the airborne sound energy. In all cases, **continuous** and **airtight** construction including sealing gaps next to other constructions and around pipes and cables is essential.

Open frameworks can also be effective. Joisted floors can have absorbent material filled on to the ceiling finish below and be given a floating and continuous **floor deck** as discussed above. Walls can consist of a pair of wood open frameworks with multi-layer plasterboard linings to their outer faces and a sound absorbent layer somewhere in between. Except for the minimum of fixings, these frameworks must be isolated from each other and from other constructions.

EXCLUDING INSECTS

It is not easy to exclude flying and crawling insects completely from a building, but once inside, they are difficult or even impossible to eradicate. Thus, the roof, the floor next to the ground, and the external walls should present a continuous barrier. This barrier should also be maintained in upper floors, internal walls and wherever ducts and pipelines pass through them.

4.4 GENERAL AND ECONOMIC ATTRIBUTES

The following attributes can apply to any part of a building.

BUILDABILITY (UK) OR CONSTRUCTABILITY (USA)

In *Faster Building for Industry* (1983) the authors commented that

> Design should incorporate 'buildability' – making it easy to plan procurement of materials and the moving of the various trades on and off the project, and minimising the possibilities of disruption by external factors. Simple designs using readily obtainable standard materials and components generally facilitate speedy construction.

Of course, simple design implies simple and straightforward construction processes of putting one substance on another. However, the design can also place constraints on the speed of erection and may indicate unavoidable delays.

For instance, where in situ (i.e. cast-in-place) concrete construction is being used, complementary operations of erecting and dismantling the formwork and placing reinforcement will be required. The **setting times** for concretes, mortars, plasters and paints will influence the rate of progress on the site. Their **hardening period**, during which they will be developing their strength, will determine the delay before succeeding processes can start.

On the other hand, building open frameworks and other lightweight constructions is speedy and can be helped by off-site fabrication. The size and mass of formed units will influence the plant required and the time for assembling them in position. (The factors which influence the productivity of construction operatives are discussed later.) Buildability will be encouraged if the designers plan their buildings and position the various structural and prefabricated parts to fit a modular grid of, say, 300 mm.

DURABILITY

Durable means being capable of resisting the effects of natural agencies and climatic forces occurring in a particular environment for at least the required life of the building. These agencies include water vapour and other gases, water and other liquids, particularly during the processes of freezing and thawing, sunlight, heat, cold, wind-blown particles, rodents, insects and fungi.

The variation in the volume of moisture when changing from liquid to solid and back again is unusual. Most physical objects get smaller as they get colder. However, from 4° C down to zero, water increases slightly in volume. Then, as the 'phase change' to ice takes place, its volume increases by about 9%. Warming reverses this process. Thus, when water within the connected pores of a substance freezes, considerable forces can be exerted both directly and also by being transmitted through any water that remains unfrozen. This will also happen to water in a pipe. Freezing and thawing can also bring about changes in the bearing capacity of some soils.

Except for decorations, all building substances should be durable for at least the planned life of the building. We shall now consider some of their properties which might put their life-span at risk, and how such risks can be minimized. Resistance to wear is regarded as a separate property.

Porous mineral substances may be disintegrated by the pressures built up by the freezing, thawing and movement of water in capillaries. The resistance of concretes can be improved by air-entraining. However, the voids in a porous substance may not always provide pathways through it, so although it may be **permeable**, it may still be **impervious**.

Concretes, particularly those made with too much water, are likely to shrink and crack with ageing, and such cracks will provide pathways for air and moisture. Water can **leach** the soluble calcium hydroxide from hardened cement paste, leaving it weakened. Carbon dioxide from the air may also react with this hydroxide, and the resulting 'carbonation' of concrete may lead to the rusting of reinforcing steel, its consequent expansion in volume, the bursting of the concrete cover and the consequent loss of bond strength. Old concrete fencing posts often demonstrate this. Check the surfaces of all exposed concrete, e.g. lintels, for this defect.

The rainwater absorbed into the pores of stone or brickwork may contain dissolved airborne chemicals which can react adversely with the substance and cause its surfaces to disintegrate, e.g. the effects of nearby industries on the historic buildings of Athens and Venice. Metals can also be affected by such chemical solutions, the resulting **corrosion** reducing their effective thicknesses. Moisture may also promote destructive electrolytic action between the unprotected surfaces of dissimilar metals and alloys, e.g. in water-pipes, and where rainwater flows from a copper to a zinc surface. Lead may be affected by the run off from lichen-covered surfaces.

Oxygen from the air will form corroding oxides on the exposed surfaces of steels. (It will also assist the setting of protective alkyd paint films, encourage the formation of protective films on the exposed surfaces of aluminium, copper, lead and zinc, and oxidize and harden the surfaces of the bituminous coatings of mineral aggregates used in road surfacings.)

Some polymers, including woods, may **degrade** by weathering, that is, by the action of rain, heat, oxygen and solar radiation. Woods are also porous, and will absorb moisture through their surfaces. The **moisture content** of a wood is its water content expressed as a percentage of its dry weight. Woods in a centrally heated building will have moisture contents of between 12% and 16%.

The spores from the fruiting bodies of **wood fungi** will germinate even in the dark on the surfaces of woods whose moisture content exceeds 20%. Such fungi will decompose parts of the structure of the wood, producing water (which makes the wood even wetter) and food for the fungus. Some species of fungi will extend their cobweb-like hyphae through the building structure in search of more food sources and will seek to reproduce themselves by releasing spores from fruiting bodies.

Woods are also a **food source** for certain insects during parts of their life cycles. Termites are a type of ant that eats wood, but needs a warmer climate than that of the UK. Here, furniture beetles are the chief culprit although the House Longhorn Beetle is active in some areas on the west side of London. Adult beetles lay their eggs in pores and crevices in wood surfaces, and the hatched larvae bore into the wood searching for the sugars it contains, finally emerging from 'flight holes' to repeat the cycle after the damage has been done.

Threats to the durability of timbers can be resisted by penetrating preservative, or by applied surface treatments, e.g. paints, stains, although these are not durable and have to be renewed at intervals. Some birds and rodents will damage soft substances, e.g. softwoods, fabrics, some plastics.

RESISTANCE TO WEAR

Except for soft, sound absorbing floor finishes, substances used for finishes should be **hard** and **tough**; that is, they should be able to withstand the impact forces and abrasion likely to be caused by the users of the building. They should also be unaffected by contact with water and other liquids used for cleaning. Floor surfaces should retain their **non-slip** properties as they become worn.

Substances used in the moving parts of hinges, locks, door handles, water taps, circulation pumps and other mechanisms should create the minimum of **friction**, so that they can continue to function for the whole of their planned life. Even so, hinges, locks, bolts, door handles etc. should be lubricated from time to time.

EASE OF MAINTENANCE AND RENEWAL

It should be possible for the surfaces of finishes to be **cleaned without damage**. The bond between a finish and its base should be adequate, but capable of being broken if necessary so that the finish can be replaced. Applied decorations should, themselves, be capable of being redecorated.

USING PRODUCTS FROM SUSTAINABLE AND READILY AVAILABLE SOURCES

Some construction products are made from minerals that are in plentiful supply, although not naturally renewable, e.g. bricks from clay, cement from clay and limestone. Clearly, supplies of some raw materials are limited, e.g. oil for plastics, lead ores, stone for road surfaces. Other products will be made from organic materials whose supply will be naturally sustainable provided it is properly managed, e.g. wood from forest trees. Suppliers may be able to certify that their products come from legal sources.

RESISTING THE EFFECT OF CHEMICALS

Resistance to chemicals is largely a property of **impermeable** substances such as laboratory glassware, lead and plastic pipes, and vitrified clay drain pipes and floor tiles, which exclude them altogether. **Permeable** substances, including those with micro-cracks created during manufacture, will offer greater surface areas for chemical action. However, they may still not allow the chemicals to pass through.

Concrete with low permeability and high resistance to chemicals can be produced by adding the minimum proportion of a suitable cement to durable aggregates. Only enough water for a workable mix should be used, and the mix should be thoroughly compacted and kept moist until cured.

When such a mix is difficult to compact, adding more cement and water (not just more water), will increase its workability. Adding pulverized fuel ash or some other potentially cementitious substance (or ***pozzolana***, named after a town not far from Mt Vesuvius) will also help. The cement used in concrete within the ground can be selected for its ability to **resist sulphate attack**.

ENERGY EFFICIENT

Large quantities of energy will be consumed in the manufacture of some products, e.g. steel, aluminium. Others may not be so demanding. Environmental pollutants may be produced as a side effect of both energy generation and manufacturing processes. Having to renew parts of a building adds to its consumption of energy. Thus, both locally and elsewhere in the world, and both initially and later, others may pay a price for the benefits enjoyed by the occupiers of a building.

4.5 OBSERVING DEFECTS

All building substances expand or contract with changes in temperature, and some also respond to changes in moisture content. This is not always important, but can lead to defects, as can failure in some of the other attributes.

You may be able to see some defects from your chair. For instance, the decorations may no longer have an acceptable appearance, or the finishes or the structure may be cracked. Look for cracking particularly above and below window openings. Cracking is often due to shrinkage, e.g. of concrete units, particularly where the cracks are midway between restraints, e.g. between buttressing walls. There may be gaps below the skirting where wood joists have shrunk.

On looking closely at wood flooring, the small 'flight holes' of mature furniture beetles may be seen. Rectangular cracking and hollowing of the surface of, say, a wood skirting to a solid ground floor, particularly where against an external wall, may indicate the dreaded 'dry' rot. Sound timbers will resist attempts to insert a sharp pointed object but, where possible, do this test where it won't show! The American Institute of Architects (AIA) in its *Handbook of Professional Practice* reports that, in the USA, termites strike five times as many homes as fires, and do more economic damage than all tornadoes, hurricanes and wind storms combined.

Internally, whitish crystalline deposits on wall surfaces may indicate the limits reached by rising damp. Other water staining, mould growth, or shadowing on internal finishes may indicate the effects of thermal bridges. It may also indicate the breakdown of the moisture exclusion system. Externally, this breakdown may show at ground level, at various roof junctions, and at the rainwater gutters and pipes.

Many defects show themselves by irregularities. For example:

- some roof slates out of line or lying in the gutter will indicate a failure of the slate nails, and the need to arrange to strip and re-slate the roof;

- a sagging ridge line will suggest that the roof is flattening and the supporting walls may no longer be vertical;
- breaks in a brick arch over a window will indicate a movement in the supporting walls;
- a door not fitting its frame or a frame with a sloping head will point to the uneven settlement of the walls;
- if the corner of a wall does not line up with the corners of other walls, either some force must be pushing it over or its foundations are failing to sustain it.

Cracks in enclosing walls etc. may give the impression that the building is partly or totally unsafe, although the cracks may be stable and not serious. On the other hand, they may be the result of the building not having withstood abnormal environmental conditions such as:

- hurricane force winds,
- earthquakes,
- changes in the ground water-table,
- differential soil shrinkage during hot dry summers, particularly where trees were demanding moisture,
- an increase in heavy road traffic, and
- mining subsidence.

Expect to find defects in parapet walls, particularly where their tops are not protected by impervious and continuous copings, and in gutters on the tops of external walls or between the roof slopes.

4.6 COMMENTARY

In spite of the efforts of those involved with producing a new or refurbished building that is fit for its purpose, its owner/occupier may misuse or neglect it, for example, by:

- not keeping it ventilated,
- not keeping the heating system on in winter,
- not renewing decorations, particularly external ones, before surfaces begin to deteriorate,
- poorly maintaining the weathershield, e.g. not repointing walls, not clearing leaves from gutters, not renewing mastic sealing round window frames,
- not replacing decayed, broken or worn parts, or parts with a limited life, e.g. a built-up felt roof covering,
- not inspecting and maintaining gas fired or other boilers,
- not cleaning flues,
- not re-washering leaking taps,
- providing opportunities for vandalism or for thieves to enter,
- heaping soil against the damp-proof courses.

4.7 SUMMARY

The chapter has been concerned with improvements that have been made to environmental standards in recent years and some of the problems that have occurred. The attributes considered were those of being weathertight, condensation-free, energy-efficient, fire-resistant, secure, sound-resistant, insect-free, buildable, safe to build, safe to occupy, durable, wear-resistant, easily maintained and renewed, chemicals-resistant, and built with products from sustainable sources. They are summarized later in Figure 7.1. For each, the physical phenomena and the natural agencies involved were outlined, terms were introduced, and suitable technical solutions were described. The chapter concluded with a consideration of common defects.

REFERENCES

American Institute of Architects. *Handbook of Professional Practice*, AIA, Washington.

Building EDC (1983) *Faster Building for Industry*, HMSO, London.

European Union (1989) Council Directive 89/106/EEC of 21 December 1988 on the approximation of the laws, regulations and administrative provisions of the Member States relating to construction products, *OJ L* **40**, 11.2.89, p. 12.

FURTHER READING

British Standards Institution. *BS. 8220 Security of Buildings against Crime*, BSI, London.

The Building Regulations Approved Documents, HMSO, London.

Building Research Station (1991) *Thermal Insulation – Avoiding Risks*, HMSO, London.

Cook, Geoffrey K. and Hinks, A. J. (1992) *Appraising Building Defects*, Longman, Harlow.

Everitt, A. (1994) *Materials*, Longman Scientific and Technical, Harlow.

Illston, J. M. (ed.) (1994) *Construction Materials, Their Nature and Behaviour*, E & FN Spon, London.

Johnson, Stuart *et al.* (1993) *Greener Buildings, the Environmental Impact of Property*, Macmillan Press Ltd, Basingstoke.

Specification, Emap Business Publications, London.

Stollard, Paul and Abrahams, John (1991) *Fire from First Principles*, E & FN Spon, London.

The publications of the Health and Safety Executive.

Construction products 5

To provide ourselves with shelter, humankind turns to whatever raw materials are available, although, from the earliest times, these always received some preliminary processing on the site. The Israelites in ancient Egypt made mud blocks from the soil in the vicinity, and villagers in Africa do the same today. The stones for medieval cathedrals usually came from the nearest suitable quarry from which it was possible to transport them, but most of them were shaped on the site. Settlers in North America built their cabins with roughly squared and end jointed logs cut when clearing their fields.

Over the years, as transportation and communications improved and industries developed, this processing has been transferred to factories capable of supplying many widely dispersed construction sites with **construction products**. Specialist producers are able to achieve higher standards with greater economy. They can also develop substitute substances and new products designed for speedy assembly. Thus, in industrialized countries, but not necessarily elsewhere, the construction process is now almost solely one of purchasing construction products and either (a) mixing and placing them or (b) preparing and assembling them.

5.1 CATEGORIES OF SUBSTANCES

This section repeats and extends earlier comments on the substances of products, i.e. the physical matter in which certain qualities or properties exist. A property of a substance is a physical characteristic which can be independently studied and possibly measured or analysed, perhaps in a laboratory. Almost all tests involve the destruction of the objects being tested, the exceptions being certain **non-destructive tests** such as the stress-grading of timbers, which measures their strength. Thus, hardly any of the substances in buildings will have actually been tested, even for their fire resistance. Instead, representative samples will have been selected and tested to give confidence that the substances have the properties claimed for them.

Building substances can be assigned to one or other of two fundamentally different categories, which we shall call A and B, depending on whether they acquired their properties after the construction process or beforehand. Note: A = after, and B = before.

CATEGORY A SUBSTANCES

Category A substances are those whose important properties develop as a result of chemical changes to some of their constituents after they have been incorporated into the building. Examples include concretes, mortars, plasters, bitumens, asphalts, paints and other liquid treatments, adhesives.

Obviously, the properties of such substances cannot be tested directly without damaging the building. However, their simpler constituent substances, e.g. cements, limes, plasters, water, sands, paint pigments, mediums or thinners, can be tested. Also, it is possible to make test samples of the substances, e.g. concrete test cubes, at the time the work is being done.

Note that the same name may be used for both the substance in its final state in a building and the building products with which it is made. For example, 'paint' may refer to the surface treatment of a door or to a liquid in a container; concrete may refer to the substance of a foundation or the wet-mix in a truck.

CATEGORY B SUBSTANCES

These substances acquired their important properties before they were incorporated into the building. Examples include fired clay (or 'ceramic'), precast concretes with a variety of aggregates, natural stones, natural woods and fibres, wood-based composites, metals, plastics, glass. Their manufacturers can be expected to have already tested them by sampling.

5.2 AN ANALYSIS OF PRODUCTS

An immense number of different kinds of products are available in any industrialized society, and any one merchant may stock between, perhaps, 5000 and 20 000 items. A glance through a builders' merchant's catalogue or a visit to their yard will give some indication of the ranges of products that are manufactured in the expectation that they can be sold. Other products will be designed and made especially for a project. In countries where domestic production is limited, the range will be smaller, and products may be imported especially for a particular project.

Instead of discussing each kind of product individually, it will be more effective if we can identify groups, or **classes**, of products that have significant properties in common and which, for certain purposes, can be regarded as being much the same. The classes must be such that any one kind of product clearly belongs to one class only, and that, together, the system of classes must provide for all kinds of products. That is, the classes must be **mutually exclusive** and **collectively exhaustive**. Our purpose will be to rationalize (a) this analysis of construction products, (b) our study of their processing on site and (c) the calculation of their quantities.

FORMLESS AND FORMED PRODUCTS

Liquid or granular products are supplied in cans, bottles, bags or some other type of container. Some of them will be labelled as being potentially hazardous to users. They are made by the **process industry** and, being measured by weight or volume, can be called **dimensional products**. The volume of a product is usually measured by the volume of its container. The construction industry tends to class them as **formless products** and, after site processing and chemical transformations, they harden and become what we are calling category A building substances.

The size, shape and rigidity of almost all the other products will already have been established during their manufacture. As their substances will be unchanged by being incorporated into a building, they belong to category B. They are made by the **manufacturing industry.** Because they can be picked up, handled and counted, they can be classed as **integral products**, although the industry tends to call them **formed products**. We have been referring to individual ones as 'units'. Thus, our two primary classes of products will be based on either their lack of form or their **form**, and will reflect the substance categories A and B discussed above.

Many products are homogeneous; that is, they have the same substance throughout, e.g. cement, paint, lightweight blocks, tiles, and each type can be used in the construction of more than one kind of part. For example, different thicknesses of walls can all be built with the same sized bricks; ordinary Portland cement can be used in mortars, renderings and concretes.

In contrast, some formed products are complex and, when fixed, will become parts with particular functions. Some will be parts of the building enclosure, e.g. doors, windows. Others will be associated with services, e.g. washbasins, taps, boilers. In such cases, the names of the products will be the same as those of the parts they will become when fixed in position. Following the international SfB classification system (see Chapter 6), we shall refer to these products as **components**. In everyday speech, simple 'units' may also be referred to as components, as they will all contribute to the composition of a whole.

Many standard-sized components are selected from the wide choice available, and are assembled on site. Components that belong together, as in the case of window frames with their casements or sashes, are likely to have been assembled in the factory together with suitable ironmongery.

MAIN AND DEPENDENT PRODUCTS SETS

Usually, but not invariably, more than one type of product will be used in the construction of each part of an element at a particular location. But only one, e.g. **bricks** for a wall, **plasterboard** to line a ceiling, will have been chosen to give the part its desired attributes. We have been calling this the **main product**, although some have called it the **dominant product.** Obviously, when making a design decision, this product has to be chosen first. In fact, this is the only real decision that has to be made. However, many main products have technical limitations, and depending on what these are, so other products must be selected, as follows.

- Where the continuity of the part cannot always be maintained by using the chosen formed main product. For example, straight pipes will be used to construct the main runs of drainage pipelines, but bends, junctions, channels and other special products will be required to enable them to fit together at changes of direction and at connections between pipelines. Not surprisingly, these are called fittings, but they must not be confused with washbasins, lighting fittings etc. (see Chapter 1). Over 50 different fittings may be available to suit the various technical problems encountered with just one diameter of a particular type of straight drain pipe and, of course, this range will not apply to any other size of pipe, or to pipes of different substances.
- The main product may be inadequate for maintaining the continuity of a particular attribute, e.g. the strength, stability and weathertightness of a masonry wall over and around a window. In such cases, special products, e.g. lintels, cavity trays, mastic pointing, cavity closure blocks, have to be introduced. Some of these may also be called fittings, e.g. wedge-shaped arch facing bricks.
- Formed products have to be joined to each other, or fixed to different ones, and this requires a range of jointing and fixing materials. Some of these materials will be formless, e.g. mortars, adhesives, while others will be formed units, e.g. nails, ties, bolts, cramps. Jointing and fixing units made for specific kinds of formed products, e.g. gutter brackets and bolts, hook bolts for sheet coverings, are sometimes called accessories.

As a range of such additional, or **dependent** products will be associated with just the one kind of **main product**, and not with others, they will together constitute an exclusive set, and similar sets will form an exclusive class. This chapter is mainly taken up with examples of classes of such sets which are introduced below. Each kind of construction process, and the operatives who carry them out, will be concerned with one or just a few of these sets of products. However, some products are used more generally, e.g. cement, water, screws and nails.

5.3 FORMLESS PRODUCT CLASSES

Formless materials are almost always used in combinations, where they react with each other to form new category A substances. Plain or 'mass' concrete is just concrete on its own. Concrete that is reinforced with steel bars is considered later.

(A1) materials for concretes and mortars

Main	Dependent
Wet mixes of constituents for concretes and mortars	Various Portland and other cements, limes, fine and coarse aggregates of various grades and substances, water, admixtures for air-entraining and other purposes, steel fabric wrapping for steel sections, pulverized fuel ash (PFA), expansion jointing

(A2) materials for thick coatings of specified thicknesses (i.e. plaster type) such as mortar floor screeds and wall renderings, granolithic paving, and coatings of plaster, asphalt and laterite soil, i.e. 'mud'.

Main	Dependent
Asphalts	Felt underlay, expanded metal
Mortar, plaster, and rendering wet-mixes	Cements and limes, plastering and rendering sands, various plasters, angle and edge beads, plaster cove, expanded metal lathing, water
Granolithic and terrazo wet mixes	Granite and marble chippings, ordinary and coloured cements, brass and plastic dividing strips, water
Artex	Surface sealers, jointing tape
Asphalts, macadams, tar surfacings	Roadstone, aggregates, stone chippings, road studs, weedkiller
Reed, straw and heather thatch	Hazel rods, iron hooks, wood battens

(A3) materials for thin coatings of implied but unspecified thicknesses (i.e. paint type) for decorating or preserving surfaces.

Main	Dependent
Paints, stains, wood treatments waterproofers, flame retardants, surface hardeners	Sealers, thinners, fillers

(A4) materials for fillings of variable depth (i.e. gravel type) must be strong, irregular in shape and size, but capable of being spread, compacted and roughly levelled.

Main	Dependent
Broken bricks, quarry waste or other stone	Sand for blinding
Various grades of aggregate	----

5.4 FORMED PRODUCT CLASSES

(B) FORMED PRODUCTS AND FIXINGS GENERALLY

Some of the properties of the substances of formed products have already been discussed. We continue with some general consideration of the shape and size of the units.

- **Shape** Almost all formed units are three-dimensional objects with flat sides at right-angles to each other; that is, they are rectangular, or **orthogonal**, in shape. This enables them to be fitted together easily so as to cover a surface, or to form a wall etc. The versatility of bricks is due to the close proportions of 2:3:6 between their dimensions.

  Units that are intended for finishes, e.g. floor tiles, are likely to have one dimension that is very much smaller than the other two. This **minor** dimension can be thick enough for the unit to be rigid, or so thin, e.g. carpet, that the unit can be rolled. It will provide the thickness of the construction while the two **major** dimensions will contribute to the extent of the surface area.

  Timbers, pipes, structural steel and other **sections** have two minor dimensions and one major one which will provide the construction with its 'extension'. These dimensional differences provide the basis of the classification system for formed materials.
- **Size** The actual size to which a formed material is manufactured is called its work size. It is quite impossible to make anything absolutely accurately, and it is normal for the maximum acceptable deviations, or 'tolerances' over or under the standard size to be included in its specification (see BS 2028). Normally, work sizes are used for the dimensioning

of drawings: for example, the thickness of a one-brick thick wall should be shown as 215 mm, not 225 mm.

FORMED PRODUCTS WITH TWO MAJOR DIMENSIONS

(B1) Small, uniform-sized rigid formed units, their fittings and fixings (i.e. tile type).

Main	Dependent
Brick and concrete paving units, stone setts	Mortar, sand
Wood flooring blocks	Adhesive
Floor and skirting tiles	Skirting angle tiles, adhesive, grout
Wall tiles	Rounded edge and angle bead fittings, adhesive, grout
Slates, shingles	Ridge tiles, lead flashings and gutters, wood battens and tilting fillets, nails, clips
Roofing tiles	Ridge, hip, valley and verge tiles, vents, lead flashings and gutters, wood battens and tilting fillets, nails, clips

(B2) Large units and their fixings used in combination (i.e. trussed rafter type).

Main	Dependent
Precast wall panels and floor beams	Mortar, concrete, reinforcement
Precast concrete flags	Mortar
Dressed stone walling units	Mortar, dowels, cramps
Trussed rafters	Nails, anchors, clips
Prefabricated wood panels	Nails, bolts
Steel or precast concrete frame members	Steel brackets, cleats, bolts
Precast piles	Driving heads and shoes

(B3) Rigid flat slabs, boards and sheets and their fixings (i.e. plywood type). Many types of manufactured boards and sheets are supplied in the old imperial 8 ft × 4 ft size (2440 × 1220 mm), which is as much as one man can handle, particularly in a high wind. Such sheets may be referred to as 'panel products'. Wood boards for floors and other load-bearing surfaces may have their long edges tongued or grooved. Wood matchboarding and flooring are manufactured to a variety of widths and thicknesses, and supplied in a variety of standard lengths, e.g. in the UK, up to 4.8 m in 300 mm intervals.

Main	Dependent
Plywood, fibreboards, hardboards, blockboards, chipboards, melamine faced boards, strandboards, insulating boards	Various kinds of nails, screws, adhesives
Wood flooring and matching	Nails
Plasterboards	Nails, scrims, jointing tapes, adhesives
Glass and polycarbonate sheets	Putties, glazing strips, mastics
Multiwall polycarbonate sheets	Glazing bars, verge and eaves pieces, flashing strips

(B4) Profiled sheets, their fittings and accessories (i.e. corrugated roofing type). Each type of profiled sheet is likely to be supplied in more than one length, but the same width. Its corrugated, or wave-like, transverse section increases its effective thickness, or depth, and thereby its strength.

Main	Dependent
Corrugated galvanized steel, rigid PVC, glass reinforced plastic, and coated steel sheets	Verge, ridge and other junctions, bolts, hook bolts and nuts, screws and washers, self-adhesive flashing strips

(B5) Flexible sheets, quilts and mesh (i.e. felt type)

Main	Dependent
Damp proof membranes	----
Building papers, vapour control and breather membranes	Galvanized clout nails, adhesives
Roofing felts	Bitumen, clout nails, adhesives
Flexible damp proof courses	----
Insulation sheets and quilts	Insulation retainers, clout nails
Wallpaper	Paste
Carpets	Underlay, edging and door strips, carpet tacks
Fencing	Posts, struts, strainers, wire, staples, bolts

(B6) Malleable sheets and strips (i.e., sheet lead type)

Main	Dependent
Lead, aluminium, copper and zinc sheets	Underlay, screws and washers, solder

FORMED PRODUCTS WITH ONE MAJOR DIMENSION

(B7) Solid sections and fixings (i.e. timbers type). Round sections are indicated by giving their diameters. Electric cables are usually described by giving the cross-sectional area of the conductors.

Main	Dependent
Sawn and surfaced timbers of various sizes and species (in North America, when not over 114 mm thick, timber is called 'dimension lumber')	Nails, screws, straps, shoes, struts, hangers, connectors, clips, anchors, roof vents, preservative
Skirtings and other joinery sections	Nails, screws
Structural steel sections	Cleats, brackets, bolts with nuts and washers, holding-down bolts, rivets
Electric cables	Clips, earthing clamps, conduits and fittings
Gutters	Fittings, brackets
Precast kerbs, channels, edgings	Quadrants, concrete, mortar

(B8) Hollow sections, their fittings and fixings (i.e. pipe type). Hollow sections may be called either tubes or pipes, depending on their type. Tubes are usually longer than pipes and made of metal. Although drain pipes of various substances are specified by giving their nominal internal bores, other hollow sections are described by giving their outside dimensions or diameters.

Main	Dependent
Various round steel tubes	Screwed, welded and flanged fittings, brackets, hangers, pipe insulation
Service, soil, waste and ventilating pipes of various substances	Fittings and brackets, taps, valves, ballvalves, traps, pipe insulation
Plastic and clayware drain pipes	Fittings, chambers, gullies, joint rings
Flue pipes	Fittings, brackets
Rainwater pipes	Fittings, brackets
Steel round, square and rectangular hollow structural sections	----

5.5 OTHER PRODUCT CLASSES

(AB) MATERIALS FOR COMPOSITE CONSTRUCTIONS

Composite constructions are made from a mixture of formless and formed materials.

(AB1) Reinforced concrete See (A1).

Main	Dependent
Concrete wet mixes, steel wire, bar (rod) and steel mesh fabric reinforcement	Cements, aggregates, admixtures, formwork, supports, releasing and retarding liquids, tying wire, spacers, chairs, prestressing anchorages

(AB2) Small structural units in mortar and associated products (i.e. block type)

Main	Dependent
Common, facing and engineering bricks, various kinds of blocks, ashlar and rubble walling stones, mortars	Masonry sands, cements, limes, wall ties, brickwork reinforcement, preformed cavity trays and soakers, lintels, sills, chimney pots, frame cramps, air vents, arch formers, damp-proof courses, wall profiles and fixings, flue linings, bends and terminals, stone quoins and dressings

(C) COMPONENTS AND FURNISHINGS

(C1) Components and their fixings. This class of products includes the fittings associated with services and installations. When fixed, they become parts with their own functions.

Main	Dependent
Doors, door frames and linings, windows	Frame cramps, dowels, hinges, sills, thresholds, draught excluders
Locks, bolts and other hardware	Screws
Staircases, stair treads, and handrail	Brackets and screws
Roof windows and lights	Flashings
WC suites, wash basins, baths, sinks, shower and other sanitary fittings, waste disposal units	Brackets, screws, toilet roll holders, grab rails, shower enclosures and trays
Cisterns, cylinders, boilers, pumps, valves, thermostats	Jointing materials, insulating jackets and lagging sets
Radiators, towel rails, unit heaters	Brackets, stays
Sealed glazing units	Spacer blocks, mastic, beads
Manhole covers and frames, step irons	Mortar, grease
Electrical switches, plug sockets, heaters, distribution units, switch gear, smoke detectors alarms, entry control units, lighting columns	Boxes, lamps and lamp holders, luminaires

(C2) Furnishings

Main	Dependent
Chairs and other furniture, cupboards, direction signs, blinds, loose floor coverings	-----

(D) SOILS, ROCKS, LANDSCAPING

(D1) Soils and rocks

Main	Dependent
Vegetable soil	----
Subsoils, rocks	Termiticides
Graded fillings	----

(D2) Landscaping

Main	Dependent
Turf, grass seed	Metal edgings
Trees, shrubs and other plants	Stakes, ties, guards, tree shelters
Seats and other furniture	Restraint straps, concrete

5.6 COMMENTARY

On a construction site, or when looking at an actual building, any particular individual product, say a door, or a length of skirting, can be identified and discussed. But before that time, whether we are making design decisions, estimating, buying or ordering, we can only deal with it as a member of a group, or **class** with common characteristics.

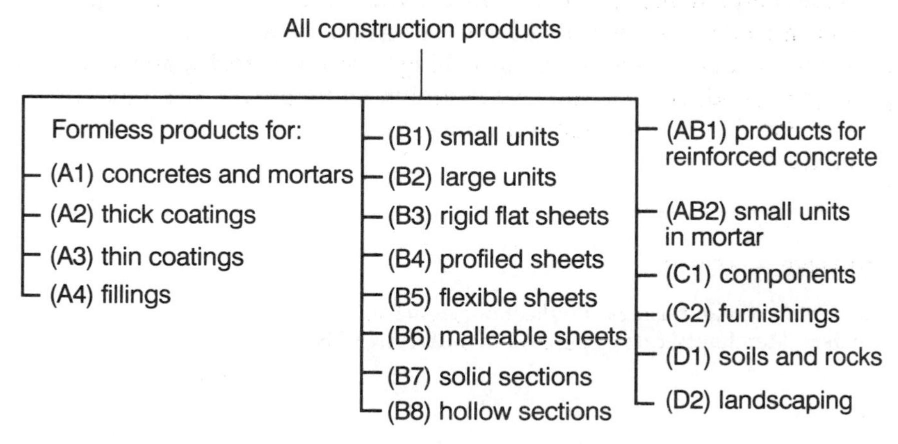

Fig. 5.1 An array showing a classification system for main construction products.

Such classes are concepts, or ideas, not actualities, and the classes of products we have been considering are also just ideas; that is, they are powerful generalizations which exist only in our minds. For convenience, and to enable us to distinguish them from the others, we give them names, or **labels**. As other people may use the same labels to mean different things, we should use them with caution when speaking or writing.

5.7 SUMMARY

The properties of the physical matter, i.e. the substances found in buildings are either:

A created by the chemical transformations of formless building products after they have been placed in position, or
B created naturally or during their manufacturing process prior to being delivered to the site.

The qualities, or **attributes** of a part of a building come from its main, or dominant product. Such products can be categorized as being **independent** as they are chosen to meet certain design criteria. Each kind of main product will have certain **dependent products** associated with it. These are:

1. the formless constituents of formless main products, and
2. those products manufactured specifically for use where their formed main product is technically inadequate.

This chapter has sought to reduce the variety of construction products to 15 classes. Fourteen are used for sets of main and dependent products used in the in-place construction of parts of buildings of whatever sizes and shapes are required by the design. The 15th class is for **components**, i.e. those formed products that are, in effect, prefabricated parts which only need to be secured in position.
Three further classes cover furnishings, soils and rocks, and landscaping products. The classes are shown on Figure 5.1. Examples of sets of main and dependent products for each of the classes of main products were given.

FURTHER READING

Specification, Emap Business Publications, London.
Builders' Merchants' catalogues and manufacturers' literature.

PART TWO
Design

INTRODUCTION

Most readers will earn their livings by working with drawings, words, numbers and other signs that represent intentions or facts about a building. That is, they will work with data about the building, and not on the building itself. The word **data** is commonly treated as singular even though it is the plural of **datum**.

Of course, this data must be communicated in such a way that it can be received through either our eyes or our ears, and interpreted in our minds. This part is concerned with:

1. the decision-making process of transforming the needs of an owner to instructions on what to do on the site,
2. some of the knowledge used by everyone involved, and
3. the ways in which such instructions are recorded and communicated.

Because these instructions will be carried out by operatives working on the site, they can be called **operative information**. Part Three deals with generating management information from these instructions so that the correct products and operatives will be directed to the site as and when required.

CONTENTS

1. Chapter 6 **Project initiation** takes the owners' point of view. It describes the analysis of their needs and the possible constraints, available options and associated risks that must be investigated before the real situation is revealed. This information will constitute the project brief and will lead to a detailed statement of what the owners will expect the project to provide, called the design brief. The specialist professionals who work in the industry and the services they offer in providing solutions to the owners' requirements are introduced.

2. Chapter 7 **Scheme design** begins by introducing the idea of building elements whose purposes are generally the same, although their technical details will be different in different buildings. The information likely to be required at this stage is then discussed, and the SfB classification system for building documentation introduced. The creative decision-making process of planning the spaces and the building layout is described, and the fresh information which should be generated is outlined.
3. Chapter 8 **Design efficiency** discusses the effects of size and shape on the relative amounts of the enclosing elements. It describes the measurement of ratios between the amounts of these elements and the floor area required by the owner, which enable comparisons to be made between alternative designs. These measurements are extended to show how the rate of heat loss from the proposed building can be calculated.
4. Chapter 9 **Detail design** considers the procedure of deciding how to construct the various elements so that they will perform appropriately. Over the years, practitioners have evolved many satisfactory ways of constructing the various elements and dealing with associated problems. These will be found in the Building Regulations, published Standards, text and reference books, and products manufacturers' literature. They are also studied in building construction courses. After some general matters, the factors that affect the detail design of each of the parts of the elements featured in Chapters 1–4 are considered.
5. Chapter 10 **Project communications** describes the various types of drawings, schedules and specifications used by designers to arrive at and record their decisions. Information systems for co-ordinating this data are introduced.

Project initiation 6

6.1 THE OWNER

A project will be triggered by some ideas concerning the need for a particular kind of building or building works, and by the decision to investigate the various possibilities. Although no site may yet be owned, it will be convenient to speak of the initiator of the project as the **owner**. This is also the term used in the USA. In the UK, members of the design team (see later) refer to the owners as their **client**. In a UK contract for construction work (see Chapter 17), they are called the **employer**.

Managers of building contracting and subcontracting organizations may also speak of 'the client', but this is misleading. The designers' client is the constructor's **customer**. The constructor's duty is to supply a product of the required quantity and quality, at the time, and for the price stipulated in their contract.

The owner may be an individual. More likely, the owner will be a body of persons legally able to act together as if they were one individual, i.e. a corporation. Particularly in the earlier stages, the owner is likely to be represented by a **commissioning team** of individuals, each with a different interest in the project. For example, if more factory space is required, the team might include the works manager, the chief accountant and the company secretary.

An organization may arrange for the physical assets that will facilitate its objectives, e.g. properties, building services, computer networks, to be managed as a whole. This function, called facilities management (FM), may include initiating new projects, and acting for the owner in setting up and leading the commissioning team. It can be provided either by an independent organization or 'in-house'.

Of course, a project may not be for the owner's own use. A private company may seek to make money by providing in advance for the needs of others, e.g. by developing estates of houses for sale, or by constructing a block of offices. A public body may have a duty to build roads, schools, social housing etc. A self-motivated individual may recognize or create an opportunity and undertake a project while accepting its financial and other risks. John Portman, who began by practising as an architect in Atlanta, Georgia, is a modern example of such an **entrepreneur**.

ANALYSING THE PROJECT

Think before you act.

The owner should first prepare what some call a **project brief**. This should include an assessment of the purpose of the project, its benefits, its risks, any legal, physical and financial constraints, and the alternative ways of bringing it to fruition, i.e. of **procuring** (delivering) it. In other words, it should answer the following questions:

1. **What** is required? The background to the owner's requirements should be described, e.g. an increase or decrease in business activity, imminent changes in manufacturing technology, increasing membership of a social club. The financial benefits of the various alternatives should be quantified, e.g. not having to pay rent for accommodation, increased production, increased efficiency. The spatial requirements could be expressed by giving the useful floor areas of the various rooms, or the number of units of a particular kind of accommodation, e.g. motel rooms, two-person flats. In the UK, these are known as **functional units**. There may be some essential features, e.g. access and parking for vehicles, not more than one storey.
2. **Where** should it be, e.g. near a motorway junction, near a bus service? Legal, practical and economic questions concerning a possible site or an existing building are listed below. A site may already be owned.
3. **When** is it wanted? The answer to this critical question may point to acquiring an existing building or to purchasing and erecting a sectional transportable building. It will certainly indicate the type of contractual arrangements that would be appropriate for a new or refurbished building (see section on 'Building procurement').
4. **How** might it be achieved and **how much** expenditure can be afforded? See the subsections 'To lease, buy or build' and 'Economic questions' below.
5. **Who** should be appointed to provide professional services? This would depend on the chosen procurement method and is also discussed in the next subsection. The job titles and functions of suitable members of the professional team will be found in the last section.

While professional help may also be required in preparing both the project brief and the design brief (see next section), such appointments should be solely for that purpose. Although a would-be owner may be tempted to appoint a project architect immediately, this should be resisted. Architects will have their own individual approach to their work, and may be personally committed to a particular style of building.

It is better to wait until after the main points of the design brief have been identified, as these can act as an agenda for meetings between the commissioning team and a number of potential design teams. From such discussions, the owner's team will learn who will suit them best, and the design teams will be better prepared to submit competitive fee bids.

TO LEASE, BUY OR BUILD?

Nowadays, anyone needing some accommodation has a choice between leasing, buying and building. The final choice will depend on the date by which the accommodation will be required, on whether suitable buildings or building sites are available, and on the likelihood of funding and its annual cost.

An agreement to have the use of a building for a stipulated period and, in return, to pay rent to the owner, is called a **lease**. Usually, the occupier will also be liable for some of the running costs, e.g. redecoration, heating, insurance, some repairs, local taxes. Except for the effects of future rent reviews, the expenditures involved are tolerably predictable. For businesses, such payments are trading expenses. Advice on available properties and their rentals should be sought from valuation surveyors (appraisers).

Buying an existing property, which actually means buying the land it stands on, is straightforward and the financial commitment is fairly certain. Its price will be known, its suitability and the costs of minor adaptations can be established, those adaptations can be done quickly, it can be occupied almost immediately, and its running costs can be evaluated. Planning permission will be required from the local authority for a change in use, e.g. from private house to health clinic. A building surveyor will give advice on the feasibility and cost of refurbishment and conversion. Funds will be required for the purchase, the building work, and for taxes, fees and expenses.

In contrast, the final cost of acquiring a site and designing and constructing a new building is uncertain, its qualities will, to some extent, be unknown, and its completion date may be months, or more likely, years away. During that time, economic circumstances, rates of interest, and the supply of and demand for properties may change.

For instance, during the recession in the early 1990s, rents for office space in London were roughly halved, as was the rate of interest. In 1995, it was estimated that the market values of perhaps one and a half million houses in the UK had declined to such an extent that their owners owed more on their mortgages than the houses would fetch if sold. (This unhappy state is sometimes called **having negative equity**.)

But, of course, constructing new buildings is what concerns us most in this book. Before proceeding very far, it is advisable to check that there are no insurmountable obstacles to doing the project, i.e. that it is **feasible**. The questions to ask will depend on the circumstances, but the following questions should be included.

LEGAL QUESTIONS

1. Is the site likely to be owned by the time construction is due to start? Design work need not be delayed while a site is acquired, provided the risks are acceptable.
2. Has anyone else a legal right to use the site? For example, are there public footpaths or bridle ways? Do telephone lines, electric cables, water mains, and sewers already cross the site or do public services undertakings have the right to install them in the future?
3. What future developments in the locality are likely to affect the usefulness or value of the site?
4. Is the site either derelict, contaminated, or filled? The owner of land may be liable for all the costs of clean-up operations, and of seepage from it.
5. What is the site used for at present, e.g. agriculture, housing, industrial, and is its proposed use different? If so, is the responsible public authority likely to approve the change in use?
6. Does any current planning permission affect the use of the site and existing buildings? Should an application for outline planning permission be made now?

The appropriate local government body must be satisfied that a development, redevelopment, or alteration is appropriate for the site. In the UK, the process is called **planning approval**. Probably the same local body will also have a duty to make sure that any new project will be safe, healthy and energy efficient throughout its life. Such powers may be based on national, regional or local government laws. In the UK, planning and the national Building Regulations are administered by local government councils. Officials are usually willing to discuss a proposal for development, and this is vital in the case of projects involving any change in the use of the site, e.g. from agricultural to industrial.

In the USA, the individual counties within each state publish and administer their own Codes. Many are based on one or other of the regional or national **model codes** produced by groups of professional

Building Officials. The Uniform Building Code has also been adopted by other countries.

ECONOMIC QUESTIONS

1. Are funds available from the owner's own resources? If not, is the financial status of the owner such that others will be willing to finance the project?
2. What are the risks to occupiers and to adjoining owners from pollution existing within the site and what might be the annual cost of insurance against claims? Is the site likely to be affected by near-by contaminated land?
3. Is the project financially worth while? That is, will it be worth more than it costs? Of course, if the owner wants it and can afford it, the question does not arise.

This last question is vital when making choices concerning the commitment of finance, land and other scarce resources for a long period. The value of a building will depend on the net annual income which it will earn for the owner or, perhaps, the annual rent for someone else's building which will not have to be paid.

A way of relating these annual amounts to their equivalent capital values is considered in Chapter 20, together with suggestions on how to control the costs of a new building. A valuation surveyor (appraiser) should know about the demand for this kind of property and its likely price if it had to be sold.

The approximate cost of constructing a building can be estimated by multiplying the required number of functional units by a suitable **functional unit cost rate**, based on experience. Information on the historical costs of various types of buildings may be found in technical publications, and in subscription services such as the Building Cost Information Service (BCIS). However, buildings of the same type will still differ in many ways, not least in the extent of their site works. Also, the market price for building work can go up or down. Thus, rather than use a single cost rate, it is wiser to think of what the upper and lower limits of such a rate might be. The costs of land, fees, finance, furnishings and equipment etc. will add to the uncertainty over the total cost.

PRACTICAL QUESTIONS

1. Can connections be made to public services such as water, electricity, gas, sewers and roads, and do these services have enough spare capacity?

2. What access to the site from the public road is available to site traffic, will it be adequate for modern vehicles, and will there be space for them to turn round?
3. Will the project fit on the site, and how many floors might be required?
4. What is the nature of the ground, what is its bearing capacity and does it contain naturally occurring sulphates or other injurious chemicals? Should a specialist organization be instructed to make a full ground investigation? At what depth is water present in the ground? (Answers will influence decisions on the structural systems and their foundations.)
5. Are there any impediments over or on top of or beneath the site, such as power lines, existing buildings, archaeological remains, deep sewers, filled waste, Radon gas, termites (but not in the UK)?
6. What are the neighbouring buildings like? Should this development be harmonized with them in any way? Might their presence restrict piling, scaffolding, deep excavations or other construction operations, or access to the site?

THE PROJECT BRIEF

All the relevant facts arising from the above questions must now be incorporated into a project brief for consideration by the owners or their commissioning team. We shall assume this leads to a decision to proceed with the construction of a new building or the conversion of an existing one. The next step is for the commissioning team to translate the owner's needs into instructions to the designers, by formulating the **design brief** (UK) or **program** (USA).

6.2 THE DESIGN BRIEF

The process of exploring what is required to satisfy the needs of the owner can be lengthy, and may require some assistance from the design team once they have been appointed. In any case, the designers can be expected to have questions of their own relating to these requirements, and this exchange of information can extend well into the design process.

CONTENTS OF THE DESIGN BRIEF

A brief can be considered under six headings, although these can be further subdivided.

1. **Requirements for useful spaces, their performance, contents and arrangements:**
   - purpose and size of each space and its location relative to other spaces;

- provision for the movement of people and things;
- doorways and other openings;
- ceiling, wall and floor finishes;
- furniture and its layout;
- furnishings;
- sound resistance between spaces;
- maintainable temperature;
- natural and artificial lighting level;
- positioning of heating or cooling units, power and lighting points;
- equipment to be accommodated;
- preserving privacy;
- possible alternative uses for the accommodation at some future date.

2. **Similar requirements for circulation and ancillary spaces** as detailed above, e.g. corridors, stairs, lifts (elevators), foyers, toilets, stores, heating and other plant rooms, fuel stores.

3. **Performance of the physical structure:**
   - the structural system, including the loads to be imposed during use;
   - fire resistance;
   - thermal, light and sound transmittance;
   - safety;
   - durability;
   - ease of cleaning and maintenance;
   - instructions to specify particular products.

4. **Fittings, services, installations and equipment:**
   - heating and ventilation system and controls;
   - sanitary fittings and hot and cold water supplies, internal drainage;
   - lifts (elevators);
   - refuse (trash) disposal;
   - telephone, television and computer cabling;
   - smoke, carbon monoxide, and fire alarms and fire-fighting equipment;
   - entry control system;
   - storage fittings;
   - room numbering and other signs.

5. **External works:**
   - access for persons and vehicles;
   - roads and pavings;
   - vehicle parking;
   - footpaths;
   - pedestrian areas;

- landscaping and furniture;
- boundary and other fencing, and gates;
- lighting;
- security and defensible spaces;
- protection of existing trees and other features;
- site clearance;
- external drainage;
- connections to sewers, and to water, gas and electricity supplies.

6. **Generally:**
   - provision for handicapped persons;
   - services to be provided directly by the owner;
   - provision of warranties and bonds;
   - appointment of structural, services and other consultants;
   - appointment of clerk of works or resident engineer;
   - limits to authority to vary the works;
   - procedure for approving design by owner;
   - financial budget;
   - use of products from renewable sources;
   - use of environmentally friendly products;
   - constraints arising from answers to legal, economic and practical questions;
   - compliance with standards set by government departments and others;
   - information relevant to health and safety on the project;
   - security;
   - contract documents to comply with the conventions of **Co-ordinated Project Information (CPI)** (see Chapter 10);
   - procurement method (see next section);
   - tendering procedures and financial approvals;
   - selection and appointment of general (or **principal** or **main** or **prime**) contractor(s) and subcontractors;
   - phasing, handing-over arrangements and dates for completion;
   - user manuals;
   - name boards and other signs;
   - quality expectations, including general appearance, ornamentation and applied decorations;
   - whole life costs (or 'costs-in-use') expectations (see Chapter 20).

Latham (1994) suggests that the owners' wishes will also normally include the following:

1. value for money;
2. pleasing to look at;
3. free from defects on completion;

4. delivered on time;
5. fit for the purpose;
6. supported by worthwhile guarantees;
7. reasonable running costs;
8. satisfactory durability.

HEALTH AND SAFETY PLAN

Under the UK's Construction (Design and Management) Regulations 1994 (CDM), all new construction work must be notified to the Health and Safety Executive (HSE) unless it will take no more than 30 days, or involve no more than 500 person days of construction work. There are some exceptions, but the Regulations always apply where five or more operatives will be working on the site at a time. The owner (called the client in the Regulations) must appoint a competent **planning supervisor** to:

1. notify the HSE;
2. prepare and maintain a pre-tender health and safety plan;
3. ensure that designers take account of health and safety risks; and
4. prepare a health and safety file for handing to the owner on completion. This will contain information on the design of the building and the facilities it contains that might affect the health and safety of occupants and workpeople in the future.

The owner must also supply the planning supervisor with any information on the site or the existing buildings and their present use that is relevant to health and safety. Later, the owner must also appoint a **principal contractor** to develop the construction phase of the plan and ensure that everyone on the site is aware of and complies with it.

6.3 BUILDING PROCUREMENT (DELIVERY)

If a suitable building can be leased, the building work is likely to be restricted to minor matters such as redecoration. Nobody invests money in someone else's building without good reason. In any case, the lease is likely to include conditions stating what can and cannot be done to the building.

The discussion that follows will apply equally to the purchase, conversion and refurbishment of an existing building and to the construction of a new building.

MINIMIZING THE OVERALL PROJECT TIME

The simplest and least risky procurement procedure is an **end-on** or linear one of first completing the design and then constructing the Works.

For various reasons, the process of designing and obtaining approvals can often take longer than the actual construction. To reduce the overall project time, these two stages can be overlapped by some **fast-track** method of **parallel working**. While earlier completion can bring considerable financial and other advantages, the increased size of the professional team enhances the risk of conflict and failure.

However, according to the NEDO publication *Faster Building for Industry* (1983), which reported on the procuring of new industrial and commercial buildings,

> The general belief that speed costs money is quite unfounded – fast building is possible without sacrificing either cost or quality. The customer must want it, and must choose a building team which will understand and share this objective. Responsibilities within the team must be clearly defined and in particular the customer must be clear as to who is the team leader.

The study also stated that

> Fast projects require:
>
> - customers with knowledge of what is achievable . . . accurately specified by dates or deadlines
> - customers well-advised about the workings of the method of contract organization and about the building team they choose
> - coherent management responsibility for the progress of the project throughout
> - overlapping pre-construction activities
> - arrangements which allow early, precise and integrated procurement and preparation for construction
> - a clear statement of design which takes into account practical aspects of organizing work on site
> - competent and adequate site management and supervision
> - control over site labour resources
> - good communications and incentive to complete on time.

CHOOSING WHICH TYPE OF CONTRACT

The factors which will determine the choice of contract type include:

1. the extent to which the owner seeks to pass the risks of construction to others;
2. the extent to which commissioning and project management skills are available within the owner's organization;
3. the availability of individuals with design, cost control and project management skills from which the professional team can be drawn;

4. the dates for starting and completing the Works,
5. the extent to which design and production information is to be completed before work starts on the site;
6. construction operative and management skills available;
7. the size, competence, financial stability, reliability and availability of suitable construction and specialist organizations;
8. the sources of investment funds and the restrictions on their use.

Chapter 17 discusses the more usual contractual arrangements although we may expect innovations in the future to suit the changing needs of owners.

OPEN AND SELECTIVE TENDERING

Normally, the design team will invite competitive tenders (bids) for the Works from potential contractors, and will supply documents that describe what is required. In effect, they will be attempting to create a market place for the Works and the owner will usually accept the lowest bid. There are two main approaches.

Open tendering is where an invitation is given to any interested party to ask for the contract documents and to submit a bid. This invitation is usually made by inserting an advertisement in the trade press. The approach is suitable for large Works contracts and, depending on their size, may be essential under European Union (EU) regulations.

The alternative is for the design team to submit a list of suitable tenderers to the owner for approval. Then, a few weeks before the documents will be ready, the selected tenderers are asked if they would be willing to submit a bid. If this does not result in a suitable number of competing tenderers, there will still be time to invite others. The procedure is described in the Code of Procedure for Single-stage Selective Tendering (1994).

6.4 THE PROFESSIONALS

In their *Handbook of Professional Practice*, the American Institute of Architects (AIA) states: 'A team of experts and specialists now divides the labor of designing and constructing our built environment'. A group of persons who claim, or 'profess', to have expert knowledge in a branch of learning, and who base their vocations on this knowledge, constitute a **profession**. Those for whom they apply these personal abilities they call their **clients**. They will have inherited their standards of ethics and of service to their clients from those earliest professionals who served in the Church, the armed services, and the law. Members of a profession are expected to put their duty to their clients before their own interests.

On the other hand, being a **professional** rather than an **amateur** may simply mean that one undertakes certain activities in order to earn one's living. Nowadays, these distinctions are becoming blurred, due to the development of consumer protection law, and the recognition that all individuals have a duty of care towards each other.

Members of a profession who are in private practice, i.e. they are self-employed, will be liable to their clients for the results of any negligence by themselves or by their partners. Traditionally, all their private possessions were at risk, but nowadays, this liability is normally covered by mandatory professional indemnity insurance. The members of any one profession are likely to be organized into a corporate body, called either an institute or an institution, which sets, maintains and advances their standards, and which, in the course of time, is recognized by the state.

In the UK, such an institute receives recognition from the State through the granting of a Charter by the monarch, acting through the Privy Council. Being allowed to add the word 'Royal' to the title of the institute is the ultimate recognition of its contribution to society. In other countries, individuals may need a licence from the state before being allowed to practise.

Characteristically, a professional person exercises independent judgement in deciding what to do to satisfy the client's expectations. Even so, much of his or her time, and the time of supporting staff, will be taken up doing technical tasks rather than original work. Until quite recently, the architect was appointed to do everything necessary to achieve a design which the owner could approve, and to arrange for competitive tenders for the Works. Such a comprehensive commitment could easily be described in quite general terms. In addition, the architect would administer the contract in accordance with its conditions.

This is still the case with relatively simple projects. However, the increasing size and complexity of some modern projects, and the importance of time and cost as well as quality, has led to increased sharing of the design team functions between a number of professionals. They may also receive advice from professional builders. When sharing occurs, it is necessary for the services required from each independent member of the team to be listed in some detail, and negligence is more easily revealed.

APPOINTING THE PROFESSIONAL TEAM

At one time, the structural engineers, building services engineers, acoustics specialists etc. who contributed to a design were regarded as simply helping the architect. Their fees were paid by the architect, and they had no direct obligations to the owner. In the UK, up to the 1950s, the surveyor's charges for providing bills of quantities to tenderers were given as an item in the bill of quantities, included in the bids, and paid by the successful contractor out of the first interim payment from the owner.

Nowadays, all the independent consultants are likely to be appointed and paid directly by the owner. In the period before the architect is selected, a **professional** adviser may be appointed solely to assist the owner:

1. when considering the need for a project;
2. to appraise the feasibility of a project, and
3. to prepare the project brief which will state its purpose and objectives.

The designer, or **architect**, may be selected as a result of either an open or a limited entry architectural competition. More likely, an individual or firm will be appointed after a satisfactory interview because of the quality of their work and the competitiveness of their fees. They are likely to be consulted before the appointment of specialist designers such as **structural engineers** and **building services engineers**. An **approved person** may be employed to certify that the proposed building will comply with the requirements of the Building Regulations regarding the conservation of fuel and power. An individual of recognized competence may be employed to calculate the Energy Rating of a dwelling in accordance with the Standard Assessment Procedure (SAP) in those Regulations. Instead of depositing full plans with the local authority for their approval, the owner may engage an **approved inspector** to certify that the design complies with the Building Regulations.

A **building surveyor** may be appointed to appraise the behaviour of the building in use and the likely expenditure on maintenance, renewals and other 'costs in use'. Ensuring that the design represents value for money and that budgets are not exceeded will, in the UK, be the responsibility of the **quantity surveyor**. Where the project is large and complex, this function may be shared with the **management contractor** who will also be concerned with the buildability (constructability) of the project. In the USA, all such functions may be provided by the **construction manager**.

A valuation surveyor may be retained to negotiate the purchase of the site, and a lawyer will be required to arrange for the purchaser to become its legal owner. In Scotland, a lawyer may do both.

The building process may be managed by a management contractor, who will be provided with production information by the others, and who will enter into Works contracts for the actual construction on behalf of the owner. A management contractor may be liable to pay damages to the owner if the project does not meet its cost and time targets. The day-by-day monitoring of the Works can be by a clerk of works, or by a resident engineer. (Chaucer was Clerk of the Works to Richard II from 1389 to 1391 – but his job was to administer the whole of the royal building programme.)

This **professional team** may be led and co-ordinated by a member of the owner's commissioning team, sometimes called in the UK the **lead manager**, or the **employer's representative**. Alternatively this person may be called the **project manager**, although this title may also be applied to the individual responsible for construction. We shall be distinguishing between the **project**, meaning all those activities necessary to achieve the owner's objectives, and the **Works**, meaning those activities occurring on the site. Even so, we must expect those responsible for managing site activities to regard these as their project.

MINIMIZING CONFLICT

The success or failure of any project will depend on the personal qualities of the owner, the members of the professional building team, and those responsible for the site operatives. Conflicts between them, and the habit of always blaming others can be minimized by developing an ethos of co-operation rather than confrontation, and by clearly defining each individual's responsibilities. A constructive attitude should be displayed at site meetings and information should be readily available.

The maxim 'Keep it simple, stupid', or 'KISS', is good advice. Obviously, the fewer the individuals involved, the less the risk of conflict and 'buck-passing', and the greater the likelihood of achieving the project objectives.

QUALITY EXPECTATIONS

Consider your own experience of recent improvements in the spatial and technical standards of buildings. For instance, compare the room sizes and the numbers of electric power points in new dwellings with those in older properties. How many dwellings or schools do you know which still have only outside toilets? How long ago was it that full central heating and/or air conditioning and fitted carpets first became the norm for new houses?

At any one time, each society will have as part of its culture a set of expectations, i.e. a **value system** concerning the qualities of new buildings. To some extent, such expectations will depend on the climate. They will certainly be influenced by notions of what can be afforded. These expectations may be incorporated into legal requirements, such as the UK's Building Regulations, although higher standards than these may often be desirable.

As some organizations will be large enough to have developed their own value systems, it is important for the owner to state what qualities the project is expected to have. This might include the designed life of the project. The design team should then make sure that what they propose will match such expectations. In other words, they must all endeavour to work within the same value system.

6.5 COMMENTARY

Much of this chapter has been concerned with those facts and other information that should be included in the project brief and the design brief. Most will be confidential to the owners and their advisers, be about specific situations, be unrelated to other facts, and be of no interest to anyone else. We shall call it **private information**.

In contrast, the professionals profess to having skill in the use of **public knowledge** acquired during their initial and continuing education and training. This they apply in the interests of their clients to the private information supplied to them. One of the objects of this book is to introduce readers to this public knowledge which consists of:

1. statements describing generally applicable **ideas**, or **concepts**;
2. **rules** for doing things; and
3. published information of various kinds.

Concepts are mental tools, abstractions which our minds have invented to organize our experiences and help us solve our problems. Even so, we can write down statements about them which we and other people can then think about and criticize. For instance:

- we have made statements about the physical properties of building substances such as their coefficient of expansion, thermal resistance and mass density;
- in Chapter 5, we used the idea of form to classify main construction products and their associated dependent products.

Rules for doing things will have been either agreed or accepted by a particular community of persons. Such **conventions** need to be revised from time to time to match changes in society, in the law (itself to some extent a set of conventions), and in technology.

Some conventions may be the work of bodies representing particular interests in construction, e.g. rules for designing a steel structural frame, or standard forms of contract, or standard classification systems for construction data. Others can have the force of law, e.g. the UK Building Regulations, the CDM Regulations. Each community, whether it is the local town or county, the country, or an international trading bloc, may arrive at a different solution to a particular problem.

Published information is likely to be mainly concerned with the promotion and regulation of trade. It will range from invitations to bid for a building contract to details of international Standards and available products.

6.6 SUMMARY

Designing and constructing a new building is a lengthy and uncertain process compared with leasing or buying an existing one. Intending owners should not just appoint an architect, but should first undertake their own appraisal of their needs, the alternative ways of satisfying them, and the risks involved. A professional adviser could be appointed solely to help in the preparation of this **project brief**.

Specific items for consideration and incorporation in this brief and also in the instructions to designers, or **design brief**, were listed. The factors that might influence the choice of procurement method were outlined.

The significance of being a member of a profession and the nature of professional institutions was considered. The functions of members of the building team responsible for designing and managing a project were identified.

REFERENCES

Latham, Sir Michael (1994) *Constructing the Team*, Department of the Environment. HMSO, London.

National Economic Development Office (1983) *Faster Building for Industry*, HMSO, London.

National Joint Consultative Council (1994) *Code of Procedure for Single-stage Selective Tendering*, from the publications departments of professional Institutions in the UK.

FURTHER READING

American Institute of Architects. *Architect's Handbook of Professional Practice* (vol. 2), AIA, Washington.

American Institute of Architects (1982) *Project Checklist*, AIA, Washington.

Latham, Sir Michael. (1994) *Constructing the Team*, Department of the Environment. HMSO, London.

Thompson, F. M. L. (1968) *Chartered Surveyors*, Routledge & Kegan Paul, London.

Scheme design 7

7.1 BUILDINGS AS ARCHITECTURE

Buildings that have been designed to affect the emotions of observers and occupiers have the special quality of being **architecture**. They, and the process of designing them, are generally regarded as the highest form of fine art. However, the term may be applied to the design of any building.

Practitioners of this art will usually call themselves architects. While social and economic conditions, scientific developments, and the availability of materials and skills will influence building styles, certain desirable architectural qualities would seem to be timeless.

For instance, two thousand years ago, Vitruvius wrote to the Roman Emperor Augustus that buildings 'must be built with due reference to **durability, convenience, and beauty**'. Or rather, this is how Morgan (1960) translates the Latin text. Alternatively, the text might have meant '**structural stability, appropriate spatial accommodation, and attractive appearance**'.

However, Vitruvius was also concerned with how the influence of the climate on occupiers could be modified by an appropriate choice of location and building style.

When, in about 1450, the Italian architect/consultant Leon Batista Alberti wrote *On the Art of Building*, he reversed the order of the first two, and put the quality of the spaces first, as we did in Chapter 1. Later, Andrea Palladio followed his example, as did Sir Henry Wotton, who began *The Elements of Architecture* (1624) with:

> In **Architecture** as in all other **Operative** Arts,
> the **end** must direct the **Operation**.
> The **end** is to build well.
> Well building hath three Conditions.
> **Commoditie, Firmnes, and Delight.**

(Later, he wrote that buildings should be '**Commodious, Firme, and Delightful**'.)

Vitruvius had stated that 'Architecture depends on Order, Arrangement, Eurythmy [i.e. proportion], Symmetry, Propriety and Economy'. The influential critic John Ruskin wrote in 1853 that 'The essential thing in a building, – its *first* virtue – is that it be strongly built, and fit for its uses.' However, for many, architecture seems to be a matter of the delight of external elevations, rather than the commodiousness of spaces.

While the characteristics of particular styles of building can be identified, analysed and illustrated, as Vitruvius and others have done over the years, every building of merit will reflect the unique creative powers of its designers.

7.2 BUILDING ELEMENTS

ENCLOSING ELEMENTS

The roof, the external walls with their windows and doorways, and the floor next to the ground, i.e. the constructions above and at the sides and bottom of the building space are often referred to jointly as the **building envelope**. Their purpose is to enclose the rooms etc. and to separate the building space from the natural environment.

Similarly, the upper floors and their internal walls are jointly called the **internal division**, their purpose being to divide the building space into rooms etc. Together, the envelope and the internal division constitute the **building enclosure**. Each floor and its external and internal walls is called a storey.

Note that each of these individual constructions will always have the same purpose, no matter what it is made of. Thus, it can be regarded as one of the **elements** of the building. (An element must be an essential, distinct and well-defined object, and different from the others in a set.) The idea of an element is widely used within the design team and is also applied to services and installations. However, as the idea can be applied in different ways, it is not always clear what is included in an element.

For instance, when thinking generally about the side of a building (which is all one can do during the earlier stages of a design), we will include windows and external doorways and internal wall finishes in the external walls element. Also, those triangular-shaped gables and other external walls that are above the highest ceiling will be included in the roof element. (Gabled ends to a pitched roof contribute to the enclosure over the top of the building, not at the sides, and are an alternative to having hipped ends.) However, in the UK Building Regulations on fire resistance, a roof is not usually treated as an **element of structure**. Some classification systems for elements are mentioned later.

THE FUNCTIONS OF ENCLOSING ELEMENTS

The purpose of each element is achieved by what the element does for other elements and for the internal spaces. These are called its **functions,** and each can be expressed as a **verb + noun** statement. For instance, we could say that one of the functions of a roof is:

- to exclude rain, snow, airborne particles, wind, birds, insects, and reptiles.

Another might be, but isn't always:

- to stabilize the tops of the external walls by acting as a brace, or collar.

Because every element has its own purposes and functions, we often call them **functional elements**. As we have seen, each function of the element is likely to be provided by one of its parts, i.e. by a layer of suitable materials in an appropriate position within the thickness of the element. Functions and their associated attributes can be classed as being either:

- **regulative** to regulate can mean either to totally exclude or to restrict the passage of some form of energy, some agency, or certain living creatures; enclosure elements can be regarded as regulating regions, within which the actual modifying process takes place;
- **supportive** of other elements and of applied finishes;
- **visual**, i.e. relating to the appearance of the element; or
- **economic**, i.e. influenced by restrictions in the supply of scarce resources, as considered at the end of Chapter 4.

The main functions of enclosing elements can be summarized as follows:

Regulative

- to regulate the passage of heat or cold, fire, smoke, light, sound, moisture, air and living creatures from one side of an element to the other;
- to resist the spread of flame across the surfaces of elements and through voids within their constructions.

Supportive

- to resist and transfer the dead, imposed, and wind loads safely to another supporting structure or to the ground;
- to provide support for their non-loadbearing parts and their finishes;
- (in the case of a floor finish) to provide a safe and level horizontal surface.

Visual

- to communicate to observers and visitors the purpose, quality and significance of the building, the priorities of the owner, and the creativity and technological skill of the designers;
- to provide an element with a suitable appearance.

Economic

- to achieve its other functions at minimum cost over the life of the building;
- to minimize the environmental pollution resulting from its construction;
- to optimize its initial and continuing demand for energy and other scarce resources and the benefits gained by its occupiers over the planned life of the building.

THE ATTRIBUTES OF ENCLOSING ELEMENTS

The element attributes mentioned in Chapters 1 to 4 can be assigned to one or other of the classes of element functions given above, as shown in Figure 7.1.

During the scheme design stage, it should be possible to prepare (probably informally) an elemental performance specification (see Chapter 10) that also stipulates their attributes.

7.3 DESIGN STAGES

The following pre-design and design stages are from the Plan of Work of the Royal Institute of British Architects (RIBA). Equivalent terms from North American practice are given in brackets. The pre-design stages A and B, which included producing the project brief and the design brief, were considered in the previous chapter on 'Project initiation'.

A – **Inception** stage (pre-design services).

B – **Feasibility** stage (pre-design services and site analysis). It may not always be practicable for the owner to prepare the design brief without the help of the design team. In any case, as more information becomes available, the design brief may need to be amended and enhanced. Members of the professional team should have been appointed before the next stage starts.

C – **Outline proposals** and D – **scheme design** stages (design – includes schematic design and design development). These are the stages where the truly creative design work is done. For smaller projects, both stages are likely to be combined into (in the UK at least) the **sketch plans** stage. On larger projects, the stages will be:

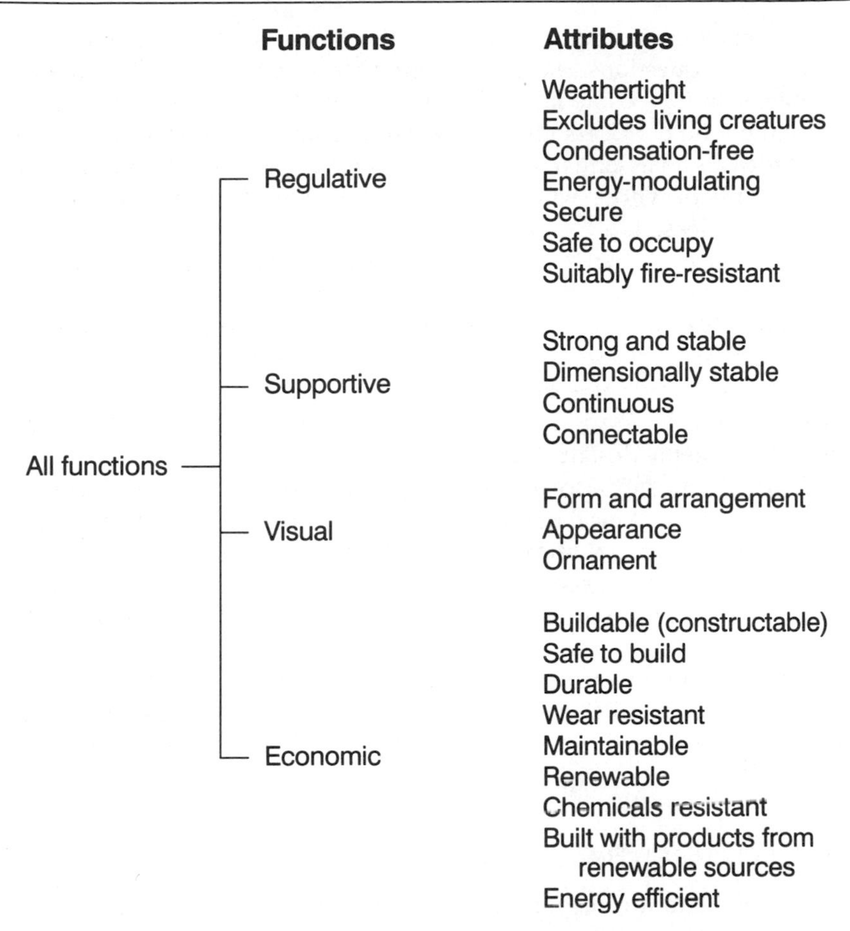

Fig. 7.1 An array of classes of element functions and their related attributes.

- C – preparing alternative outline (schematic) proposals for consideration by the owner, and
- D – developing the chosen option.
- E – **Detail design**, or 'working drawings' stage (design implementation).
- F – **Production information** stage (production documents).

These stages offer what might be called an ideal, or **normative** analysis of the design process. In reality, the overall process is never a sequence of distinct and separate procedures, where each one is completed before the next is started. Instead, each is likely to be repeated as earlier thoughts and decisions regarding the design are modified and refined. In a word, the design process is somewhat **iterative.**

It begins with getting to grips with the owner's requirements, asking constructive questions, and discovering and absorbing relevant facts. Creative solutions come later. The discovery of an unexpected fact may throw the whole process into confusion. Discarding the results of superseded ideas is an essential aspect of the creative process, although design team members whose work will appear to have been wasted are not likely to be pleased.

Even after work has started on the site, the owner might wish to make changes to the brief. These may be the result of, say, changes in the requirements of the project, or changes in the arrangements for finance. Some changes may be simply capricious. But 'He who pays the piper calls the tune', and there may be an agreement to reimburse the designers' extra costs. Even so, changes to the brief ought not be made once the end of the scheme design stage has been passed.

Major upsets can also be caused by the requirements of regulative authorities, the technical decisions of specialist designers concerning, for example, the structural frame or the building services, and by errors discovered in previous decisions. Incorporating significant changes to the design can involve returning to an earlier stage, and could disrupt the whole process.

Even so, having recognizable stages does enable the outputs from each to be listed and the work still to be done before that stage is complete to be identified. Meanwhile, work may have started on the next stage(s).

THE CHICKEN AND EGG QUESTION

Individuals who are responsible for setting tasks for themselves or for other people are constantly wondering what should be done next. The parts of the design process, and the parts of the buildings themselves, can be described as being interdependent, i.e. the parts depend on each other. Even so, in any project, a start must be made somewhere. The question is, where? In spite of what has just been said about interdependence, there are dependences in the design process. That is, certain decisions must await others. Only when the preceding items have been dealt with will it be sensible to proceed.

For example: no sensible person will undertake work before agreeing terms for payment; choosing a facing brick comes after deciding to have a brick wall; oil storage will be required only if oil has been chosen as the energy source for heating. When tempted to make a start on a particular aspect of the design, first ask the question 'what does it depend on?'.

Designers are likely to start by collecting as many facts as possible about the owner's requirements and on the limits, or **constraints** to their freedom to make decisions. This analysis of the problem is then followed by the investigation of possible design solutions. Occasionally, the opposite will apply, and the designers will seek to apply a particular building solution to the owner's requirements, e.g. when extending an existing building, or when a building is to be in a particular style.

7.4 GATHERING PRIVATE INFORMATION

If answers to all the relevant legal, economic and practical questions listed in the previous chapter have not been given in either the project brief or the design brief, they must be considered now. The requirements of the design brief will lead to further questions and investigations.

INFORMATION ON THE SITE

1. Survey the site and plot its features, levels and boundaries.
2. Take photographs of the site and neighbouring buildings. What does the site suggest to you about its potential? Time spent talking to neighbours, looking into local excavations, watching the movement of persons and vehicles, viewing the site from different vantage points, and generally absorbing the atmosphere of the neighbourhood will not be wasted.
3. Investigate the nature of the ground beneath the site, its natural drainage and its ability to resist loads.

INFORMATION ON THE ACTIVITIES TO BE ACCOMMODATED

1. What is the initial purpose of the building and what other purpose might it serve in the future?
2. What activities are to be accommodated, and are any incompatible?
3. If these activities are already taking place elsewhere, analyse and record the movement or circulation of people and objects.
4. Are there any requirements for the reception, processing, storage and dispatch of materials and products?

CONSTRAINTS

The following will be among the constraints on a design:

1. requirements of the design brief;
2. decisions already taken on e.g. the functional systems;
3. requirements of local planning and building control authorities and, through them, the Fire Authority;
4. providing for wheelchair users and the ambulant disabled;
5. any legal requirements, including the UK's Building Regulations;

6. the need for security;
7. the size and shape of the site;
8. the area required for car parking and other site works;
9. the locations, heights and external treatments of neighbouring buildings;
10. preserving sightlines from existing buildings;
11. locations of connections to public services;
12. locations of access points from public roads;
13. maximum safe bearing capacity of the ground at different depths;
14. availability of products, operatives and plant;
15. quality standards to be achieved;
16. maximum dimensional deviations to be tolerated for the various parts;
17. completion date;
18. cost limit of the Works and the gross floor area that can be afforded (see below and Chapters 8 and 20);
19. the absolute need to preserve the health and safety of operatives and the public during construction;
20. the absolute need to preserve the health and safety of occupiers of the completed building;
21. safe demolition.

HEALTH AND SAFETY

The owner should have already appointed a competent planning supervisor, possibly from within the design team. If not, this must be done before designers start their work. All those who produce production information are regarded as designers under the CDM Regulations. They must give due regard to health and safety in their work and ensure that information on any health and safety risks is included in the pre-tender stage health and safety plan.

7.5 STORING AND RETRIEVING PUBLIC KNOWLEDGE

DATA STORAGE AND RETRIEVAL

To retrieve a document, we must know where we put it. Some people manage to keep their personal papers in order. Others maintain one or more shapeless heaps and have difficulty in finding what they are looking for. The uniqueness of every building project, and the many specialists involved, calls for good communications and an efficient referencing and retrieval system.

An industrialized trading community produces a wide variety of different products, and publishes technical documents about all of them.

Once we have obtained the technical data we need, and have read it, we must file the documents where we can find them easily, either to read again or to remove for recycling. Trade literature is likely to have a life of only one or two years. Of course, we are also interested in filing and retrieving drawings, correspondence, and other data relating to whole buildings, their elements and parts.

CLASSIFYING GENERALLY

The advantage of subdividing a collection of objects with the same general purpose, e.g. all construction products, into classes was introduced in Chapter 5. Members of any one class will have certain properties in common and knowledge of these properties can be applied to any member of the class. As the members of each of the other classes will have their own special properties, the classes are **mutually exclusive**. Also, together, the classes must accommodate all possible members, i.e. they must be **collectively exhaustive**. Each class will have its own reference or **index** to point to it, identify and label it, and distinguish it from the others. Classes that are, themselves, grouped into more general classes, or **categories**, may be referred to as **basic classes.**

THE SFB SYSTEM

SfB stands for *Samarbetskommittén för Byggnadsfrågor*, which was the name of the committee that, in 1950, in Sweden, pioneered the idea of having a classification system for the co-ordinating of construction industry information. It is published by CIB (1973) and the UK variant is called CI/SfB, the Construction Indexing Manual (1976). Its purpose is to provide an indexing system for sorting and labelling documents so that they can be easily retrieved from store.

Any set of classes based on a particular property will be associated with an indexing **notation** system, and may be referred to as a **facet**. All notation systems are based on the capital, or **upper case** letters A–Z, the small, or **lower case** letters a–z, and the numbers 0–9. The SfB notation system with examples is shown in Figure 7.2 and an example of its application is shown in Figure 7.3.

An index based on one aspect of an object, that is, on a single facet, may not be sufficient. For example, where there are many trade catalogues on the same form of product, e.g. pipes, it may be helpful to have a set of parallel and equally important classes for the various kinds of substance, e.g. cast-iron, fired clay, plastic.

Then, when indexing, say, catalogues for drainage pipes and fittings, the appropriate references for form and for substance can be placed side

No.	Table for	Notation	CI/SfB box compartment	Index	Refers to
0	Purpose groups	Up to three digits	1st	712	Primary schools
1	Functional elements	Two digits within brackets	2nd	(27)	Roof
2	Constructions	A capital letter	3rd	N	Rigid sheet overlap work
3	Materials and substances	A small letter and a digit	3rd	e5	Slate
4	Activities	Capital letter and two small letters	4th	Ajt	BRE digests

Fig. 7.2 The CI/SfB tables and their notation system.

by side to form a **co-ordinate** index for each catalogue. This is not the same as a hierarchy, as the individual indexes have equal significance.

In the UK, the index of an item of trade literature is often given in a CI/SfB classification box (see Figure 7.3). This will be found at the top right hand corner of the front page, and contains compartments for indexes from up to four different facets. Indexes for form and substance from Tables 2 and 3 are combined and given in the third compartment. Thus, the index for slate roofing would be written in compartments 2 and 3 as (27) Ne5.

COST INFORMATION

The cost limit for the Works as a whole is a major constraint on the design. Either this limit will have been given in the design brief or it must be calculated from the total of the required useful areas (see also Chapter 20). Once agreed, its feasibility must be maintained as the

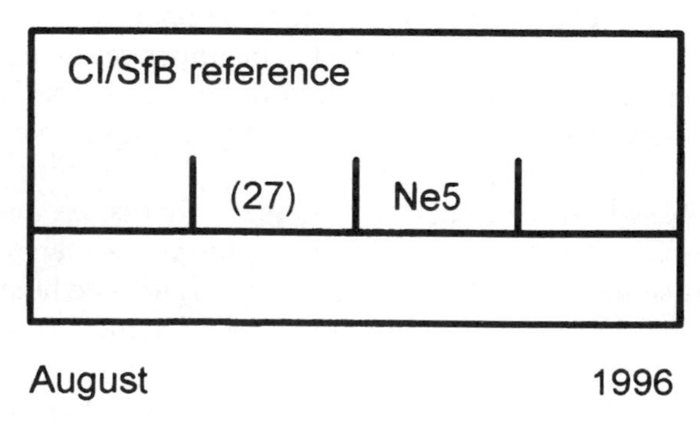

Fig. 7.3 An example of a CI/SfB classification box.]

design is developed. At this time, the best sources of cost information will be that on similar projects built recently in similar circumstances.

In the UK, the Building Cost Information Service uses the *Standard Form of Cost Analysis* (1969). The purpose of its detailed set of elements is to 'provide data which allows comparisons to be made between the cost of achieving various building functions in one project with that of achieving equivalent functions in other projects'. A similar list, called the *National Public Works Conference List of Elements and Sub-elements*, is used in Australia. Sub-elements are their equivalents to our parts. In these classification systems, wall foundations and the lowest floor are all brought into the substructure element. Rainwater drainage installations are usually treated as services. Planning and controlling the price of a building are discussed in Chapter 20.

INDEXING BOOKS

Much of our knowledge will be found in books. If our collection of books is small, it should not be difficult to search for and find the correct title. But if the collection is large, or if we only know what the subject matter is, and not which book it is in, then we have a **retrieval** problem.

In a library, this problem is dealt with by classifying and indexing. That is, a comprehensive list of subjects (the classes) is prepared, and each class is given a unique reference or tag within a comprehensive indexing notation.

Libraries in the UK are likely to use the Dewey Decimal Classification. Melvil Dewey (1851–1931) was a US librarian. Its three-digit notation can be extended by giving numerals to the right of a decimal point, and, possibly the first three letters of the author's name. Any one class might contain only a few, or even just one book.

Then, if the books are shelved in index order, any particular book should not be difficult to find, once its index is known. This index would be revealed by either a manual or an electronic search in the library catalogue for the name of either the author or the subject or, possibly, for one or more 'keywords' on the subject. The index range for construction starts at 690. That for architecture at 720. The rather similar Universal Decimal Classification system provides an alternative to 'Dewey'. Libraries in the USA may use the classification system developed by the Library of Congress.

PLANNING INFORMATION

Sources of dimensional data on human beings (i.e. **anthropometric data**) and their needs for space when undertaking various activities will be found in 'Further reading'. A home-grown example is given in Figure 7.4. The requirements of the UK's Building Regulations for means of escape in case of fire, access for disabled people, sanitary accommodation, and other health and safety matters are given in Approved Documents B, E, F, G, H, K, L, M.

OTHER INFORMATION REQUIREMENTS AND SOURCES

The members of the professional team and their staff will all be specialists in their own areas of public knowledge, and will have their own sources of information. Specialist areas will include acoustics, interior lighting, heating, ventilating and air conditioning, site investigations, interior design, structural design, conservation of fuel and power. Books and articles in periodicals will give planning details of individual buildings and those of a particular purpose group. In this introduction, we can only note the complexity of building design and hope that readers who are encouraged to become specialists will have gained some awareness of the contributions of others and of the need for the design team to work together.

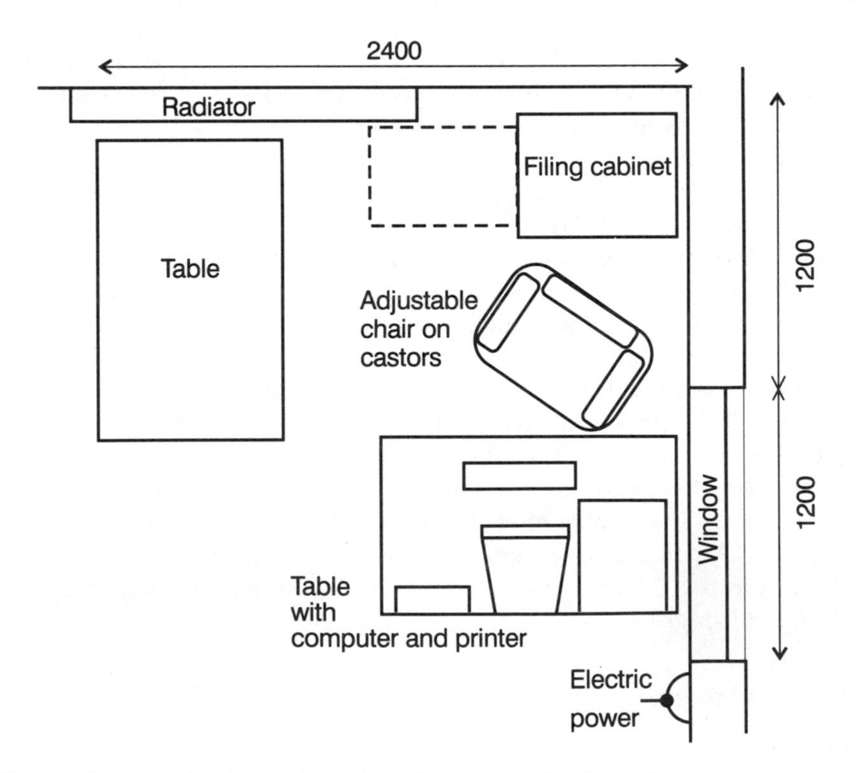

Fig. 7.4 An example of spatial needs – plan of the author's workspace.

Information on products and their use in technical solutions for the various elements and parts is more likely to be required during the detail design stage (see Chapter 9). Sources of information will include products manufacturers' data, the Approved Documents relating to the Building Regulations, British and European Standards, publications from the British Board of Agrément, the Building Research Establishment and various specialists. In the USA, see publications from the American Society for Testing and Materials (ASTM) and Products Directories from Underwriters Laboratories Inc. (UL).

Standard specifications will offer opportunities for constructing a Works specification by inserting and deleting text, either manually or electronically. The Architectural Liason Officer at your local police HQ can advise about security.

ELECTRONIC STORAGE AND RETRIEVAL

Developments in computers and peripherals, digital storage and communications, and the technology associated with the Internet are likely to encourage the growth of specialist information service providers to the construction industry. Individuals will then have ready access to up-to-date public knowledge on almost everything, and be able to communicate simply, speedily and cheaply with each other.

Using one or more keywords to retrieve electronically stored information avoids having to index and store the physical documents. The long-term effects on the structure and procedures of the industry and on education and training could be profound.

7.6 PLANNING THE BUILDING

PLANNING THE USEFUL SPACES

The design brief will either list the rooms etc. required and their sizes, or state what activities each space must accommodate. The floor areas of such spaces are called the **required useful areas** (sometimes called usable areas). The commissioning team may require the design team to think about and add the following:

- sanitary accommodation, cleaner's stores, cloakrooms and other **ancillary areas**; and
- boiler, lift and tank rooms and other **mechanical and electrical services areas**.

To plan a useful space,

1. list the activities that are to take place and the objects to be accommodated;

2. quantify their needs for space and for movement between them, and establish their interrelations;
3. assemble these spatial needs into a rational plan shape while allowing for access space;
4. determine the size of the space, i.e. its 'spatial' dimensions of length, breadth and height;
5. list the thermal, ventilation, sound and natural and artificial lighting levels to be achieved;
6. describe the visual quality of the environment to be achieved and the durability of its surfaces;
7. state the need for privacy and security;
8. decide on window sizes and their orientation;
9. identify or determine the maximum floor loading;
10. position the heating or cooling units, sanitary conveniences, washbasins and other appliances, lighting, power and other service outlets, furniture and fittings;
11. list the maximum number of persons to be accommodated and the maximum distances to spaces with related activities, e.g. to stores, toilets, library, restaurant and to exits from the building;
12. check the above with the requirements of the brief.

THE LAYOUT STRATEGY

1. consider the possible apportionment of the site to buildings and other uses;
2. consider the modelling of the building, e.g. shape, number of storeys;
3. appraise the security risks;
4. decide the orientation of the building;
5. locate the various activities within the building and segregate incompatible activities;
6. decide on a suitable horizontal and vertical dimensional planning grid;
7. decide the strategy for the provision of sanitary accommodation;
8. decide the strategy for dealing with a fire, including staircases, escape corridors, fire doors and exits, and how a fire will be contained. How many occupants must be catered for? Will any of them be less than fully mobile, and what is the greatest distance they should travel before reaching safety?

PLANNING THE LAYOUT

Plan the layout and the circulation areas within the constraints and strategy considered earlier. Obviously, the various useful spaces must fit together in an acceptable way, and this will probably mean modifying earlier thoughts on their shape and spatial dimensions. Also, in a multi-storey building with loadbearing external and internal walls, those of upper storeys should have similar walls directly below them, i.e. a space on any upper floor should have a corresponding space on the floor below.

1. outline possible alternative floor plans, elevations and structural systems and consider their impact on observers;
2. for each alternative, consider possible horizontal and vertical circulations, lobbies, foyers, sanitary accommodation, stairways and lifts (elevators);
3. for each alternative, decide suitable storey heights and consider possible sizes and shapes for the building envelope(s);
4. identify possible locations for drainage and service pipelines, boundary walls and fencing, seating, planting, paving, and other external works;
5. allocate movement and parking areas;
6. locate the entry and exit points for people and vehicles;
7. in collaboration with specialist designers, consider the requirements for the structural system, and for mechanical, electrical, and other services, and how to accommodate them within the envelope;
8. refine the chosen proposals in collaboration with the structural and services designers.

TESTING AFFORDABILITY

The relations between quantity, quality and cost are explored in Chapter 20.

FUNCTIONAL SYSTEMS

Where some of the parts in adjoining elements have one or more attributes in common, they can be thought of as forming a **functional system**. For example, some parts in the elements of the envelope function together as a **thermal regulating system**. Others will regulate the passage of fire, or of light, sound, air, water, smoke, water vapour and other gases, dust and living creatures. Similarly, the various structural parts will constitute the **strength and stability system**. In each case, the appropriate part(s) in each element must be joined to those in their neighbouring elements so that the system is continuous. Of course, most parts will contribute to more than one function.

Strategic decisions on the various functional systems will include how to achieve:

1. vertical and horizontal movement;
2. regulative systems for light, heat, sound, for excluding the weather, for protecting against fire, and for security;
3. supportive systems to provide strength and stability;

4. heating/cooling, hot and cold water, electric light and power, emergency lighting, telephones and other services;
5. sanitary accommodation and waste disposal installations.

7.7 DESIGN PROCEDURES

At one time, the designer would first design the building enclosure, and then pass the drawings to structural and services engineers and other consultants. The structural engineer would decide what was necessary to make it strong and stable, and the mechanical and electrical services engineers would decide what was required to keep it warm etc. Nowadays, an interior designer might decide what the building will look like on the inside. Each is likely to have a different opinion of what is important in a building.

While this step-by-step approach may still be adopted, there is increasing recognition that decisions made concerning one element can affect other elements as well. In other words, the elements are **interdependent**. For example, the height of the external walls might depend on the minimum headroom beneath the beams supporting an upper floor, the depth of these beams and the depth of the floor. The floor may have to accommodate heating mains, air conditioning ducts and other services.

This interdependence encourages the various specialist designers to seek to integrate their contributions, even though it may cause them to repeat many procedures. Every designer and design team will have their own approach to their work. The following is offered as an indication of what information has to be considered and produced. Drawings, specifications and schedules are discussed in Chapter 10.

OUTLINE STAGE

1. Gather all the private information on the project, such as the items listed earlier. This will point to the public information that will also be required, e.g. planning, technical and cost data, parts of the Building Regulations.
2. Identify the constraints on the project.
3. Prepare a timetable for the stages at which reports will be made to the owner.
4. Outline a reasonable number of possible alternative solutions to the brief. This corresponds to stage C of the Plan of Work. The number will depend on the size of the building and, in spite of the expense of making them, these outline proposals should be sufficiently developed for their relative advantages and disadvantages to be described to the owner. They could, for instance, be based on 1, 2, 3 and 7 of **planning useful spaces**, on the **layout strategy, on planning the lay-**

out, and on the strategic decisions on functional systems that have been agreed with specialist designers. The substances which would be seen from the outside should be listed.

5. Calculate the design efficiency ratios, or indexes, for the external walls girth, the gross floor area, and the internal walls girth for each alternative proposal (see Chapter 8). These are obtained by dividing their quantities by the net useful floor area.
6. Prepare a report and presentation drawings for the owner, and discuss and amend as required. Even when alternatives are not being offered, a progress report should still be made to the owner. This should always include a financial appraisal. Obtain approval to proceed with the chosen design.

SCHEME DESIGN STAGE

1. Locate the temporary bench mark and assign its notional level. Decide the finished levels of the tops of ground and upper floors, external paving and other horizontal planes relative to that of the benchmark.
2. Prepare multiplane orthographic general location drawings suitable for specialist designers and regulatory agencies. Identify the subdivision of the building into compartments and the escape routes in case of fire. Show windows, doorways and other alternative technical solutions. Calculate overall spatial dimensions by adding the dimensions of the rooms etc. and wall thicknesses.
3. Check the design efficiency indexes.
4. If a dwelling, assess its likely SAP (Standard Assessment Procedure) Energy Rating (see Building Regulations Approved Document L).
5. Prepare a performance specification for the enclosure elements and decide on the kinds of products to use for visible parts.
6. Have specialist designers prepare design drawings for the building structure, the interior, mechanical, plumbing, electrical and other services.
7. Identify the visual attributes to be provided.
8. Prepare a site plan showing the location of the building and the external works.
9. Prepare a block plan showing the site and its location.
10. Apply for full planning permission from the local Planning Authority or other regulatory agency.
11. Prepare a financial appraisal for the project as a whole (see Chapter 20).
12. Check that the proposals meet the requirements of the brief and will accord with the owner's value system. Do they also comply with the answers to questions made during the project initiation or pre-design

stages? Will they give an appropriate impression, as discussed in Chapter 2? Amend the proposals as appropriate.

13. Prepare a final report and presentation drawings for the owner, and discuss and amend as required. Obtain approval to proceed with the chosen scheme.

7.8 SUMMARY

After looking at some historical views on architecture, the notion of functional elements was introduced. It was suggested that element functions and attributes were either regulative, supportive, visual or economic. The internationally recognized stages in a building project were reviewed.

This chapter considered the outline and scheme design stages. The private information which the designers would need, the constraints on their freedom to act, and the public knowledge they might employ were discussed. The SfB classification system was introduced. Matters to be considered when planning the useful spaces and the overall layout were listed. It concluded with a suggested procedure for arriving at alternative outline designs and their functional specifications, the development of the chosen scheme and its approval by the owner.

REFERENCES

Alberti, Leon Batista (1991) *On the Art of Building in Ten Books*, translated by Joseph Rykwert, Robert Tavenor and Neil Leach, MIT Press, Cambridge, Mass.

Palladio, Andrea (1965) *The Four Books of Architecture*, facsimile republication of the work originally published by Isaac Ware in 1738, Dover, New York and Constable, UK.

Vitruvius (1960) *The Ten Books of Architecture*, translated by Morris Hicky Morgan, Dover, New York and Constable, UK.

Wotton, Sir Henry (1970) *The Elements of Architecture*, facsimile reprint of the first (1624) edition, Theatrum Orbis Terrarum Ltd, Amsterdam and Da Capo Press, New York.

FURTHER READING

The design process

Baker, G. H. (1989) *Design Strategies in Architecture*, E & FN Spon, London.

Broadbent, G. H. (1988) *Design in Architecture*, David Fulton, London.

BS 5606, Accuracy in Building, The British Standards Institution, London.

Durant, David N. (1992) *The Handbook of British Architectural Styles*, Barrie & Jenkins, London.

Gordon, D. E. and Stubbs, S. (1991) *How Architecture Works*, Van Nostrand Reinhold, New York.
MacDonald, A. J. (1994) *Structure and Architecture*, Butterworth Architecture, Oxford.
Stollard, Paul and Abrahams, John (1991) *Fire from First Principles*, E & FN Spon, London.

Planning data

Neufert, Ernst (1980) *Architects' Data*, Granada, London.
Tutt, Patricia and Adler, David (eds) (1979) *New Metric Handbook, Planning and Design Data*, Butterworth Architecture, Oxford.

General reference

'The art of architecture', in *The New Encylopaedia Britannica*, 15th edn (rev.), Vol 13.
Brett, P. (1993) *Building Terminology*, Newnes, Oxford.

Design efficiency 8

8.1 ELEMENT RATIOS

Buildings with the same purpose can differ considerably in both their size and their shape. They will also differ in the extent to which they provide useful accommodation rather than just space, and in the relative amounts of their functional elements. That is, they will differ in their **design efficiency**. The study of the physical form of buildings and of the quantitative relations between their enclosing elements is sometimes called **building morphology**.

As seen from your chair, the building effectively stops at the finish to the external walls. The thickness of those walls is of little interest to the occupier and, in any case, other buildings may have walls of different thicknesses.

When comparing the size and shape of different buildings, at least in the UK, the convention is to measure within the inner faces of their external walls. Every **storey**, i.e. each floor and its walls, is separately measured. The amount of useful accommodation is usually measured by its floor area, and we shall return to this later. To make comparisons between buildings, each must be represented by a single quantity.

We shall start by looking at the proportion, or ratio, between (a) the sum of the girths of each of its storeys, and (b) the overall or **gross area** of all the floors in a building. We shall also consider how this **element ratio** is affected by its size and shape. In all cases, we shall express the ratio as the amount of (a) for each unit of (b). That is, the ratio will be the quotient resulting from dividing (a) by (b).

However, these ratios only help us to understand the differences between various sizes and shapes of buildings. They do not tell us how efficient the designers have been in achieving the owner's requirements. This we shall investigate in the next section by considering the ratios between the gross floor area, the area of the external walls, the internal walls areas and the **useful floor areas** (or those required by the owner, if these are less).

THE RATIO BETWEEN THE PERIMETER AND THE GROSS FLOOR AREA

The ratio will depend on the plan shape, the number of storeys, and the size of the building. For simplicity in demonstrating these relations, we shall work in single-storey **building units** where each unit is a square on plan. Figure 8.1 represents a single unit square building with sides of length P and an area of A. Figure 8.2 represents a similar square building with sides of length $2P$ and a floor area of $4A$.

In Figure 8.1, the ratio between the girth of its perimeter and the area is $4P/A$. As we are only considering the numerical ratios, its P/A ratio will be 4. Similarly, the P/A ratio of Figure 8.2 is 8/4, i.e. 2, only half that of Figure 8.1. Thus, increasing the size but keeping the shape will proportionately reduce the ratio, and reducing the size will increase it.

Figure 8.3 represents another shape of four-unit building. Its external walls girth has increased from $8P$ to $10P$, and its P/A ratio to 2.5. Figure

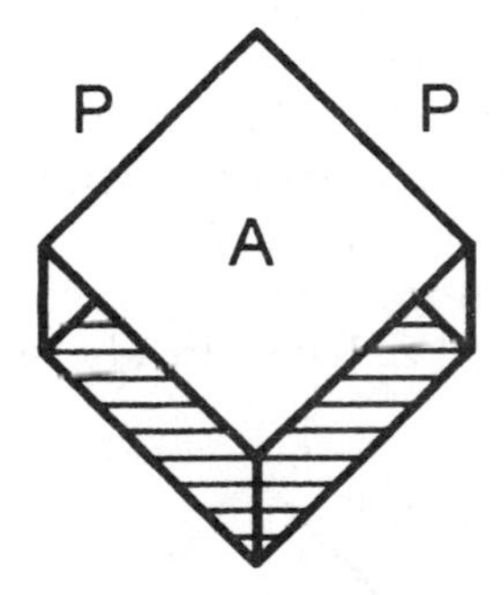

Fig. 8.1 Planometric schema of a single-unit building

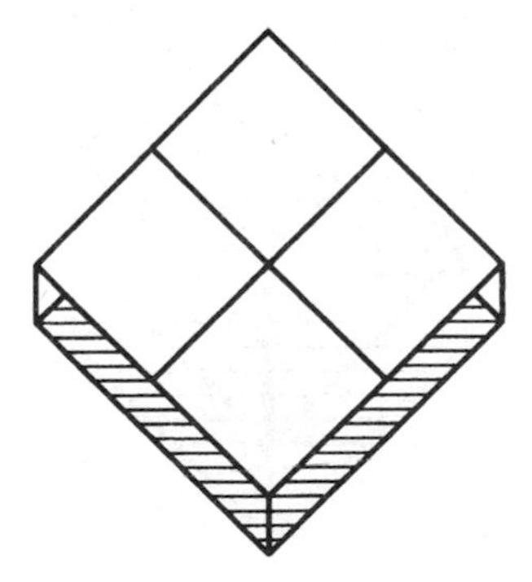

Fig. 8.2 Planometric schema of a four-unit building

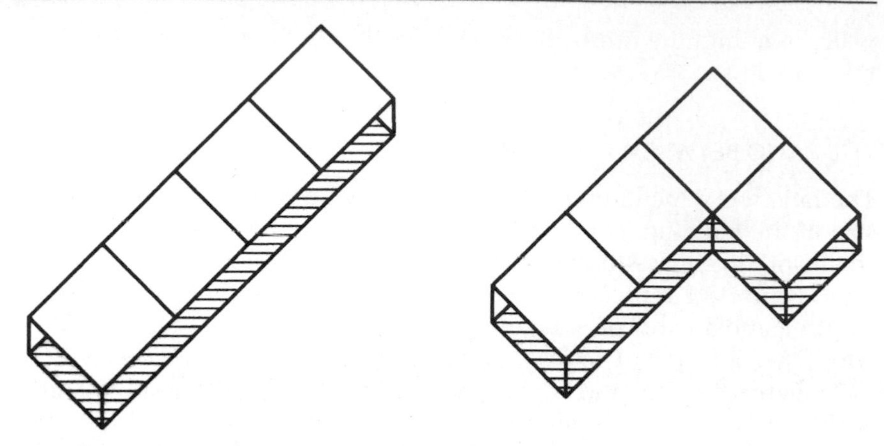

Fig. 8.3 and **8.4** Schemas of non-square four-unit buildings.

8.4 shows that the external girth remains the same when the units are rearranged into a 'dogleg' shape. However, an irregular outline affects the buildability of the design and increases the cost of construction. Note that this girth could also enclose a simple rectangle containing six units with a *P/A* ratio of 10/6, or 1.66.

Consider what happens when the shape on plan and the gross floor area remain the same, but the area is distributed over more than one floor. Figure 8.5 shows four single-unit Figure 8.1 storeys one on top of the other. As we are measuring each storey, its *P/A* ratio will be the same as that of Figure 8.1, i.e. 4. We have already seen that this ratio is twice

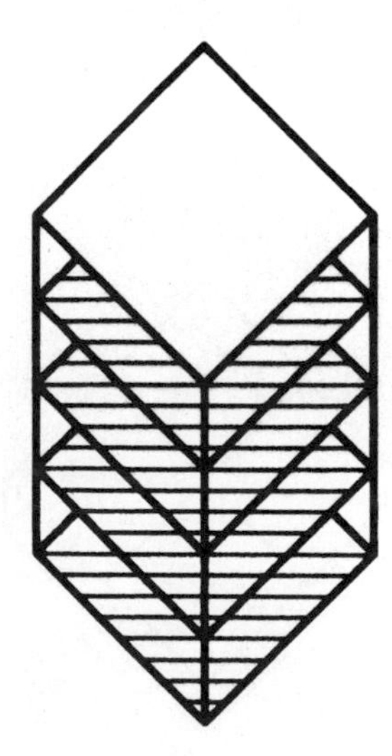

Fig. 8.5 Schema of a multi-storey (story) single unit building.

what it would have been had all the floor area been provided in a single storey.

Similarly, the *P/A* ratio for a nine-storey building would be three times as much as that of a single-storey building of equivalent area. The rule is that the ratio varies with the square root of the number of storeys (n).

Notice how changing from single-storey to n-storeys also affects the horizontal elements. The area of the lowest floor is now only $1/n$ of what it was, the remainder being replaced by upper floors (which will cost more). The roof area, excluding any overhangs, is always the same as that of the floor next to the ground, even when one portion of the building is higher than the rest.

As we have seen, the lowest possible external girth can be achieved by having buildings that are square on plan. Not only will this minimize the initial cost of the external walls, it will also minimize the cost of keeping it warm or cool.

Examples of square on plan buildings include some college libraries and warehouses. A drawback may be that opportunities for having windows are also minimized, and natural light may only dimly penetrate into their interiors. Users of libraries will need a high level of artificial lighting in any case. The natural lighting of a single-storey warehouse is likely to come mainly from roof lights.

Alpine chalets are often square on plan, although this is disguised by their having gable-ended roofs with wide overhanging eaves and verges. Low levels of natural lighting may appeal to farmers who spend their days in their fields. However, in most buildings it is important for natural lighting and ventilation to reach most of the rooms. The plan shapes of Figures 8.3 and 8.4 remind us of blocks of apartments, hotels, schools and hospitals.

8.2 MEASURING DESIGN EFFICIENCY

FLOOR AREA ANALYSIS

In Chapter 7, we identified four different kinds of spaces, which we called:

1. the required useful areas,
2. the ancillary areas,
3. the mechanical and electrical services areas, and
4. the circulation areas.

These four kinds of horizontal areas, together with the area occupied by internal walls, partitions, chimney stacks and other **internal divisions**, will add up to the **gross floor area**. In the UK at least, this is measured within the enclosing external walls. In North America, the horizontal area occupied by the external walls is also included to give the **architectural area**. In other words, this is measured to the outsides of the external walls.

In the USA, the floor area available for each occupant of a block of flats or some other shared building may be called the **net assignable area**. The sum of these areas, plus the **common parts area** and the **internal divisions** will equal the gross floor area.

USEFUL FLOOR AREA INDEXES

Most of the decisions made early on in the design process have a greater influence on the project than those made later. What is required at the earliest possible time are pointers, or **indexes**, that will indicate whether the initial and the running costs of the emerging design are likely to be economical or not. The higher the index, the greater the relative amount of the element, and the greater its demands for resources, including finance.

This analysis can (and should) be carried out as soon as the horizontal dimensions of the rooms etc. are known. Design efficiency indexes are the ratios calculated by dividing the measured quantity of each enclosure element, including the gross floor area by the **useful floor area**. In North America, these indexes are actually called ratios.

It can be argued that the useful area used in all these calculations ought to be the total required by the owner rather than that provided in the design, as these will never be quite the same. But, in any case, where required room areas are given in the brief, the ratio between these and the areas actually provided by the design, i.e. the **actualization index**, should be produced.

Each index can be compared with those of other designs or actual buildings of the same type. However, as we have seen, the extent of any one element will depend on the overall size, shape and number of storeys of the building. Having more of one usually means needing less of another, e.g. the more non-square, i.e. the narrower the building is, the greater the extent of the external walls, and the less the need for internal walls.

Thus, where possible, it is advisable to compare buildings of similar size and shape. At this early stage in the design process, we can only look for pointers to design efficiency or otherwise. Even so, eliminating uneconomical schemes now can save much time, money and heartache later.

We will now consider how to obtain the quantities for calculating the following indexes:

1. **external walls girth index**, i.e. the sum of the perimeters of the storeys divided by the useful floor area (we adopt the spreadsheet convention

of the oblique line, or 'slash' to indicate 'divided by'; in the example given later, this is 44.40/83.75 = 0.53);
2. **gross floor area index,** i.e. the gross floor area divided by the useful floor area (96.40/83.75 or 1.15);
3. **internal walls girth index,** i.e. the sum of the lengths of the internal walls and partitions in every storey divided by the useful floor area (20.20/83.75 = 0.24).

8.3 CALCULATING GIRTHS AND AREAS

Most buildings are composed of plane constructions with horizontal floors and ceilings and vertical walls with right-angled junctions. We can describe these constructions and the spaces they enclose as being **rectangular**, or **orthogonal,** and we shall limit our discussion to such buildings.

Again, we shall begin with an existing building, although, in practice, the ideas would be applied mainly to the designs for new projects. Should the reader feel encouraged to investigate the design efficiency of, say, a dwelling, and dimensioned drawings are not available, a dimensioned plan of the rooms and the outline of the building envelope should first be sketched as in Figure 8.10.

Start in a room that is a simple rectangular box in shape. Wherever you are in the room, the horizontal length of the space between the shorter opposite walls will be the same. A wood lath of slightly less than that length (to allow for irregularities) should fit between the walls wherever it is held.

This lath can be imagined as a horizontal line with an 'open' arrow head at each end, and the room can be imagined as being filled with such arrowed dimension lines, all identical and parallel to one another (see Figure 8.6). Those against the ceiling, the two longer walls, and the floor will indicate the extent of those constructions in that direction.

In the same way, the space within the room can also be imagined as being filled with arrowed lines between the longer opposite walls, or between the floor and the ceiling. On a floor plan or a vertical section drawing, one such arrow in each direction will represent them all.

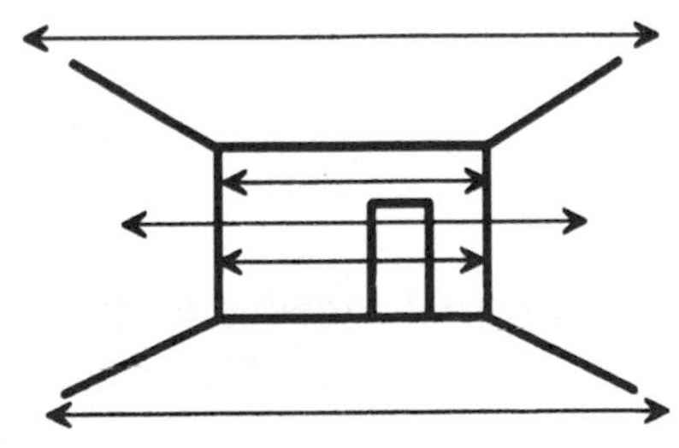

Fig. 8.6 Room perspective with arrowed dimension lines

Dimensions are rather like the scalars of vectors and, to be consistent in our consideration of Works data, we shall adopt the '*i*, *j*, *k*' vector notation in future. '*i*' will refer to dimensions 'across' floor plan drawings, '*j*' will refer to those 'up and down' the drawing, and '*k*' will refer to heights or depths.

To distinguish the dimensions of these spaces from, say, wall thicknesses, we shall be calling them **spatial dimensions**, or '**S-dimensions**'. A more detailed discussion of S-dimensions and the relations between them is included in Chapter 13. Note that the length or width of a floor or ceiling can be obtained just as precisely by measuring the lengths of the walls instead.

DIMENSIONAL DATA FORMATS

Calculating any quantity involves the following two stages:

1. obtaining and recording the dimensions;
2. carrying out arithmetical operations on these dimensions.

In computer jargon, these two stages might be described as (1) data capture, and (2) data processing.

In the north of England and in Scotland, the custom has been to set down the data on identical objects as a single row using a format similar to that in Figure 8.7. (The contents are in imperial units and have no relevance to building morphology.) This format provides columns for:

1. the number of identical objects,
2. up to three variable dimensions '*L*' (length), '*B*' (breadth or width) and '*H*' (height or depth),
3. the resulting quantity, and
4. the reference or location of the object(s) being measured.

(1)	(2L)	(2B)	(2H)	(3)	(4)
	120.0	3.0	4.6	1620.0	18" walls
2	24.0	2.6	3.0	360.0	9" walls
				27) 1980.0	
				73.3	

Excavate surface trenches not exceeding 5'0" deep and cart away
73 cubic yards

Fig. 8.7 Historical measurement format – northern UK.

(1)	(2)	(3)	(4)
2/	24.0		Excavate surface trenches not
	2.6		exceeding 5'0" deep and
	3.0	360.0	cart away 9" walls

Fig. 8.8 Historical measurement format – remainder of UK.

The description of the object is written below the rows of numbers and across the full width of the format. This arrangement also appears to be in use in North America.

In the remainder of the UK, the format has usually been compressed by putting the dimensions of each object one under the other in a single column, the last in each set being underlined. Column (4) is widened to accommodate the description and the format is shown in Figure 8.8. This arrangement is more suitable for the 'half-width' pocket-sized dimension book used by surveyors when measuring on site after the building had been built.

We shall base our own approach on the arrangement in Figure 8.7. Being tabular, the data can be set out in the same way whether the entries and calculations will be made manually or on a computer spreadsheet. Note that in Figures 8.11 and 8.13, the *i* and *j* dimensions are given first, and followed by the number of identical spaces, as this is the order in which the data will be captured. The symbols used in spreadsheets, e.g. + (plus), – (minus), * (multiply), / (divide) and ^ (to the power of), will be used to indicate arithmetical operations.

The table, or the equivalent screen display of a computer spreadsheet, offers a rectangular array of spaces, called **cells,** with lettered columns and numbered rows. A cell is a kind of box for storing and displaying either a numerical value, some text, or a mathematical formula and the results. The address of a cell containing a value to be used in a formula is its column capital letter and row number.

ROOM GIRTHS AND AREAS

In the approach described below, the floor girths and areas will be calculated by the formulas $(i + j) * 2 * n$ and $i * j * n$. Readers who are more comfortable with imperial units should measure either in inches, or in feet to two places of decimals. The number of identical rooms or other spaces (n) is always given although, conventionally, this number would be stated only if there was more than one of that size.

Let us assume that plan (a) on the left in Figure 8.9 represents a rectangular room whose horizontal dimensions are (i) 5.50 m and (j) 4.50 m.

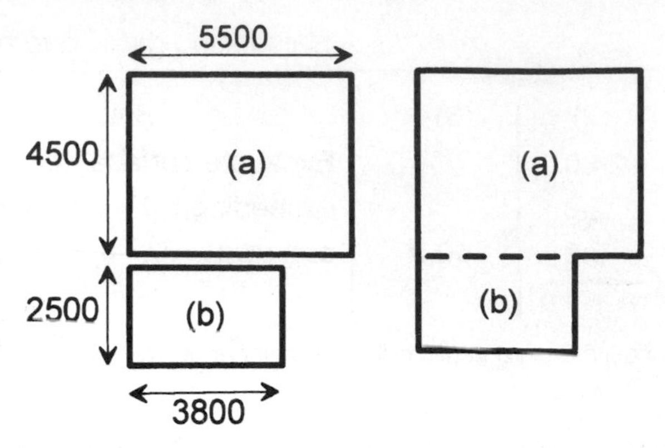

Fig. 8.9 Floor plan of the room used to illustrate the 'spaces' approach to calculating room girths and areas.

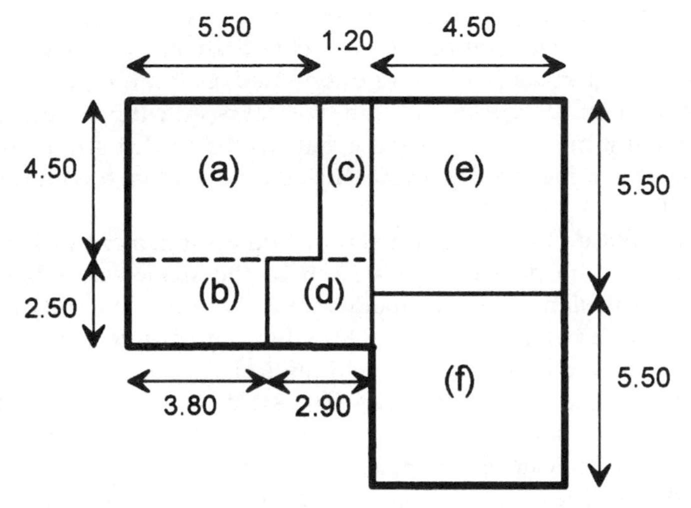

Fig. 8.10 Floor plan of building used in example of calculating the useful floor areas.

	A	B	C	D	E	F	G
1	*Ref.*	*(i)*	*(j)*	*Number*	*Girth*	*Horiz. area*	*Useful area*
2	(a)	5.50	4.50	1.00	20.00	24.75	24.75
3	(b)	3.80	2.50	1.00	12.60	9.50	9.50
4	(a-b)	3.80	0.00	-1.00	-7.60	0.00	
5	(c)	1.20	4.50	1.00	11.40	5.40	
6	(d)	2.90	2.50	1.00	10.80	7.25	
7	(c-d)	1.20	0.00	-1.00	-2.40	0.00	
8	(e,f)	4.50	5.50	2.00	40.00	49.50	49.50
9	Totals	17.40	25.00		84.80	96.40	83.75

Fig. 8.11 Capturing rooms dimensions and calculating the useful floor area.

This data and the calculations can be recorded as shown in columns B–F of row 2 of Figure 8.11.

When dealing with a room that, although orthogonal, is 'L' shaped or otherwise complex on plan, it should be analysed into simple rectangular spaces. There is always some choice in their arrangement.

Let us imagine that the space (a) above is one constituent of the 'L' shaped room also shown in Figure 8.9. Space (b) is the other constituent, with (*i*) = 3.80 m and (*j*) = 2.50 m. This data and their quantities are shown in row 3 of Figure 8.11.

The boundary between these two constituent spaces is indicated on Figure 8.9 by a broken line, and is within the room and not against the walls. Thus, to obtain the correct girth of the complex space, twice the length of this boundary must be deducted from the sum of the girths of the constituent spaces. However, as the formula for the girth includes for the doubling of the dimensions, we need only enter the length of this **adjustment line** once, with –1 as the number to give it a negative value. To maintain consistency, it is given a width of zero and labelled (a–b) as shown on row 4 of Figure 8.11.

In Figures 8.10 and 8.11, this room is incorporated into a larger example with circulation spaces (c) and (d) and rooms (e) and (f). The quantities for each floor of a multi-storey building should be separately totalled, and the useful areas extracted. As this exercise would be done during the outline stage of the design process, we can expect the walls to be shown as single lines on the plan of the building, and we shall ignore their thicknesses.

BUILDING ENVELOPE HORIZONTAL AREA AND GIRTH

First, the total space enclosed by the external walls must be analysed into an arrangement of simple rectangular spaces (see Figure 8.12). The dimensions of these constituent spaces and their adjustment lines can usually be built up from the spatial dimensions of the individual rooms.

Enter the horizontal dimensions and numbers of identical spaces in columns B, C and D and calculate the girth and the horizontal area in columns E and F as shown on Figure 8.13. Separately measure each storey of a multi-storey building. The indexes are obtained by dividing the girth and the horizontal area by the useful floor area from Figure 8.11. Both Figures 8.11 and 8.13 give the same total horizontal area, thus confirming the accuracy of our data.

CALCULATING THE GIRTH OF INTERNAL WALLS

At least one wall of the room you are in is likely to be an external wall,

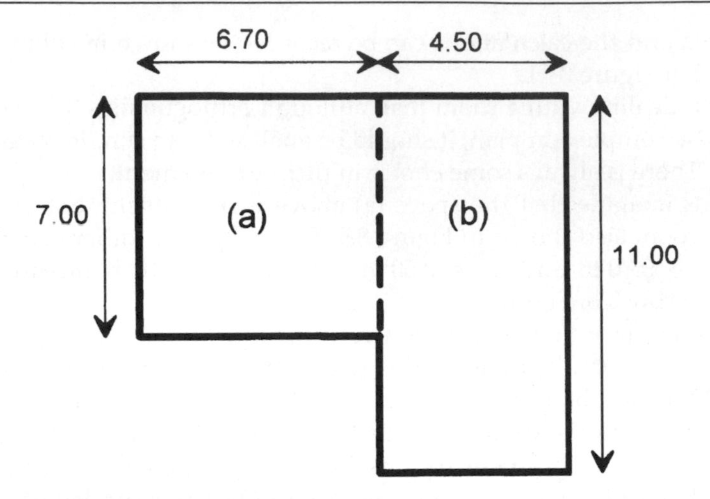

Fig. 8.12 Outline plan of building used in the example of the 'spaces' approach to calculating the girth and area of the envelope, its design efficiency indexes, and the rate of heat loss.

	A	B	C	D	E	F
1	*Ref.*	*(i)*	*(j)*	*Number*	*Girth*	*Horiz. area*
2	(a)	6.70	7.00	1.00	27.40	46.90
3	(b)	4.50	11.00	1.00	31.00	49.50
4	(a-b)	0.00	7.00	-1.00	-14.00	0.00
5	Totals				44.40	96.40
6	Indexes				0.53	1.15

Fig. 8.13 Capturing envelope dimensions and calculating the external girth and horizontal area indexes.

the remainder being internal walls. In our example, we know the following for that floor:

- the sum of the girths of all the rooms etc., i.e. 84.80, and
- the total girth of the inner face of the external walls, i.e. 44.40.

When the latter is deducted from the former, we are left with the total girth of the faces on both sides of the internal walls of 40.40. Half of this is the girth of the walls themselves, i.e. 20.20. This can be checked by reference to Figure 8.10. Divide this length by the useful floor area to obtain the index, i.e. 0.24.

8.4 CALCULATING THE RATE OF HEAT LOSS

As soon as decisions have been made on floor-to-ceiling heights, it becomes possible to calculate wall areas. Columns are inserted in the table for these '*k*' values and for vertical areas and the enclosure volume, as shown on Figure 8.14. As a result, we are able to obtain the areas of all those elements that make up the envelope and subdivide it into floors and rooms etc.

The areas of the finishes to the ceilings, walls and floors can be calculated in the same way. By inserting columns for their likely average costs per unit and for extensions and totals, we could begin to build up an estimated cost for the Works. This version of the dimensional model is sometimes called a **cost model**.

However, at this stage we are concerned with predicting design efficiency and we shall use the model to calculate the rate of heat loss of the building shown on Figure 8.10. This **elemental method** can be modified to suit whichever calculations are required to satisfy Building Regulation L1, Conservation of fuel and power.

Predicting the thermal transmission of an element was considered in Chapter 4. This U-value is expressed in watts for each m^2 of the element, when the difference in temperature between the external and internal faces is 1° Celsius.

At this early stage in the design process, the rate of heat loss from a building can only be estimated from the proposed U-values for the envelope elements, based on expectations regarding the relative amounts of each technical solution. For example, where 35% of the external walls

	A	B	C	D	E	F	G	H	I
1	*Ref.*	*(i)*	*(j)*	*(k)*	*Number*	*Girth*	*Vert. area*	*Horiz. area*	*Volume*
2	(a)	6.70	7.00	2.60	1.00	27.40	71.24	46.90	121.94
3	(b)	4.50	11.00	2.60	1.00	31.00	80.60	49.50	128.70
4	(a-b)	0.00	7.00	2.60	-1.00	-14.00	-36.40	0.00	0.00
5	Totals					44.40	115.44	96.40	250.64
6	Thermal transmittance co-efficients etc., say						1.31	0.70	
7	Total thermal transmittance						151.23	67.48	
8	Rate of heat loss per degree (W/K) =							218.71	

Fig. 8.14 Calculating the rate of heat loss.

area are likely to consist of double-glazed wood windows or doors (U-value 3.0), and the U-value for the exposed walls will be 0.40, the average will be 1.31. The rate of heat loss for a 1° difference in temperature can be calculated by the formula used in Figure 8.14, i.e.

1. (external walls area * average U-value) +
2. (horizontal area * (U-value for roof + U-value for ground floor)).

A heat loss index can be calculated by dividing the total heat loss by the useful floor area. The aggregate volume of the various constituent spaces is also shown in case the energy required to heat incoming air is also to be calculated.

The dimensions model can be amended as the design is developed and, particularly if on a spreadsheet, will continue to provide up-to-date indexes.

8.5 COMMENTARY

In the UK, buildings are usually classified according to their purpose, using Table 0 of the CI/SfB system mentioned in Chapter 7. It is arranged in the form of a **hierarchy**, or **inverted tree-structure** with a number of levels, as in Figure 8.15.

This table assigns three numerals to each class of building to act both as a pointer, or **index**, and as a label for data. The first digit refers to one of the very general categories, e.g. Education, Scientific and Information = 7. The second digit refers to one of its sub-categories, e.g. Schools = 1. The third digit refers to one of the basic classes, e.g. Secondary and high

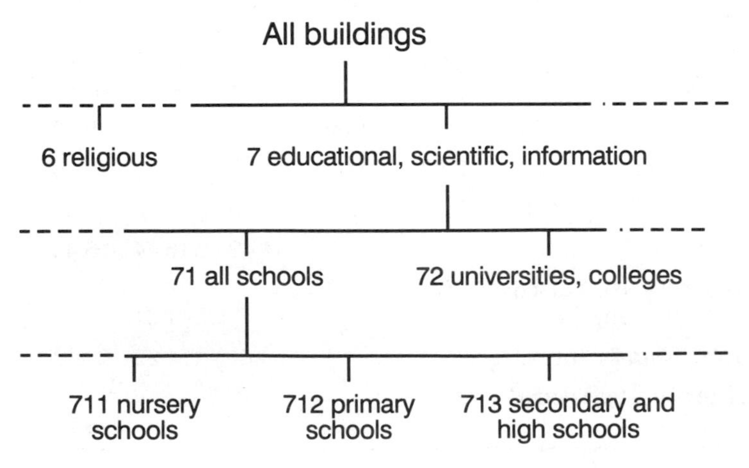

Fig. 8.15 Extract from Table 0 of CI/SfB – occupancy purpose groups.

schools = 3. Thus, the index for this class is 713. The nine general **occupancy classes** used in the USA are rather different, and both differ from those in the Dewey Decimal Classification which is used in libraries in the UK.

8.6 SUMMARY

The chapter has been concerned with predicting the economy (or otherwise) of a design at the outline, or schematic design stage. The useful floor area in a design is what matters most. The less the construction works necessary to achieve this, the less the cost is likely to be. We explored the regularities of form to be found in orthogonal buildings, and the relative amounts of the enclosing elements of buildings with differing shapes, sizes and numbers of storeys. Ratios that point to the efficiency of one design compared with another we called **indexes**. They were obtained by dividing the following quantities by the useful floor area:

1. the external walls girth,
2. the gross floor area;
3. the internal walls girth.

The quantities and indexes were calculated from the lengths, breadths and heights of the rooms etc. and of the building envelope, using a tabular format similar to that of a computer spreadsheet. These spatial dimensions we called 'S-dimensions'. A method of generating internal wall girths was described. Complex orthogonal spaces were treated as sets of simple constituent spaces with negative S-dimensions at their common boundaries. The format was also used to demonstrate how to calculate the rate of heat loss from a building.

REFERENCE

UK Building Regulations, *Approved Document L*, HMSO, London.

Detail design

9

9.1 TECHNICAL SOLUTIONS IN GENERAL

The information on the scheme design discussed in Chapter 7 was mainly about the locations of the various elements, their functions and attributes, and the dimensions of the spaces they enclosed. These decisions would have been recorded on location drawings and site and block plans (see Chapter 10) which would also show where these functions and attributes differed, e.g. at doorways and windows. Strategic decisions would also have been made about the classes of main products to use, particularly those that would be visible.

FEATURES OF TECHNICAL SOLUTIONS

During this next stage, which corresponds with stage E of the Plan of Work, decisions must be made on what attributes the elements must have, and on the technical solutions that will provide them. Details of these attributes are likely to remain in the designers' heads, only the details of the technical solutions being recorded.

Part of the stock-in-trade of the design team consists of a working knowledge of a wide variety of these solutions. Regional variations are to be expected. Skill in detailing, however, lies in selecting those element technical solutions that are compatible with one another and that can be joined together satisfactorily. The process will inevitably involve some iteration.

The subsections of this chapter follow the order of the subsections in Chapters 1, 2 and 3. They are not concerned with offering specific solutions, which should always reflect the requirements of a particular scheme. Instead, they are intended to provide a background of decision-making factors which, when applied to the circumstances of that scheme, should point to possible solutions. For each part of an element, decisions must be made on:

1. its relative position within the element;
2. its substance;
3. acceptable dimensional tolerances;
4. any dimensions other than those based on the already decided dimensions of the spaces, i.e. the width and/or thickness of a part constructed in place, and all three dimensions of a door or other part constructed by fixing a component; we shall be calling them technical, or T-dimensions;
5. how gaps between adjoining parts will be covered or sealed;
6. what main and dependent products to use in its construction;
7. whether to stipulate its construction method or rely on custom and practice;
8. its surface finish.

Decisions will be recorded on assembly drawings or given on schedules and elaborated in the specification. When the detail design is split between two organizations for the reasons discussed in Chapter 17, these decisions may be made in the following two stages:

1. Decide on the attributes of the parts and specify (1) to (5) above. This is called a descriptive specification.
2. Decide on (6) to (8) above to match the requirements in (1) to (5). This is called a prescriptive specification.
   These specifications are described in the next chapter.

SOURCES OF TECHNICAL INFORMATION

Although most buildings have a unique size and shape, usually their technical solutions will have been well tested in earlier buildings. Descriptions of these 'portable' technical solutions will be found in product manufacturers' literature, text books, and research and other publications. They also feature in the UK's Building Regulations. Students learn about them during courses in building construction.

However, four important changes have been taking place in the UK in recent years.

1. Codes of practice have been increasing in number. Each relates to a specific aspect of the design and production of building and civil engineering construction and is the work of specialists within the industry. Codes of practice are published as British Standards by the British Standards Institution and are being harmonized with those in other EU countries.
2. The approach adopted in the Building Regulations has changed. Earlier versions consisted of lengthy performance requirements interspersed with 'deemed-to-satisfy' descriptive and prescriptive technical

solutions. The current Regulations (1991) are concerned solely with health, safety, energy conservation and the welfare and convenience of disabled people. Acceptable technical solutions in certain circumstances are illustrated in separate **Approved Documents** with capital letter identifiers. Where these apply, designers are given the option of either adopting them, or of demonstrating to those responsible for building control that either the rules in the relevant codes of practice have been followed, or that some other approach is appropriate. Their requirements are not repeated in this book as they are always subject to revision. However, the letter of the related Document is given at the start of each subsection where it might apply.

3. The increasing influence of the EU, including the acceptance of the certification standards of other nations when specifying construction products.
4. The manufacture of products and the provision of professional and other services of an assured and consistent quality (but not necessarily to the requirements of a related British Standard) through the application of BS 5750 – Quality Systems. This standard has now been revised and renumbered as a series beginning with BS EN ISO 9000.

GENERAL DEPENDENCES

Chapter 7 introduced the idea that almost all scheme design decisions depend on what had been decided earlier. Similar dependence relations exist between the scheme and detail design procedures. The following are some of the more general dependences which constrain the detail design sequence.

- For many parts of elements, the choice of technical solution will depend on whether strength and stability will be provided by loadbearing walls or by a structural frame. This decision should have been made during the scheme design stage.
- Decisions on the substances of many of the visible parts of elements should have been approved by the owner at the end of the scheme design. These decisions will determine the classes of construction products from which the final choice must now be made. For instance, an earlier decision to paint an external rendering must now be made specific by choosing the brand of paint and its colour.
- Decisions on specific main products will also point to the kinds of dependent products that will also be needed.
- The widths and/or thicknesses, i.e. the technical or T-dimensions, of the parts of an element (see above) will depend mainly on their substances, their loadings, and their height, span, or some other spatial, or S-dimension(s). This distinction between S- and T-dimensions is considered further in Chapter 13.

- The products used at a junction between two different technical solutions will depend on those of the parts being joined.
- The exposed sides of an open framework must be given a continuous lining.

Examples of such decision sequences are:

- an earlier decision to use a composite class AB substance such as reinforced concrete must now be converted to decisions on the T-dimension(s) of the members (see above), and on the strength of the concrete and the diameters, lengths, shapes and positions of the steel reinforcing bars;
- where a formed class B main product is chosen, the manufacturer will have already decided most of its fittings, accessories, fixings and methods of jointing. These must now be identified and specified.

DETAIL DESIGN SEQUENCE

The detail design sequence is, to some extent, a matter of personal choice, although some iteration may be necessary to ensure that everything fits. One approach is to deal with each element in turn, particularly when using one of the extended lists of elements mentioned in Chapter 7. For consistency however, we shall adopt the sequence in Chapters 1, 2 and 3 although only for the parts of the building enclosure elements. Additional insulation will be treated as a part of the structure. Services and installations will be somewhat dependent on these and so must come afterwards.

Where appropriate, each subsection will start with the following:

- **dependences**, i.e. other design decisions and factors on which decisions regarding the technical solution for the part being considered are likely to depend;
- **special attributes**, i.e. those attributes that are the main reasons for including that particular part in the element;
- **Approved Documents** The letter references of those Building Regulations Documents (if any) approved by the Secretary of State for the Environment and relevant to the subsection. The Approved Document on Materials and workmanship (Regulation 7) will be assumed to apply generally. A list of related Standards and other publications is given at the end of every Document.

The following more general attributes will be left to the reader to apply as appropriate:

- continuous;
- connectable;

- buildable;
- safe to build;
- durable;
- dimensionally stable;
- suitably fire-resistant;
- maintainable;
- renewable;
- chemicals resistant;
- safe to occupy;
- built with products from renewable sources;
- energy efficient.

Types of specifications are discussed in the next chapter.

CONSTRAINTS ON TECHNICAL FREEDOM

The Building Regulations and other rules in the public domain limit or **constrain** the freedom of action of the designers. Other constraints will include the express wishes and the private knowledge of the owner about the way the building is to be used. It is always possible that some of these may have been overlooked when the design brief was being prepared. While the designers profess to having appropriate public knowledge, they will only know what they have been told concerning the owner's private information. With experience, they learn what questions to put to the owner's project team. Constraints will include:

- the express desires of the owner;
- the purpose of each space and any adverse agencies involved in its use, e.g. oil, grease, mobile equipment, heavy or concentrated applied loadings;
- general and exceptional sound levels likely to be generated by occupiers;
- the sound levels acceptable by occupiers of adjacent spaces;
- acceptable intervals for renewing the various finishes and external and internal decorations;
- the maximum size of objects to be taken through doorways;
- the owner's value system regarding, e.g. energy conservation, substances from renewable sources;
- possible alternative uses for the building that would call for higher standards of, say, strength, stability, or fire resistance in the future; it is more economic and satisfactory to build to these higher standards in the first place, rather than to have to modify the building later;
- the obligation under the CDM Regulations to ensure the health and safety of those who will construct, maintain and occupy the building.

PRIME COST AND PROVISIONAL SUMS

An expenditure that can be directly attributable to a particular service is called its **prime cost** (see Chapter 16). Prime cost (PC) sums are the amounts of money to be included in a contractor's bid to pay for certain services from others. They will be stipulated in the production documents and will be for:

- the purchase of specialist products from suppliers nominated by the design team, and
- the execution of work by specialist subcontractors similarly nominated.

Often, the extent of works to be undertaken will be uncertain, e.g. when remedying the effects of wood rot. A budget for the actual expenditure can be created by instructing bidders to include a stipulated **provisional sum** of money.

9.2 PARTS VISIBLE FROM THE INSIDE (SEE CHAPTER 1)

APPLIED INTERNAL FINISHES

- **Dependences** Whether the background structure is continuous or an open framework.
- **Special attributes** Attractive appearance, light reflecting, wear resistant, easily cleaned, conducive to a quiet environment.
- **Approved Documents** B, E.

Types of finishes were considered in Chapter 1. Specific main and dependent products must now be prescribed for:

1. ceiling, wall and floor finishes, any cornices and skirtings at their junctions, and mat wells at entrances;
2. finishes to staircases (steps and any nosings must be non-slip and highly wear-resistant);
3. stairwells, balustrades and lift entrances;
4. the surfaces of exposed structural members or their casings.

A greater thickness of floor screed may be necessary to accommodate, say, cable ducts within the screeds.

Seal concealed spaces above suspended ceilings and subdivide larger spaces with cavity barriers. Finishes to ceilings and to compartment floors and walls must resist the spread of flame over their surfaces and also over the backs of their bases where these are exposed to the air, e.g. above a suspended ceiling. Design decisions can conveniently be recorded on schedules.

FURNISHINGS

- **Dependences** The substances of the internal finishes and their backgrounds, the size and location of windows, the backgrounds of signs.
- **Special attributes** Attractive appearance, wear resistant, legibility of signs.

Include curtains, blinds, room identifiers or descriptors, and direction, escape route and other signs within the building.

DOORWAYS

- **Dependences** The purpose of the doorway and if in an escape route. The thickness, strength and stability of the surrounding wall, and the type of wall finish. Whether the total of the door areas (plus those of windows) is within the permitted percentage of the floor area.
- **Special attributes** Secure, strong, stable, dimensionally stable, energy efficient.
- **Approved Documents** B, L, M, N.

The choice or design of a door and its hardware, its frame and its fixings, will depend on its purpose. Use safety glazing in vision panels. The choice of architrave will depend on the frame, the wall finish and, possibly, the skirting.

The chosen substance of a door and its frame or lining and architraves will determine suitable kinds of surface treatment.

Identify fire doors and give their minimum integrity period in minutes and if restricted smoke leakage at the ambient temperature is required. Provide each fire door with a self-closing device or a suitable mechanism with an automatic release to keep it open where this is allowed by the Building Regulations.

External doors and their frames and fixings should be able to resist physical attack. Decisions can be recorded on schedules.

WINDOWS

- **Dependences** The thickness, strength and stability of the surrounding wall and the type of wall finish. Whether the total of the window areas (plus those of doors) is within the permitted percentage of the floor area. Their locations and heights above the ground.
- **Special attributes** Transparent or translucent, energy efficient, sound resistant, secure.
- **Approved Documents** B, F, L, N.

The shapes and configurations of windows, dormer windows, and roof lights make a major contribution to the appearance of a building. Their choice or design depends on the need for natural light, visibility, privacy, security and ventilation. Factors in the design of windows etc. will include:

1. the substance, shape and overall size of the frame;
2. the need for privacy;
3. glazing or multiple glazing, and whether with double casements or double-glazing units;
4. the substance of the glazing material and its ability to break safely;
5. any large areas of uninterrupted glazing which should be 'manifested';
6. glazing method, e.g. mastic putty, a drained and ventilated system (see Figure 1.5);
7. the sizes and configuration of the fixed and opening panes and framing;
8. the exposure to the wind;
9. the ease of operation of the opening and closing mechanisms;
10. whether it must be possible to clean both sides of the glazing and to reglaze from inside the building;
11. whether it must be possible to redecorate the outsides of the window from inside the building;
12. hardware, weatherstrips and seals, trickle ventilator, and insect screen;
13. their surface treatment, if any;
14. the structural element that will provide strength and stability, and the method of fixing to it;
15. the blinds or other means of regulating the passage of light;
16. maintaining the safety of occupiers, particularly small children;
17. if to be used as a means of escape;
18. security.

The thickness of the glazing will depend on the pane size and the exposure to the wind, while double or treble glazing will affect the transmission of light, heat and sound. The substance of the window units will determine the kind of protective treatment, if any. Schedules are a convenient way of recording decisions.

STAIRWAYS

- **Dependences** The technical solutions for the upper floors, the floor to floor S-dimension, the maximum number of users.
- **Special attributes** Fire resisting, strong and stable.
- **Approved Documents** B, E, K, M.

A staircase should provide its users with:

1. a progression of regular and level steps at suitable intervals, and wide enough for their purpose; the rise of each step and to a landing must be the same throughout, and treads must all have the same going; avoid tapered treads, i.e. winders;

2. square, i.e. 'quarter-space', landings to limit the number of risers in a straight flight and to enable users to pause for rest; an equivalent space should be kept clear at the bottom and the top of the staircase;
3. handrails for guidance and support to suit the needs of users, including the disabled; balusters should be sufficiently close that small children will be unable to get even partly between them;
4. adequate headroom (at least 2 m) for people and the objects they might be carrying;
5. a safe and smoke-free means of escape in case of fire.

When designing the stairways it must be assumed that lifts will not be used during a fire. The exposed surfaces of wooden staircases should be suitably decorated and are likely to be carpeted. The finish to concrete steps and any nosings must be non-slip and highly wear-resistant. A loft-ladder may be required to reach the roof void through a trap-door.

LIFTS

- **Dependences** Strong, stable, fire-resisting and smoke-free shaft. Internal finishes. Electric power supply. Accessible space for operating gear.
- **Special attributes** Quiet, robust, low maintenance, and having a smooth suitable speed of operation.
- **Approved Documents** B, M.

The maximum rate at which persons will be expected to enter or leave the building or move between floors will indicate the number, capacities and speeds of passenger lifts. Provide sufficient floor space in the lift car for a wheelchair and attendant. Identify the maximum size and weight of any object to be carried by a goods lift.

Lifts are normally chosen from those available from specialists who will both manufacture and install them, and will specify the electrical services required and their locations. Depending on the design, they will require shafts with suitably sized openings and machine rooms, all protected against fire and smoke.

Integrate landing doors, their frames, architraves and other finishes with the rest of the internal finishes. Noise from the operation of the doors, and from motors, signalling devices and passengers, should be appropriately suppressed, depending on the use of the building. Air must be allowed to move freely away from moving lift cars.

FITTINGS

- **Dependences** The sizes of recesses for built-in fittings.

Cupboards and other built in joinery fittings should be chosen or designed, and may need hardware and decoration. Sanitary fittings will be considered with services.

9.3 PARTS VISIBLE FROM THE OUTSIDE (SEE CHAPTER 2)

ROOF COVERING

- **Dependences** The shape and slope of the roof and its exposure. The locations of any abutments against higher walls, and the type of covering approved at the scheme design stage. The locations and sizes of dormer windows and roof lights. Whether on an open framework or a continuous construction.
- **Special attributes** Attractive appearance, weathertight, exclude insects and other living creatures, durable.
- **Approved Documents** A, B, C.

The type of covering and the main product will probably have been chosen as part of the scheme design. While it may be adequate to specify an asphalt covering by referring to the appropriate British Standard, you will probably wish to choose the manufacturer, style and colour of a single lap tile.

Decide the overlap between roofing units and thus the spacing of their longitudinal supports. The roofing unit manufacturer will probably recommend a suitable size for these supports. The choice of main product will also point to those dependent products that will be required to maintain a continuous weathertight covering at the junctions between different slopes and at eaves, verges and abutments against higher constructions. Provide continuous supports for lead valley and other gutters. Are there any junctions requiring flashings?

EAVES AND VERGES

- **Dependences** The shape and slope of the roof covering, the roof structure, the products used in the roof covering and in the exposed surfaces of the external walls.
- **Special attributes** Weathertight, durable, excludes living creatures.

Eaves and verges are the technical solutions to the junctions between the roof covering and structure and the external wall finish and structure. Being interdependent, they are probably best designed together. Projecting the roof covering at eaves and verges will protect the upper faces of the external walls, and their fascias, bargeboards and soffits will enclose the roof structure. The fascia will provide support for rainwater gutters. The widths of fascias and bargeboards will partially depend on the overall thickness of the roof structure.

Ensure that the bottom edges of the units of a tiled sloping roof covering will be supported by the projecting top edge of the fascia and stiffen this with a triangular fillet. Use raised curbs, possibly with metal edge trims, to restrain surface water on flat roofs except at gutters or outlets.

RAINWATER INSTALLATION

- **Dependences** The positions of eaves, the area of the roof, the peak rainfall rate.
- **Special attributes** Weathertight, durable.
- **Approved Document** H.

The roof slopes should direct the water away from the building as quickly as possible and into gutters fixed to the eaves. Then, any overflow due to exceptional rainfall or a blockage will discharge where it will do least harm. Avoid internal gutters if at all possible. Size the gutters and downpipes to suit the likely extremes of rainfall intensity and duration, the roof surface area, and the distance between outlets. They may require decoration.

If downpipes must be internal, avoid joints where possible. Choose products with the most durable substances and joints, as these parts are the most difficult to maintain, and deterioration can go unnoticed until major damage, e.g. wood rot, has occurred.

THE FACES OF EXTERNAL WALLS

- **Dependences** The technical solutions for the structural walls, including the main products to be used in their construction.
- **Special attributes** Attractive appearance, weathertight, exclude insects and other living creatures, durable.
- **Approved Documents** A, B, C.

If a structural wall is to be self-finished, the type(s) of main product(s) will have been approved during the scheme design stage. The technical solution will include the name of their manufacturer, their identifier, colour and texture, the bond, the mortar and the jointing. Brick walls one-brick thick or more can be made of facing bricks backed with common bricks. Are any patterns, e.g. soldier courses, bands, to be created by using bricks of different colour and/or texture?

Examples of applied finishes that might be used to enhance the appearance of a structural wall are given in Chapter 2. Alternatively, the wall can be provided with a rain screen, possibly consisting of an impervious cladding with an attractive appearance, an insulating layer and a vapour control layer, all securely fixed to the structural wall and with a drained and ventilated cavity in between.

EXTERNAL WORKS

- **Dependences** The natural levels of the site and the layout of the new buildings. The activities to be accommodated, including car parking. Access from the highway. Ownership of boundaries. See also the later subsection on 'Clearing and preparing the site'.
- **Special attributes** Safe to use, durable, secure.
- **Approved Documents** B, K.

Specify:

- gates, walls and fencing,
- roads and paving and their curbs and drainage,
- parking barriers,
- site planting, security and other lighting.

VISITOR INFORMATION

- **Dependences** The use of the building.
- **Approved Document** B.

Identify the postal address of the property and, possibly, the name of the future occupier, as instructed by the owner. Position legible car parking direction and exit signs. Should these be illuminated? Arrange for site display boards giving information on the project. Include the names of the owners, the professional team and, possibly, the funder. Suitably restrict the freedom of the constructors to advertise.

9.4 THE BUILDING STRUCTURE (SEE CHAPTER 3)

ROOF STRUCTURE

- **Dependences** The shape and S-dimensions of the plan shape to be covered, the locations of strong and stable supports, the locations and sizes of dormer windows and roof lights, and the substance of the roof structure approved in the scheme design. The dead load of the covering, the expected imposed load of snow and people, and the maximum force of the wind.
- **Special attributes** Strong, stable, energy efficient, fire resisting.
- **Approved Documents** A, B, E, F, L.

The substance, the T-dimensions and the configuration of the roof structure will depend on:

1. whether the roof is to be pitched or flat;
2. the roof covering and its supporting battens or deck;
3. if pitched, the loads to be supported by ceiling joists, e.g. storage cisterns full of water;
4. the habitable or storage spaces to be accommodated and the supporting structure required for their finishes;
5. if a trap-door (i.e. a horizontal door) is to be provided for access;
6. the type of main product to be used to finish the ceiling below;
7. the maximum snow and other imposed loads;
8. the maximum exposure to wind;
9. the clear span; and
10. the level of fire resistance to be achieved.

There is obviously some interdependence between the loads on the battens or deck, the spacing of rafters or joists and the T-dimensions of these members.

For single family houses, suitable T-dimensions for traditional roof members are given in Approved Document A. Ensure that the roof structure is securely connected to its supports. These can be either:

1. the walls, provided they are able to resist the loads and are capable of transferring them safely to the ground; or
2. an independent structure.

Locate the position of extra thermal insulation (see Figure 4.1). The stages in the elemental method of calculating the minimum thickness of extra insulation are:

1. decide the required U-value to be achieved by the design;
2. choose the thermal insulation product;
3. identify its thermal conductivity;
4. calculate the minimum thickness, i.e. the base thickness required by the Regulations, ignoring the contribution from other parts;
5. calculate the equivalent thickness contributed by those parts;
6. deduct (5) from (4) to give the thickness of extra insulation.

Why not use more insulation than the minimum? In the long run, its extra cost may be less than the saving in the size of the heating installation and in energy costs. Include a vapour control layer below the insulation.

Provide cross ventilation to remove humid air and minimize condensation within any void within the roof by inserting insect-proof air vents in the soffits of projecting eaves or, alternatively, in the roof coverings. Include fire resisting sealed cavity barriers to roof voids unless the compartment walls are sufficient.

UPPER FLOOR AND STAIRCASE STRUCTURES

- **Dependences** Imposed loads and dead loads including those of non-loadbearing partitions. Whether a compartment floor. The types and substances of upper floor structures included in the scheme design.
- **Special attributes** Strong, stable, fire resistant, sound resistant.
- **Approved Documents** A, B, E, K, M.

Decisions on the thicknesses and other details of the floors will depend on:

1. the spans;
2. the maximum dead and applied floor loadings;
3. the pipes, ducts, and other services conductors to be accommodated, and their outlets;
4. the dead loads of non-loadbearing walls (or **partitions**); and
5. the desired level of fire and sound resistance.

Hoisting and fixing precast concrete units is speedier than constructing a cast-in-place concrete floor or staircase, although using mortar or wet-mix concrete to connect the individual beams and join them to other structures will cause some delay while it hardens.

A wooden open framework of floor joists will require some kind of continuous floor surface or deck that will unite the various structural members. Stiffen the wooden joists with solid or herringbone strutting and connect them to supporting walls.

Reduce impact sound transmission through the structural floor by isolating the finish with some 'flabby' substance and a secondary, probably wooden, open framework floor structure (see Figure 4.2). A similar structure will provide a space for service pipes and cabling. These will form concealed spaces which may require barriers to the spread of flame, insects, sounds etc. through the voids.

LOADBEARING EXTERNAL WALLS

- **Dependences** The technical solutions for roof and floor structures and their loads. Decisions made earlier on 'The faces of external walls'.
- **Special attributes** Strong, stable, dimensionally stable, sound resisting.
- **Approved Documents** A, B, C, D, E, L, N.

A wall that is to sustain loads from other elements must be capable of conducting them safely to either the ground or to another loadbearing element. The thickness of a loadbearing wall will depend on:

1. the dead and imposed loads from the roof and upper floors;

2. the direction of such loads, e.g. vertical, sloping outwards;
3. the effects of the wind;
4. its height and length between lateral, i.e. buttressing walls;
5. its substance; and
6. whether it is of continuous or open framework construction.

Where the outer leaf of a cavity wall provides the rainscreen, the loadbearing inner leaf might be of masonry or a wooden open framework. Provide lateral support wherever possible by designing intersecting walls to act as buttressing walls. If necessary, stiffen their corners with larger masonry units. A masonry wall should have as high a thermal resistance as possible, consistent with its need to be strong, stable and dimensionally stable. Additional insulation (see Figure 4.1) can be provided:

- as partial or complete cavity fill between two masonry leaves;
- behind the impervious external cladding to a rainscreen;
- within an open framework;
- as a lining to a plasterboard internal wall finish.

Its thickness can be calculated by the elemental method as described in 'roof structure'. Avoid interstitial condensation by putting a vapour control layer as near the warm side of the insulation layer as possible. Provide cavity barriers where compartment walls occur. Allow for dimensional movements in larger constructions. Maintain stability over openings for windows and doorways by selecting suitable lintels. Avoid thermal bridges.

Where the roofs of a building are at different levels, some walls that were internal ones at lower levels become external ones at higher levels and their attributes will change. Thermal insulation will be required, and moisture must be excluded both generally and at the junctions where the change occurs.

OTHER LOADBEARING WALLS

Much of the above will also apply to other loadbearing walls. Take compartment walls through roof spaces and provide fire-stops at their junctions with the coverings.

PARAPETS

- **Dependences** Technical solution of external wall. Eaves gutters, rainwater pipes and outlets. Roof structure.
- **Special attributes** Attractive appearance, weathertight.
- **Approved Documents** A, C.

Avoid them if you can (see Figure 2.8). Parapet walls which project above the level of the roof covering are particularly exposed to the natural elements. Their copings should be strong with throated projecting edges, be well anchored to the walls, and have the minimum of vertical joints. Use some highly durable substance for these joints. Prevent moisture in the parapet wall from penetrating into the internal portions of the walls below by a suitable system of dpcs.

STRUCTURAL FRAME

- **Dependences** The dead, imposed and wind loadings. The substances for structural elements included in the scheme design. Notional foundation and floor levels, and the bearing capacity of the ground. The strategy adopted for dealing with a fire.
- **Special attributes** Strong, stable, dimensionally stable, continuous, connectable, buildable, safe to build, durable, suitably fire-resistant.
- **Approved Documents** A, B, E.

Detailing the technical solution of a structural frame and its foundations will be the responsibility of either its manufacturer or a consulting engineer, who will need information on the loadings to be sustained, the constructions to be supported, and the bearing capacity of the ground at various levels.

A steel frame will be erected speedily by its supplier but must be protected from fire by some form of casing. If not precast, reinforced concrete frames will be cast in place, floor by floor, with delays while the concrete hardens, but will need no additional protection. Decorate exposed surfaces (see internal finishes).

Decide how to connect the other elements to the structural frame. Column base foundations are likely to be deeper than those for walls, and may be cast first.

NON-LOADBEARING WALLS

- **Dependences** Finishes to both sides.
- **Special attributes** Stable, dimensionally stable, sound resistant.
- **Approved Documents** B, E, N.

The choice between continuous and open framework construction and decisions on their substance and thickness will depend on:

1. the required thermal and sound resistance;
2. the height of the wall and its length between lateral supports;
3. the substances of its finishes; and
4. their buildability.

These decisions will involve some iteration in the design process. The choice will be limited by the maximum dead load that can be resisted by the sustaining construction.

FLOORS NEXT TO THE GROUND

- **Dependences** The levels and slope of the surface of the ground after the vegetable soil has been removed. The total thermal resistance to be achieved. The perimeter/area ratio of the floor, the dead and applied loads, and the nature of ground contaminants, if any.
- **Special attributes** The exclusion of living creatures, moisture, water vapour, and, possibly, Radon gas. Providing a level support for floor finishes, and transferring loads to the ground.
- **Approved Documents** C, L.

Where the surface of the ground is almost horizontal, and the finished floor level can be just above the natural ground level, a ground supported floor is the obvious technical solution (see Figure 3.7). This will consist of a hardcore bed and a concrete ground slab with a damp-proof membrane either below or above the slab. All damp-proof membranes must be continuous with the external and internal walls damp-proof courses.

The top of the floor will normally be level with the dpc to the external walls (see below). Under a basement floor, use a concrete blinding layer instead of hardcore to support the membrane. Ducts may be required below the lowest floor to accommodate heating mains and other services. These will require access for maintenance and renewals.

On a sloping site, or where the floor level will be rather high for it to be supported off the ground by hardcore filling, a suspended floor will be more appropriate. This can be made of wood, precast concrete units or in-place concrete, with intermediate supports provided by sleeper walls (see Figure 3.7). Cover the ground below with a damp-proof membrane and a layer of concrete. This can be either level or sloping, but must allow any moisture to drain away, and can be supported on hardcore filling. The space in between should be ventilated, although living creatures must be kept out.

Thermal insulation can be introduced under the finishes or within any air space, and round the external edges of the structure.

FOUNDATIONS

- **Dependences** Dead and imposed loads and wind loads from supported elements. The type and bearing capacity of the subsoil and its depth below the surface. The presence of sulphates in the soil.
- **Special attributes** Strong, stable, durable.
- **Approved Documents** A, C, E.

By 'foundations' we shall mean just the lowest parts of the building which finally transfer all the building loads on to the ground. However, the term can also include the foundation walls that connect the building enclosure walls to those foundations.

Begin by calculating the dead, imposed and wind loads of each element, but include the floor next to the ground only if it is suspended. Apportion these to each linear metre of the continuous external and internal loadbearing walls and to each individual support, e.g. a structural column or masonry pier.

Next, identify the nature of the subsoil and the notional level at which it should provide adequate support and be unaffected by frost. Then decide what its safe bearing capacity per square metre can be assumed to be (see BS 8004 Foundations).

Calculate the width of a continuous strip foundation by dividing its total load per linear metre by this safe bearing capacity. Alternatively, refer to published tables, e.g. in Approved Document A. For stability, strip foundations must be positioned centrally under the walls and widened under supporting piers, chimney breasts etc. to maintain the same projection. On sloping sites, decide where strip foundations will be stepped.

Calculate the areas of individual foundations by dividing their point loads by the safe bearing capacity. The projecting portions of foundations are, in effect, cantilevers which tend to be bent upwards by the resistance of the ground. Mass concrete foundations must, therefore, be thick enough to resist this bending. As a rule, this thickness will be the same as the projection, but with a minimum of 150 mm. Alternatively, add steel reinforcement.

Extend the enclosing walls downwards to their strip foundations with durable foundation walls. Alternatively, extend the foundations up to just below ground level by filling the trenches with concrete (this is called 'trench fill'), and finish with a low foundation wall. Structural columns will normally extend downwards to the tops of their individual foundations. Where structural columns are close to or within loadbearing walls, all their foundations will be interlinked. In such cases, if the tops of their foundations can all finish at the same level this will help their buildability.

CLEARING AND PREPARING THE SITE

- **Dependences** Existing buildings, fences, power lines, sewers, trees, vegetation, and other obstructions to the development. The previous use of the land. The presence of toxic waste, landfill or other solid, liquid or gaseous contaminants that might endanger health and safety. The depth of the vegetable soil and the type of subsoil. The extent of the new Works, pavings, entrances etc. and their foundations.

- **Special attributes** The strength and stability of soils, safe for construction processes.
- **Approved Document** C.

Should a specialist contractor be employed to demolish the existing buildings and clear the site while the design is being completed? If you suspect that the site is contaminated, are you seeking specialist advice? Have you told the owner, as it may be advisable to insure against the long-term risks? Are any existing cables or pipelines to be diverted or sealed off? Site preparation procedures are listed in Chapter 12.

9.5 SERVICES (SEE CHAPTER 1)

SANITARY FITTINGS AND WATER SERVICES

- **Dependences** The number of potential users. The sizes, shapes and locations of the spaces allocated to sanitary accommodation in the scheme design. The height and size of the building and whether it is to have a sprinkler system.
- **Special attributes** Hygienic, safe to use, suitable for the disabled.
- **Approved Documents** G, H, L, M.

Types of sanitary fitting include water closets, urinals (both often referred to as sanitary conveniences), wash basins, baths, showers, laboratory sinks. Their numbers will depend on the numbers of persons likely to be present at any one time in each part of a building. The position of each fitting should now be considered three-dimensionally. Bear in mind:

- whether to be suitable for disabled persons;
- the need for wall fixings and ducts;
- partitioning between fittings;
- the runs of hot and cold water pipelines;
- the extent of the wall area below the windows; and
- the opening back of doors.

The numbers of persons using these fittings will determine the demand for hot and cold water. This demand will determine the pipe diameters and pipe runs, the outputs required from the hot water heating and/or storage appliance(s) and the capacities of cold feed cisterns that will replenish the store of hot water.

Insulate storage vessels and provide overflow warning pipes to feed and flushing cisterns. Will the hot water storage system be unvented or vented to the atmosphere? Ducts or casings may be required for the pipe runs, together with openings through floors and walls. These must be sealed to exclude water vapour, sound and insects. (Internal drainage is considered in the section on installations.)

HEATING AND COOLING SERVICES

- **Dependences** Choice of fuel. The owner's energy efficiency policy, including the U-values of the enclosing elements.
- **Special attributes** Safe to use, energy efficient.
- **Approved Documents** J, L.

The calculated total demand for energy will depend on:

1. the thermal resistance of the envelope;
2. the incidental and planned rate at which external and internal air is to be exchanged;
3. heat gain from electric lights and other appliances;
4. solar heat gain;
5. incidental heat gains including body heat from occupants;
6. the assumed extreme external temperature;
7. the desired internal temperature; and
8. the efficiency with which fuel will be converted into useful energy.

Some excess capacity is desirable.

The requirements of individual spaces will determine the capacity, size and position of self-contained appliances or individual emitters. Consider their positions three-dimensionally, bearing in mind the wall space below windows and elsewhere, the locations of chimney breasts, if any, the positions of electric power outlets and the layout of furniture, appliances and furnishings.

An individual fuel-burning appliance will require a hearth and a chimney that extends well above the roof and has a suitable flue lining. It should have some type of air temperature control. Choose the appliance and the fireplace surround together and integrate with the internal finishes.

Link a centralized boiler plant to emitters by a circulation system of suitably sized conductors and controls and provide feed and expansion cisterns. Enable the air temperature in each space to be controlled, e.g. use thermostatically controlled radiator valves. Flues and chimneys, rooms for plant, meters, time controls and electrical switchgear, and ducts and other voids and openings for pipelines, will depend on the choice of fuel and the detail design of the installation. Insulate plant and pipelines.

Ensure that every fuel burning appliance will have a sufficient supply of external air. A continuous fuel supply may be available as a public service, e.g. electricity, gas, a district heating scheme. Alternatively, access roads and storage will be needed for intermittent deliveries of oil, coal, or liquefied petroleum gas (LPG).

ELECTRICAL AND ASSOCIATED SERVICES

- **Dependences** The numbers, positions and ratings of electrical appliances and other services, the lighting levels to be achieved, including externally.
- **Special attributes** Safe to use, energy efficient.
- **Approved Documents** B, L.

The numbers, sizes and positions of internal and external electric lights, power points, power supplies to lifts, heaters, industrial plant, boiler and radiator controls, fans, pumps etc. should also be considered three-dimensionally. The owner may require a 'clean' power supply to computers, and the ability to use cheaper off-peak electricity.

Other services needing electrical power may include entryphone systems, emergency lighting, telephones and fax machines, smoke, carbon monoxide and fire alarms, and parking control barriers. TV aerials and cabling and computer terminal cabling may also be required. These can all be included in the performance specification for the electrical services. Trenches may be required for service cables, with ducts through the substructure. Secure accommodation will be needed for meters, switchboards and distribution boards.

The height and exposure of the building will determine what kind of lightning conductors (if any) are required.

VENTILATION SYSTEMS

- **Dependences** The existence of openable windows, devices for providing background ventilation, and open-flued heating appliances. The rate at which the air within an enclosed space might become excessively humid or polluted by, say, car exhausts, tobacco smoke, or concentrations of people.
- **Approved Documents** F.

Mechanical extractors will be required in sanitary accommodation, kitchens, workplaces, car parks and other enclosed spaces where the air must be replaced at a rate that will limit the build-up of contaminants including tobacco smoke. Provide sufficient inlets for the volume of air to be removed to be easily replaced by an equivalent supply of fresh air. Ensure that a mechanical extractor will not cause an open-flued heating appliance to spill flue gases. Provide any passive stack ventilation system with ducts from ceiling level to outlets at high level, preferably at the ridge if the roof is pitched.

SERVICE CONNECTIONS

Services undertakings must have sufficient spare capacity to supply the building. Prepare a timetable for making the connections, including any needed by the constructors, e.g. water, electricity, telephone, and seek the commitment of the service providers to those dates. Each provider should state their requirements regarding the service connection from the public highway to its entry into the building. The electrical supply cables may be brought to the building either overhead on poles or in a duct in a trench and through the substructure.

9.6 INSTALLATIONS (SEE CHAPTERS 1 AND 2)

INTERNAL DRAINAGE INSTALLATIONS

- **Dependences** The extent of the sanitary installation and water supplies (see above), the positions of internal ducts and ventilation stacks to the external drainage system.
- **Special attributes** Quiet in operation.
- **Approved Documents** B, H.

The internal drainage installation will include traps and branch and vertical discharge stacks. Its pipe sizes and their runs and gradients will depend on the need to preserve water seals in traps and on the peak volume of waste water. The transmission of sounds associated with the use of sanitary fittings is a major cause of noise pollution and stress which affects residents of flats, apartments, hotel bedrooms etc. Lightweight pipework is the main culprit. Seal and firestop where pipes penetrate walls and floors.

EXTERNAL DRAINAGE INSTALLATIONS

- **Dependences** Whether to be a separated or a combined system. The location of each sewer and its invert level. The peak discharge flow rate of waste water and rainwater. The finished ground levels. See also the layout dependencies given below.
- **Special attributes** Durable, accessible.
- **Approved Documents** C, H.

Sewage treatment will normally be the responsibility of a public service organization. They will decide whether or not the storm water (which does not require treatment) can be combined with the **foul** water and sewage. Calculate the overall gradient of the installation from the following:

1. the vertical distance, or **fall** from the bottom, i.e. the **invert level** of the farthest inlet to that of the existing sewer;
2. the length of the pipeline that joins that farthest inlet to the sewer.

Assuming the levels of branch drains will allow an even fall, the ratio between (1) and (2) is the average overall gradient. If this is more than the minimum permitted fall, the installation would seem to be feasible, and the design can proceed.

The dendritic drainage layout will depend on:

- the outline of the building;
- the positions of the outlets from the internal drainage and rainwater installations;
- the maximum flow from each of these outlets;
- the levels and the configuration of the site;
- the positions of road and paving gullies; and
- the location, size and depth of the public sewers.

Design the drainage layout as a set of straight pipelines connecting the inlets, the inspection chambers or manholes, and the sewer. Put an access fitting or rodding eye at the head of each drain run and also where bends and junctions outside the inspection chambers are unavoidable. List the accumulated maximum flows expected at the junctions between pipelines and size the ongoing pipes accordingly. Decide the thicknesses and substances of beds and/or surrounds to pipelines.

Where necessary, adjust the gradients of pipelines to match the chosen pipe sizes, calculate their individual falls, and accumulate them to give chamber and manhole invert levels. Their cover levels must match those of their surroundings.

Use a schedule for recording their identifiers (usually numerals) and their internal dimensions and depths to invert, your choice of products to be used in their construction including the covers (which will depend on the traffic loads), and the pipeline connections. Draw outline sketches of the bottoms of chambers and manholes showing the positions, plan shapes and diameters of the main and branch channel pipes.

REFUSE DISPOSAL

- **Dependences** The types and capacities of containers and the frequency of waste removal by the collection service. The number of households being served, or the type of waste and the volume to be stored. The location of the vehicle access.
- **Approved Document** H.

Detail the construction of refuse chutes and their access doors and hard standings for containers.

9.7 DESIGN APPRAISAL

BUILDABILITY

Buildability was introduced in Chapter 4. Its main requirements are:

1. it should be easy to plan the procurement of construction products;
2. the design and related construction processes must be simple and straightforward;
3. the various trades should be able to move on to and off the site with ease.

Simplicity is the key to buildability. The maxim 'Keep it simple, stupid', or 'KISS', mentioned previously suggests the following.

1. The fewer the number of different technical solutions, the fewer the number of different trades, and the less they are likely to get in each other's way.
2. All the specified products should be readily available.
3. Where there is a choice, that class B formed products and components should be used rather than class A or composite class AB constituent products. This will eliminate setting and drying time delays.
4. Construction processes should involve the simple addition of materials to those already in place.
5. The work of the various trade gangs should be separated from each other in both space and time.
6. The sanitary services should be concentrated into certain areas.
7. The design should be appraised in three dimensions to see if any one part is likely to interfere with another. For instance, can the doors open fully? is it possible to get furniture through the doors and up the stairs? is there enough wall surface below windows for the radiators? are the internal and external drains clear of the structural frame and its foundations? When designing a process plant such as an oil refinery, a major concern is to avoid having two or more pipelines planned for the same space.
8. It should be possible to complete the weathershield quickly, thus enabling finishes and services to be started.

APPRAISING THE DETAIL DESIGN

One approach to building appraisal is to take each attribute system in turn, and check its continuity and effectiveness throughout the whole building and under all likely conditions. Pay particular attention to the way the building is likely to perform in extreme conditions of rain, snow, frost, hail and wind, flooding, heat or cold, concentrations of insects and internal or external noise. Consider any implied risk to the health and safety of operatives and ensure that these are mentioned in the pre-tender stage health and safety plan.

Although many of the technical solutions available to us have been tested over the years, the incorporation of new products, the quest for higher environmental standards, and the construction of larger and higher buildings, can bring unforeseen difficulties. For instance, structural problems can arise when provision is not made for differential dimensional movement in long or high buildings. This can show as cracking in masonry walls due to the excessive thermal movement of a steel frame, and the spalling of external tiling due to the creep of a concrete frame when loaded. The durability of main products must be matched by that of their fixing and jointing materials, e.g. roofing nails, mortar joints, mastic and weather-excluding seals.

Check that the building will be energy efficient, including its space heating, hot water and lighting installations. Will the technical solutions for the envelope, including windows and doorways, provide at least the required U-values? Look for possible thermal bridges and air leakage points. Choose an appropriate method and prepare calculations to demonstrate that the thermal performance of the building fabric will meet the requirements of the Building Regulations.

The noise transmission between dwellings, especially in multi-storey buildings, can be very stressful for residents. (Designers seldom occupy the buildings they have designed!) Obviously, an understanding of how weaknesses in design have been revealed in other buildings can be helpful during the detail design stage of a new project.

A **maintenance profile** will indicate where the detail design will commit the owner to future expenditures on maintenance that might not represent value for money. This profile consists of a list of the likely expenditures on maintenance, renewals, and major refurbishments. Summarize these for each interval of, say, five years throughout the life of the building. An estimate, or even a guarantee of the likely annual expenditure on fuels, water, cleaning, and other occupancy costs may be required by the commissioning team.

PRODUCTION DOCUMENTS

These will probably consist of a set of multiplane orthographic assembly and location drawings, site and block plans, schedules, and a comprehensive specification (see the next chapter). The documents will represent the final form of the design, that is, they will define what is to be provided on the site for the owner. Before that, they will be used (a) to inform those who will be bidding for the Works contract, and (b) to demonstrate that the design conforms with the local building code, e.g. the UK's Building Regulations.

While the technical solutions for roofs, floors, windows and doorways are usually more or less independent of each other, they and those for the walls are interdependent. Thus, vertical section drawings showing details of their junctions with the walls will be particularly informative.

Later, some of this data may be re-expressed pictorially for the benefit of operatives on the site, e.g. planometric projection drawings of the foundations, the walls, or the roof construction.

9.8 SUMMARY

The detail design process is one of deciding what the parts of the elements are to be, what products are to be used in their construction, and their sizes and positions within the elements, i.e. deciding on the technical solutions. This introduction to the process mainly followed the order in which those parts were introduced in Chapters 1 to 4. It began with a general discussion on technical solutions, their dependence on decisions already included in the scheme design, on sources of technical information, and on a suitable procedure. For each element and its parts, their particular dependences and specially desirable attributes were identified and some of the factors to be considered and procedures to be followed were discussed.

FURTHER READING

The Building Regulations, HMSO, London.

Building Research Establishment (1994) *Thermal Insulation: Avoiding Risks*, HMSO, London.

Relevant British Standards, British Standards Institution, London.

Cook, Geoffrey K. and Hinks, A. J. (1992) *Appraising Building Defects*, Longman, Harlow.

Griffith, A. and Sidwell, A. C. (1995) *Constructability in Building and Civil Engineering Projects*, Macmillan Press Ltd, Basingstoke.

Hall, F. (1988) *Essential Building Services and Equipment*, Newnes, Oxford.

Levy, Matthys and Salvadori, Mario (1992) *Why Buildings Fall Down*, W. W. Norton & Co Ltd, London and New York.

Specification, Emap Business Publications, London.

Stephenson, J. (1993) *Building Regulations Explained*, E & FN Spon, London.

Communicating the design

10

The task of the design team is to develop the intentions of the owner, and to transform them into instructions regarding what is to be constructed on the site. They communicate their decisions mainly through the medium of drawings, specifications, and schedules. This chapter is about these documents and the information they carry.

Individual members of the team will take information from others, including the owner, apply their expert skills and knowledge to it, and produce the fresh information they are contracted to supply. In passing, we should note that only this information becomes available to the team as a whole. Each group of specialists is likely to keep to themselves the supporting information generated while they were producing it.

Each kind of input–process–output, i.e. what each group does, is sometimes referred to as a **function**, e.g. the functions of architect, structural engineer, services engineer, cost consultant. Each function may be carried out by a different and independent firm of professionals. Alternatively, a sequence of functions can be carried out within a single organization, although this only makes economic sense if the individuals concerned can be provided with a continuous supply of work.

In any case, there is a sequence that must be followed if decisions are to be made logically. Irrespective of where the work is done, who does it, and what their job titles are, the same procedures are logically necessary and the same documents must be produced if the intentions of the designers are to be understood by others.

10.1 DRAWINGS

As we grew up, we learned not only to distinguish between familiar objects, but also to intuitively allow for the foreshortening effects of angle and distance, or **perspective**. Thus, our minds automatically transform the shapes of the objects as we see them, so that we comprehend them as actual three-dimensional (3-D) forms (see Figure 10.1).

Fig. 10.1 View through two doorways showing the foreshortening effects of distance.

A graphical representation of the projection of the boundaries of one or more objects on to a plane surface is called a **drawing**. These boundaries, or 'interfaces', may be between a solid object and either the air or another object, or between one face of the object and another. The amount of information that can be conveyed by just a few lines is quite extraordinary.

This information will probably be extended by words, numbers and other graphic symbols. A drawing should be prepared according to a well-known set of conventions, so that readers who are familiar with the conventions can be confident they understand what its originator wishes to tell them. Although some artists may deliberately explore the effects of breaking conventions, designers should be concerned with following them.

The ratio between any extension measurement made on a drawing and the actual extension when measured on the object, e.g. 1:50, is called its **scale**. The size of the representation when compared with the actual object will influence how we describe it. E.g. a drawing to a scale of 1:5 will be described as **large-scale**, whereas one to a scale of 1:2500 would be a **small-scale** drawing.

A variety of different ways of viewing an object can be depicted on drawings, each having its advantages and disadvantages. **Pictorial** drawings (see Figures 10.1–10.4) attempt to portray objects as they would actually be seen from particular viewpoints, although many will appear distorted. None will be entirely to scale. In contrast, representations on **multiplane orthographic**, or 'technical' drawings are true to shape and scale, but each can only portray the details of one plane of an object (see Figure 10.5). Such drawings are likely to be drawn to generally recognized preferred scales to minimize the number of different scale rules required by everyone involved.

WHO ARE DRAWINGS FOR?

The earliest drawings will be those which designers make for their own benefit, to discover, develop and record what is in their minds. Most drawings will communicate Works information between members of the design team and between the design team and the owner, public authorities, and the constructors on the site. Others may be produced for product manufacturers.

PERSPECTIVE PROJECTIONS

A perspective drawing attempts to portray a 3-D object as it might be seen in actuality or in a photograph. It is as if the object had been seen through a window, and its outlines had been marked on the glass. The location of the window glass, the frame, photographic film, or any other (usually imaginary) plane on to which the boundaries etc. of the objects have been projected and recorded is called the **plane of projection**.

Nothing portrays an object better than a perspective projection. In the case of a cuboid, it will show realistic and connected details of three of its six sides, and may be the only kind of drawing with any meaning for the lay person, whether owner or operative.

Vertical edges will appear as verticals. Other parallel boundary lines should, if extended, converge to vanishing points, horizontal ones meeting at the horizon. One limitation, however, is that dimensions cannot be scaled from a perspective because the line lengths are not in proportion.

A parallel perspective projection based on Figure 10.1 would have some of its faces parallel to the plane of projection, and one vanishing point. An angular perspective has its vertical faces inclined to the plane of projection and two vanishing points. Figure 10.2 shows one of the ways in which an angular perspective projection can be set out and created.

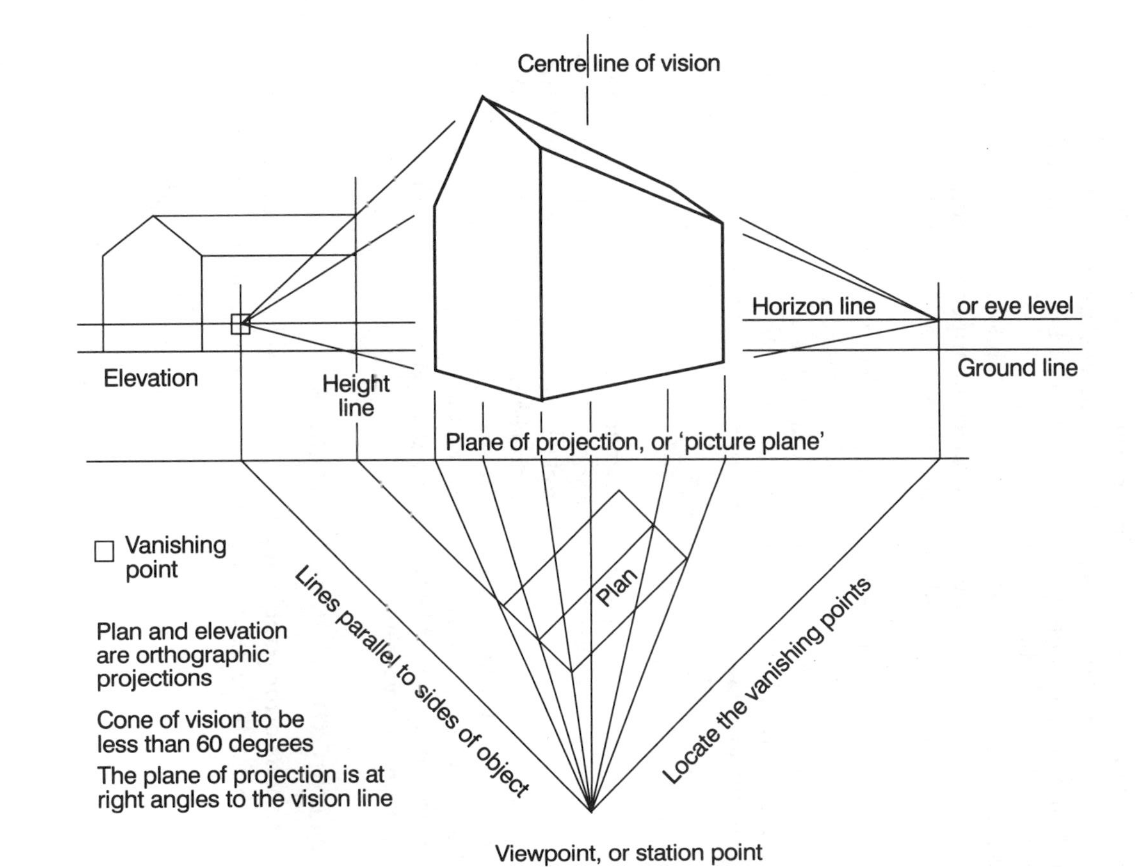

Fig. 10.2 One way of setting out an angular perspective projection.

OBLIQUE PROJECTIONS

Where one face of an object is drawn true to shape, with the other faces at 45° to it, and where all lines parallel to an axis are drawn to scale, the projection is called an **oblique** (see Figure 10.3). Lines on the other faces will only be to scale where parallel to an axis.

The front face of a **cavalier** oblique projection is true to shape and the same scale is used on all axes. This projection appears distorted. A **cabinet** projection is similar, but the receding edges are drawn to half scale, which looks more realistic.

On a **planometric** projection, the top horizontal face is the true one and the same scale is used throughout. This projection is obviously less realistic than the isometric shown in Figure 10.4 as there is no evidence of the gable ends. However, it is particularly useful for showing interiors and pipeline layouts.

AXONOMETRIC PROJECTIONS

By giving up some realism, an **axonometric projection** offers a proportionate, although distorted, 3-D view of an object where all three axes (length, width, height) are at an angle to the plane of projection. The most common form is the **isometric** projection (iso = equal), where the dimensions on all three axes are to the same scale, as on Figure 10.4. In a **dimetric** projection, the scale on one of the axes is usually half of the scale of the other two. Measurements parallel to an axis can be scaled from an axonometric drawing.

MULTIPLANE ORTHOGRAPHIC PROJECTIONS

On **multiplane orthographic** drawings, six different views at right angles to each other will be required to portray the outsides of a simple cubic

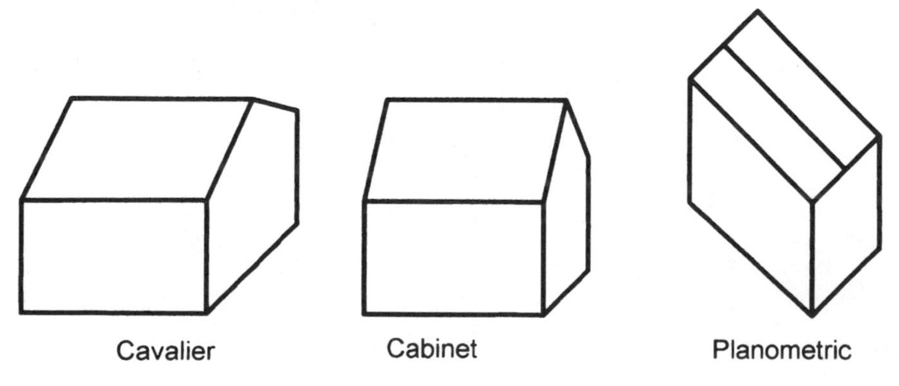

Fig. 10.3 Schemas of oblique projections.

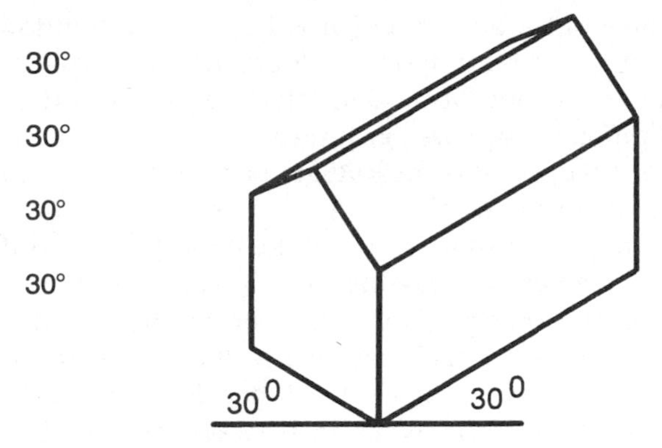

Fig. 10.4 Schema of an isometric projection.

object. The plane of projection of each drawing will be parallel to one of the faces of the object, and the lines on the drawing and the angles at which they intersect will truly represent it. They provide somewhat unnatural, disconnected views of an object, and our minds have to learn to integrate these views before we can 'read' them with confidence.

An **elevation** shows the outlines of one face of the building as seen from the outside, together with interface lines between different types of constructions and different planes. For instance, lines will indicate the boundaries between the roof and the walls, and between walls and windows and doorways. The direction of its viewpoint can be shown on the ground floor plan, or by stating the direction in which the elevation faces, e.g. south-west elevation.

A **section** is a representation of the internal structure of an object, seen as if it had been cut through by the plane of projection in a specified position. When a section shows the boundaries of constructions that will be seen beyond the section plane it is called a **cut**. E.g. a vertical cut through a window in a wall will also show the faces of the frame and the wall beyond.

Each vertical section or cut should be identified with a capital letter. Its plane of projection and identifier will be marked on the floor plans, with open arrow pointers indicating the direction of view. Vertical section planes may be staggered so that they pass through locations where information is required.

A **plan** depicts a view from above, although **plans** may refer to a set of drawings of various kinds. A roof plan will show the intersections of the different roof slopes, the gutters, and any junctions with higher walls, chimneys, dormer windows etc.

The plans of the different floors of a building convey relatively more information than the other graphics, as they are really horizontal cuts through the walls, taken at a height which will show the window and door openings as well as the boundaries of floor surfaces. 'Broken' lines are used to indicate the location of hidden interfaces, such as the edges of foundations and pipelines below a ground floor. A foundations plan will show the different widths and thicknesses of pad and strip foundations, and their top and bottom levels.

The distances, i.e. the **dimensions** between the boundaries of different constructions should be given on drawings as related sets rather than as isolated facts. For example, on the plan of each storey there should be a set of horizontal dimensions for each external wall as measured on the outside above window sill level. Each set will consist of the widths of windows, doorways and the solid walls in between and at the ends, together with the overall external length. An equivalent internal set will show the dimensions of the rooms and the internal wall thicknesses. Each overall external length must be consistent with the sums of both the external and the internal dimensions. Dimensions are considered again in Chapter 13.

Each view of a large building may occupy a separate sheet of paper, but where two or more views are to be placed on a single sheet, a decision must be made on their relative positions. In North America, construction drawings are likely to follow the convention that the views face the viewer. Starting with the front view, the view from above is placed above it and underneath it is drawn the view from below (if any). One side view is placed at that side, and the other side view and the back view are placed on the other side. (Engineering drawings in the UK may also follow this convention.)

In the UK and other European countries, elevations may follow this convention, or the placing of the side and back views may be reversed (see Figure 10.5). Plans are always placed below the elevations.

THE PRODUCTION DRAWINGS SET

Of course, different kinds of drawings are produced at different stages in the design process and for different purposes. The Works will be constructed from a set of production drawings. Normally, this set will include some of each of the following drawing types, where each drawing is related to at least one other at a higher level and smaller scale, by their having some boundary lines in common. That is, the drawings will form a hierarchy.

- At the lowest level are the extra-large-scale **component drawings.** These show the designer's requirements for components that have to be manufactured specially for the Works. They will be incorporated into the manufacturer's production drawings. Examples include window units, wood and precast concrete staircases, roof trusses, and

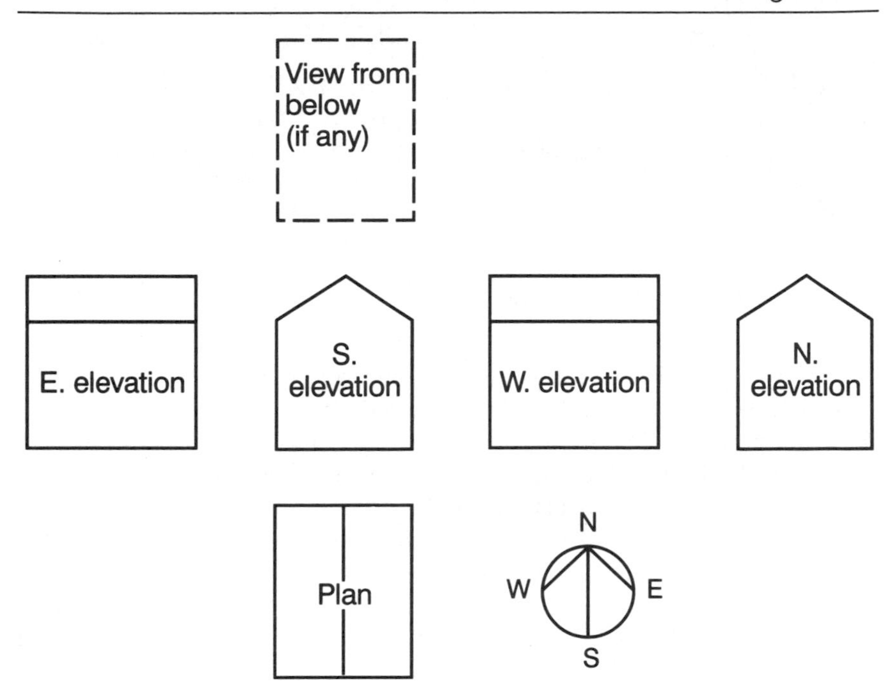

Fig. 10.5 Placing of views in European or first angle multi-plane orthographic projection.

cupboards. Their details will include their size and shape, and the arrangement of the members.

- Details of the junctions between the various technical solutions, including how components fit into the building fabric, are shown on large-scale **assembly drawings**. Such drawings are the only ones that might be usable on more than one project. That is, they can be **portable**, and, in any one design office, may be regarded as **standard details**. They are most effectively shown on large-scale pictorial assembly drawings where the details of the adjoining elements can be shown as well.
- Smaller-scale plans, elevations and sections that show the boundaries of the physical constructions of a building, the locations of any visual detail, the positions of doorways and windows, and the runs of underground pipelines, are called **general location drawings**. These boundaries also indicate the size and arrangement of the spaces within the building. Many of the drawings will include the same outlines of walls etc. and can be produced from transparent copies, or **overlays**, of the original drawings, or from their electronic equivalents.
- A **site plan** will show the positions of the external walls with respect to the site boundaries or to some fixed points on it, together with

Drawing details	Notation	Index	Refers to
Source of drawing	Capital letter	A	Architect
Type of drawing	Capital letter	L	Location drawing
Element group	(Digits)	(27)	Roof*
Sheet number	Digit	6	
			* from Table 1 of CI/SfB

Fig. 10.6 A notation system for indexing drawings.

underground pipelines and their connections, external works, existing structures to be demolished, and access to the site.

- A **block plan** will show the site in relation to its environment, including the locations of public sewers, services, neighbouring sites and adjoining roads. It may also show the outlines of the proposed Works. Some of the boundary lines on this block plan should be identifiable on a map available in the public domain, such as a UK Ordnance Survey map. Maps showing the site in the centre of the sheet and giving land parcel references and ground areas can now be produced on request by local Ordnance Survey Agents.

Drawings have their own system of classification consisting of two capital letters and two digits within brackets, as in Figure 10.6. This is followed by the sheet number within the class, e.g. AL (27) 6.

10.2 SPECIFICATIONS

TYPES OF SPECIFICATIONS

A statement describing some of the qualities of one or more existing or proposed objects is called a **specification**. In our case, the objects will be those resulting from site construction processes. The type of specification which is possible and appropriate at any one time during the design process will depend on the stage that has been reached. Decisions that are not made by members of the design team acting for the owner will have to be made by designers or others acting for the construction organization or by their suppliers.

What is regarded as the normal division of responsibilities varies from country to country. In the UK, under the influence of the EU Regulations, it is likely that more and more decisions will be left to the constructors and their suppliers. We can identify three different types of specification:

1. a description of the attributes required of the elements (see Chapter 7), called a **performance specification**;

2. a list of the properties of the substances of the parts of the elements, called a **descriptive specification**;
3. a statement of the construction products to be used in the construction processes (see Chapter 5), called a **prescriptive specification**.

For example, in the case of a concrete strip foundation to an external wall, its performance specification might require the foundation to:

- resist the loads from the wall and transmit them safely to the subsoil of a given bearing capacity;
- be continuous and durable, including being unaffected by sulphates in the soil.

The constructor would decide both the size of the foundations and their substance.

Its descriptive specification might:

- state that foundations of particular sectional sizes are to be made of concrete;
- give the minimum strength to be achieved by concrete test cubes 28 days after casting; and
- require that sulphate resisting cement be used.

Where ready-mixed concrete will be used, the decision on the mix design may be left to the concrete supplier.

Its prescriptive specification would:

- give the sizes of the foundations if they were not shown on the drawings;
- give the concrete mix design, e.g. the cement and aggregates to be used and the proportions of the mix, including the water/cement ratio; and
- state what is and is not to happen to the concrete on the site.

PERFORMANCE SPECIFICATIONS

Once the enclosure elements have been outlined on schematic or outline drawings, it is possible to think about their functions and attributes, including those whose values will vary, such as thermal transmittance or maximum floor loadings. A list of these can be called a **performance specification**.

This kind of specification is desirable when bids are being invited for the detail design as well as the construction of the Works, e.g. bids from system builders. In this case either the design team will be split, some working for the owner and others for the constructor, or the contracts with the designers will be transferred from the owner to the builder. If a performance specification is not provided, the bids will not necessarily be based on the same quality standards, and will be difficult to evaluate and compare.

It is relatively easy to stipulate the required technical attributes, as these are likely to be described in Building Codes, Standards etc. In contrast, it is probably impossible to state what the visual attributes are to be. This may be why the results of **design and build** procurement methods can be unsatisfactory. Of course, it will be in the builder's interest for the design to be buildable, but the other economic attributes (see Chapter 4) should also be stipulated.

Generally, though, this list of required qualities is likely to exist only in the minds of designers, and will be written down only when it is necessary to inform other members of the team.

DESCRIPTIVE SPECIFICATIONS

The detail design stage, which comes after the scheme design has been approved by the owner, is one of deciding how to achieve these attributes. The problems are technical, and involve deciding what substances have suitable properties, how much of each is required, and where they should be placed in relation to the other substances. This was considered in Chapter 9.

Just as we give names to the various elements, so we find it helpful to give names to their parts. Each name serves to gather up and label all our thoughts about the part and its substance, where it is, what its purpose is, and what special qualities it has that caused it to be chosen. It also acts as a verbal link between its written details and the lines on a drawing which show its boundaries. A statement about a fact is sometimes called a **descriptor**, the one naming the subject being known as its **identifier**.

A list of parts and the properties required of their substances can be called a **descriptive specification**. Unacceptable properties may also be given. This kind of specification is used when seeking suitable construction products, or as a bidding document where the constructor will be responsible for deciding what products to use.

PRESCRIPTIVE SPECIFICATIONS

Thus far, we have been concerned with specifying the qualities of the substances that will give a building its desired attributes. The final stage in this sequence of decision-making is to prescribe (a) construction products having suitable substances and (b) their site processing. This involves:

1. naming the classes of construction products to be brought to the site;
2. describing those aspects of the final assembly processes where there is a choice;

3. stating what qualities the parts, i.e. the assembled products should be given by those processes, e.g. brick jointing, floor -level tolerances.

(2) and (3) are usually called descriptions of **workmanship**. The various Parts of BS 8000: Workmanship on Building Sites cover many construction processes although much reliance still has to be placed on generally acceptable standards of specialist operative skills and on expert supervision.

A record of these final decisions is called a **prescriptive specification**. Most specifications from the design team will be a combination of descriptive and prescriptive statements, depending on the circumstances.

Some products may be specified in two stages: the first giving sufficient information to prepare the bid, and the second providing the remainder of the details in time for the work to be done. For instance, the colours of paints and other decorative products are often chosen, perhaps by the owner, when the Works are almost complete.

Sooner or later, though, all the products must be prescribed in sufficient detail for a merchant to know what to supply. Decisions that are not made by the design team will, of necessity, have to be made by the construction team.

Almost all construction products are made in anticipation of demand, i.e. they are 'standard' products, and each kind will have an identifier given to them by their manufacturers. This identifier may have been invented by the manufacturer, or it could be the one in common use in the industry. When prescribing a product we must give its full identifier so that everyone will know what is required.

Notice that we are not concerned with the actual products that will be used in the works. They may not have been made yet. Instead, we state the identifier of the required class of products to which they must belong. Their properties will be described in the manufacturer's technical literature, which may be supported by the details in national or international prescriptive standards.

STANDARD SPECIFICATIONS FOR PRODUCTS AND PROCESSES

In the UK, standards relating to building products are issued by the British Standards Institution (BSI). Their equivalent in the USA is the American Society for Testing and Materials (ASTM).

Agrément Certificates for products approved by the British Board of Agrément also stipulate how the products are to be incorporated into the works. Other product manufacturers may do the same. Specifications of good design and site practices have often been prepared by groups of experienced professionals and issued by authoritative bodies.

In the UK, such information is given in Codes of Practice issued as British Standards by the British Standards Institution. In the USA, a number of Institutes share this task, including the American Concrete

Institute (ACI), the Concrete Reinforcing Steel Institute, and the Building Stone Institute.

In Europe, starting in 1993, national specifications are gradually being superseded by Harmonised European Standards and European Technical Approvals (ETAs) issued under the Construction Products Directive (CPD). Then, the EU Conformity Mark on a product will be an indication that its manufacturer is warranting them as complying with the CPD.

CONTENTS AND ARRANGEMENTS WITHIN SPECIFICATIONS

The arrangements within a specification should reflect the way the industry works. Bids from main contractors will be based largely on subbids from specialist subcontractors, and each should have their own section of the specification. These can then be copied for the use of the specialists when bidding. They will provide a basis for their subcontracts, and will define what each specialist is to do on the site. The quality of the bidding documents can influence the amounts of the bids as they will indicate the level of competence of the designers.

In the past, the design team has produced descriptive specifications of the Works where each section consists of the work of a single **trade**. Even so, the order of these sections has been somewhat arbitrary. For instance, in Skyring's *List of Builders' Prices* (1816), foundation digging and other Bricklayers' work comes after Carpenters' work! In the main, though, the order of Skyring's sections did match the order in which the trades would have been required on the site.

Historically, a specification was written as prose. It would usually consist of a set of instructions to 'lay foundations', 'build walls', 'pave floors', 'cover roofs', 'finish walls' etc. Included in the sentences would be the materials or products to use, and the workmanship expected. Thus, the facts being communicated were held in a kind of matrix of conjunctions, prepositions etc.

However, as the making of building products has become more industrialized, so specifications have changed. Nowadays, the specification might be based on one of the almost comprehensive model specifications now available on computer disk. These are updated at regular intervals, can be modified to suit organizational practices, and may follow the conventions of Co-ordinated Project Information (see later). Of course, any published work is incomplete and partially out of date by the time it becomes available, and skill and care will still be needed when using one. Some titles are given at the end of the chapter.

LINKING PRESCRIPTIVE SPECIFICATIONS TO GRAPHICS

There is a wide variation in the practice of recording specifications and linking this written data to drawings. It is not usually necessary to have a separate specification document for small works. Instead, the identifiers of the products to be used, and any choice of process, can be written on the general location and assembly drawings. This data can then be arrowed to where the parts are delineated on the drawings.

Alternatively, the names of the parts can be written both there and against the product data. As little will be said about workmanship, selecting a reliable construction organization with proper quality control procedures is even more important than usual.

When a separate specification is provided, its clauses can be linked to locations by writing their references on the drawings and arrowing these into position. This is sometimes referred to as **annotation**.

10.3 SCHEDULES

A feature of construction work is the contrast between the relatively small number of different products used for the structural building enclosure and the large variety of products used in the internal finishes. For example, the whole of the roof will be covered by the same kind of roofing tiles, but a different kind of floor tile might be used in each room.

Of course, this is because we think of the building envelope as a single object, with its own attributes and technical solutions, but we regard each room as also being a separate object. Thus, there will be many locations within the building with the same elements and parts, although of different sizes and made with different kinds of products. Rather than write these details on the various locations on the drawings, we can give each room, window, doorway and other object its own code reference, and enter the data on a **schedule** instead. An example is shown in Figure 10.7.

A computer spreadsheet is ideal for doing this, as a schedule is also an array of columns and rows. The locations, e.g. the room numbers, are given on one axis (usually vertically), and the part names are given across the other axis. Clause references from the specification, the product identifiers or other details are written in the cells at the appropriate intersections.

Room no.	Room	Ceilings	Walls	Woodwork
301	office 21	white	zephr	white
302	office 22	magnolia	apple	white
303	toilet A	white	primrose	bluebell
304	reception	lemon	aconite	white

Fig. 10.7 An example of a colours schedule.

The following are some of the parts of elements whose locations, dimensions, and other details can conveniently be listed on schedules:

- structural columns and beams, column foundations, steel bar reinforcement;
- internal finishes and decorations;
- the parts of windows and doorways;
- details of inspection chambers, including their top and invert levels;
- electrical power and lighting points, heating radiators, sanitary fittings, fire alarms, cupboards and other components in connection with fittings, services, installations and furnishings.

10.4 INFORMATION SYSTEMS

CO-ORDINATED PROJECT INFORMATION

In the UK, production documents from the design team may be prepared using the conventions of *Co-ordinated Project Information* (1987). This claims that 'More complete, timely, relevant and conveniently arranged information should enable contractors to estimate, plan and control building work more efficiently to produce buildings of good quality, on time and within cost.'

One of its conventions is the 'Common arrangement of work sections' (CAWS). This lists about 300 different classes of work, each likely to be carried out by a different group of operatives, and categorizes them at two levels, called 'group' and 'subgroup', as shown in Figure 10.8. The notation consists of a capital letter and two digits, each lettered group having a somewhat arbitrary array of subgroups numbered 1–9, and with each subgroup having up to nine basic classes.

Thus, the index for the work section for natural slating is H62. Within any one work section of a CPI Works specification, each clause is given its own three- or four-digit tag or label. As with most notation systems, gaps are left in the sequences for later additions.

A comprehensive and specific system such as this avoids the need for other facets. Bills of quantities (see Chapter 17) can also be arranged according to CAWS, and individual items of data on drawings, schedules and bills of quantities can be annotated with the related clause tags from

Level	Notation	Index	Refers to
Group	Capital letter	H	Cladding/coverings
Subgroup	Single digit	6	
Work section	Single digit	2	Natural slating

Fig. 10.8 An example of an index from the Common Arrangement of Work Sections (CAWS) indexing system.

01 General	05 Metals	09 Finishes	13 Special constructions
02 Sitework	06 Wood and plastics	10 Specialties	14 Conveying systems
03 Concrete	07 Thermal and moisture protection	11 Equipment	15 Mechanical
04 Masonry	08 Doors and windows	12 Furnishings	16 Electrical

Fig. 10.9 The MASTERFORMAT divisions and their notation.

the specification, e.g. H62/214. However, readers should not expect everyone to follow these conventions.

THE MASTERFORMAT

In North America, the sixteen divisions of the **MASTERFORMAT** are used to organize Works specifications, construction products literature and cost data. Three more digits are added to the right of the division number (see Figure 10.9) to provide indexes for the various sections, e.g. 05210 – Steel joists. The joint authors of the system are the Construction Specifications Institute (USA) and Construction Specifications Canada.

10.5 COMMENTARY

In this chapter we have been considering the types of drawings, schedules and specifications that can be used to record and communicate information on what is to be constructed on the site. Because the work will be executed by building **operatives**, it is convenient to call it **operative information**. Equally, we shall regard all those who work on the site, whether employees of the main contractor, subcontractors or suppliers, as constituting the **operative system**.

Before work can start, various technical and legal documents relating to the directing of resources and products to the site must be prepared. Their contents, which are the main topics of the next part, we shall call **managing**, or **directive information**. The documents will be be prepared by members of the management team within each of the organizations involved, i.e. by their **managing system**. This arrangement is illustrated on Figure 10.10 although this only shows the managing system of the main contractor. In practice, the next stage will be to seek competitive tenders (bids) to construct the Works from suitable general building or other construction organizations and to appoint the most suitable one. However, such tenders will be based on predictions of what the site processes will be, which is why we must look at those first. Chapters on estimating and tendering are in Part Four.

An international unified classification is proposed that will harmonize existing information systems.

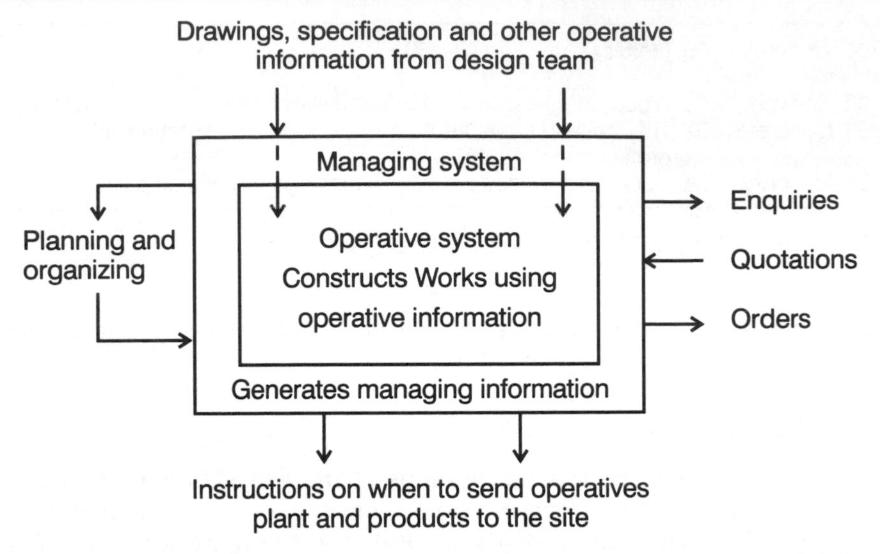

Fig. 10.10 Representation of the operative and managing systems and their information flows.

10.6 SUMMARY

Drawings, specifications and schedules are used to communicate the intentions of the design team to the operatives on the site. Thus, they form part of the operative information system.

Drawings were classified as being either pictorial, or multiplane orthographic. The former includes perspective (or 'as seen'), oblique (one true face) and axonometric (lines are to scale) projections. Variants of each type were described and illustrated.

While pictorial projections are more readily understood by the inexpert, only multiplane orthographic 'technical' drawings can be drawn true to shape and scale. Elevations, plans, sections and cuts were discussed. The purposes of component, assembly, location, site and block drawings were described. Where drawings of the various planes can all be put on a single sheet, their likely arrangements were outlined.

Performance, descriptive and prescriptive specifications were outlined in the context of the design process. The last two are linked to the detail design, being aspects of the element technical solutions. Their co-ordination with drawings was described. The trend towards the use of published standards for products and processes was discussed. Where the same parts have different technical solutions at different locations, data on them can conveniently be expressed on schedules.

REFERENCES

BS 8000: Workmanship on Building Sites, British Standards Institution, London.

FURTHER READING

British Standard 1192:1984–1990 (1990) *Building Drawing Practice*, British Standards Institution, London.

Ching, Frank (1985) *Architectural Graphics*, Van Nostrand Reinhold, New York.

Gill, Robert (1974) *Basic Perspective*, Thames & Hudson, London.

MASTERSPEC, Construction Specification Institute, 601 Madison St., Alexandria, Virginia 22314.

National Building Specification, NBS Ltd, Mansion House Chambers, The Close, Newcastle upon Tyne NE1 3RE.

National Engineering Specification, NES Ltd, 222 Balham High Road, London SW12 9BS.

Porter, T. and Goodman, S. (1991) *Design Drawing Techniques*, Butterworth Architecture, Oxford.

PART THREE
Construction

INTRODUCTION

We shall follow the example of BS EN ISO 9000 – Quality Assurance, where every 'company, corporation, firm, enterprise or association or part thereof, whether incorporated or not, public or private, that has its own function(s) and administration' is called an **organization**.

Construction organizations have to organize unique Works contracts. That is, they have to prepare the site, make arrangements for the timely direction of products, skilled operatives and plant, and construct something that is different from any other building. This involves:

1. analysing the Works into self-contained activities which will be carried out at designated locations and involve particular sets of products and/or processes;
2. for each activity, calculating the quantities of the products required;
3. deciding what operative skills will be required for each activity;
4. predicting how long each activity will take;
5. planning how these activities will fit together;
6. arranging for the provision of products, operatives and plant; and
7. directing and controlling their arrival at the site and the movement of operatives from location to location.

Members of the organization's managing system will do this by working with the operative information discussed in the previous Chapter 10. From it they will produce fresh data of various kinds to convey what can be called **managing** or **directive information**.

CONTENTS

1. Chapter 11 looks at construction organizations and **the total building process** as this differs significantly from manufacturing processes. In

manufacturing, the product passes from one process to another in a factory owned by the manufacturer. In building, operatives move round the building being constructed on a site owned by the customer. The evolution of the industry is outlined and the total building process is analysed so that readers can recognize what is taking place on a site.

2. Chapter 12 **Managing projects** considers the analysis of the building process into activities, the graphic methods available to planners, the data they use, and the control that this offers to contracts managers who have to deal with very complex situations. Their task is magnified by the actions (or, more likely, inactions) of independent subcontractors and suppliers which can, at times, delay completion until well beyond the contract date. Subcontractors have similar difficulties. The effects of such delays on the financial wellbeing of any organization are discussed.
3. Chapter 13 **Dimensional relations** This chapter demonstrates the relations between the relatively few design dimensions and the large amount of measured data that has to be generated for construction management purposes. It distinguishes between the dimensions of elements that are based on those of the spaces and decided during the scheme design stage (see Chapter 7) and those that are decided during the detail design stage (see Chapter 9).
4. Chapter 14 **Products quantities** describes how to generate product quantities from the measurements considered in Chapter 13, and is based on the product classes in Chapter 15. Readers are encouraged to calculate these for themselves from basic facts rather than to use published tables.
5. Chapter 15 **Process durations** discusses the factors that affect productivity and the difficulties of obtaining performance data by observing site processes. Constraints on productivity with each of the product classes and likely performance rates are suggested in Appendix B. This data will be applied in construction planning (Chapter 12), and in estimating (Chapters 18 and 19).

The total building process 11

Itte es ordayned by ye Chapitre of ye kirk of Saint Petyr of York yat all ye Masonns yt sall wyrke ... in ye loge ... sall be in ye forsayde loge at yaire werke atte ye son risyng, and stande yar trewly and bysily wyrkande ... all ye day ... untill itte be namare space yan time of a mileway before ye son sette ...*
(The Fabric Rolls of York Minster, 1370)

11.1 THE INDUSTRY AND ITS RESOURCES

CONSTRUCTION CONTRASTED WITH MANUFACTURING

In a manufacturing industry such as one making formed building products, materials are processed by an established and organized force of operatives and supervisory and management staff, using the manufacturer's own plant. **Plant** is usually taken to include buildings. The employees, the plant and the working capital are sometimes referred to as the manufacturer's **resources** as they remain under the control of the organization.

Each stage in the overall process is carried out by a relatively small group of people working with plant. Generally, each group stays where it is and the products are moved from group to group. The weather has little influence on working conditions. Quality control is straightforward, and there will be opportunities for economies of scale. Over a period, a number of identical or different products are manufactured in the hope that they can be sold, and are distributed to potential customers mainly through a network of outlets. Until they are sold, or, by agreement, until they are paid for, they remain in the ownership of the manufacturer.

When constructing a building, the opposite arrangements apply. A single, unique product, which we shall sometimes refer to as the **Works**, is to be constructed on land owned by the customer. (We have been calling this customer the **owner**, although others may speak of the **client**.) The manufacturing plant and buildings are brought to the site, erected, maintained and, at the end, dismantled and removed. The production

* Time to walk one mile

resources are supplied by a variety of organizations who are hardly ever accustomed to working together. Operatives construct the various parts as they move from one location to another in the Works, and their performance is affected very much by the weather. Production control and quality control are difficult to manage, and there are no opportunities for economies of scale.

Many construction products are both heavy and bulky, and receive little or no processing on site. They have to be placed in various locations and at different heights. Under UK law, when products are incorporated into the Works and thereby become attached to the site, they become the property of the owner.

AN EVOLVING INDUSTRY

The owner of a medieval castle would have brought about, i.e. **procured** its construction, extension or repair by appointing certain master craftsmen to design and supervise their portions of the Works and by providing them with operatives and materials. Political and social changes and economic developments over the centuries have resulted in an increase in the demand for a variety of different types of buildings.

Some of these would have been designed by the more able master craftsmen, who would also have managed their construction. Better educated, creative individuals also undertook these tasks. Although some were called **architects**, others were called **surveyors**, as their duty was to take a wide view of the project, and to supervise it. Sir Christopher Wren's official title was 'Surveyor-General to the Sovereign'.

During the 1600s, foreign travel became somewhat safer and easier. This enabled designers to study the classical buildings of Greece and Italy and those built more recently in Northern Italy during the Renaissance period. On their return, they applied these styles and modes of thought to their own work.

In contrast, due to the difficulty and expense of travelling and communications, individual workmen would have sought employment locally. This encouraged both masters and operatives to acquire more than one skill. For example, bricklayers would also do plastering, and others would combine plumbing, glazing and painting.

Designers also learned how to prepare drawings and specifications in advance of any construction work. This enabled individual master craftsmen to provide the owner with an **estimate** of what the work would be likely to cost. In those days, they would not have been bound by that figure. Nowadays, an estimate is an offer, or bid, which, when accepted, becomes a legally binding contract to do the specified work for a **lump sum**, or in the USA, a **stipulated sum**.

The demands of owners for more certainty in the costs of their buildings led to the monopolizing, or **engrossment**, of the various building functions into single **general building** organizations. The American Revolution of 1776 and the Napoleonic Wars increased the need in the UK for barracks to house the army, earlier practice having been to billet soldiers in private houses. For their construction, the government insisted on contracts with single organizations which then became accountable to them for the whole of the Works.

During and after the Industrial Revolution, businessmen also recognized the commercial advantages of contracting for their factories **in gross** with **general building**, or **main**, or **principal** contractors. This development required that the design functions should be exercised by independent professional architects and engineers. At the same time, the more enterprising architect-builders purchased land and built houses (including the elegant squares in central London) for sale or renting to those who were migrating from the country into the towns.

To sum up: in earlier times, construction was carried out using what we would now describe as a **directly employed labour organization**, or DLO. Nowadays, professionals acting on behalf of a would-be owner will attempt to create a market for the Works, perhaps by inviting suitable organizations to submit bids in competition. (Contractual arrangements to match the needs of owners and their projects are discussed in Chapter 17.)

ECONOMIC SURVIVAL AND THE STRUCTURE OF THE INDUSTRY

Our economic survival as individuals depends on our spending no more than our incomes. Most of us earn our livings by providing a service for someone else, and our primary concern will be to keep ourselves in work. Either we are self-employed, and have to find our own customers, or our employer does this for us. The question is, how best to provide this continuity of work.

In any area, there will always be some demand for the services of individual craftsmen to carry out maintenance and small improvements. They will work together as the need arises. Some craftsmen may extend their businesses by employing operatives with other kinds of skills, perhaps in order to meet a demand for small extensions, refurbishments and individual houses.

The latest information on the industry will be found in the current edition of the Department of the Environment's quarterly *Housing and Construction Statistics*. Most public reference libraries will have this. Latham (1994) reports that, in 1992, the construction industry 'contains 200,000 contracting firms, of which 95,000 are private individuals or one person firms ... only 12,000 contracting firms employ more than 7 people'. Seven is about the practical limit to the span of control of an employer who has to supervise the administration of the office in addition to getting work, organizing and managing it, and getting paid for it. Many are family businesses, and a few have grown, over the generations, to become regional, national or even international in their operations.

Until recently, most of these would have directly employed those operatives who could be provided with a continuity of work, including general and trade labourers, bricklayers, carpenters, joiners and painters. An insight into what it was like to be a building operative in Edwardian times can be gained from Robert Tressell's book (1965). Specialist contractors (often referred to as **domestic subcontractors**) would have supplied plumbers, electricians, roofers, floor layers and others as and when required. The reputation of any building business would have been based on the quality of their work, and many businesses took pride in maintaining this quality by apprentice training. We shall call a business with a stable labour force a **stable organization**. Recent changes in the structure of the industry have been influenced by social and economic changes, including:

- recessions and the resulting increase in competition;
- the universal availability of the privately owned motor car;
- the telephone and fax;
- changes in taxation law;
- extended credit available for the purchase of construction products;
- a fluctuating demand for its services;
- the unwillingness of young people to accept the pay and conditions of apprentices up to their 21st birthday.

Both general building contractors and housing estate developers need to be able to respond whenever and wherever their services are required. To provide this flexibility of response, many organizations have shed most, if not all their directly employed labour forces. Instead, these operatives have either become self-employed or they work for specialist subcontractors. Even the site management staff may not be regularly employed. We shall be referring to contractors with fluctuating labour forces as **unstable organizations**.

The subcontractors who will carry out the Works may not be chosen until just before they are needed, and then only after much haggling over their prices. Thus, an owner who expects to achieve a certain quality of building by carefully selecting the general contractor risks being disappointed.

Most Works contracts include provisions for periodic payments by the owner to the contractor for completed work. At present, few contracts ensure that the main contractor will, in turn, promptly pay subcontractors and suppliers. In developing countries, the owner may even make an initial payment to the main contractor to provide working capital for the project. As little capital is required to set up a building business, few have any real financial stability.

11.2 BUILDING OPERATIVES

OPERATIVES AND THEIR SKILLS

Construction processes are carried out by individuals called 'operatives', or, possibly, 'operators', and, in the past, 'artificers' or 'artisans'. Historically, young people acquired skill in a **trade**, or **craft**, by being apprenticed to a master craftsman. During this apprenticeship, which normally lasted until their 21st birthday, apprentices learned how to use the 'tools of the trade' and to work with those materials that were the prerogative of their craft, and from which the craft gained its name. At the same time, they acquired a particular place in society. Nowadays, as few general builders are able to offer apprenticeships, entrants must go elsewhere for their training, perhaps on a course backed by the Construction Industry Training Board (CITB).

In his 'London and Country Builder's VADE MECUM, or The COMPLEAT and UNIVERSAL ARCHITECT'S ASSISTANT', Salmon (1748) gives prices for the works of Bricklayers, Masons, Carpenters, Joyners, Glasiers, Plumbers, Slaters, Plaisterers, Painters, Paviours, Carvers, Smiths and Thatchers. Even so, individuals sometimes acquired skill in more than one trade, particularly where work was scarce, as in rural areas.

Elsom, in his *Practical Builder's Perpetual Price Book* (1825) comments that 'bricklayers in the country are identified not only as bricklayers, but as masons, and sometimes as plasterers and slaters'. He also reports that foundation trenches were still being dug as the first task in the process of bricklaying!

The division of building skills into **trades** or other **affinity groups**, and the designations of the operatives, are likely to vary to some extent from region to region. For example, an operative who works with wood products or their substitutes on site may be called a carpenter, or a joiner, or a carpenter and joiner. When a UK bricklayer visits the USA, he should describe himself as a mason. (In the UK, it is becoming more common to classify working with either bricks, blocks or stone as **masonry**.)

In 1969, Nelson listed 116 main trades and major specializations. Changes in building technology, and the marketing of new products encourages existing skilled operatives to develop fresh skills. Past examples include the formwork carpenter and the dry-liner. New types of plant will require skilled operators and may lead to new specialisms such as that of the ground-worker.

Specialist operatives are essential for the efficient construction of large new projects. In contrast, complex refurbishment and renewal work calls for the services of operatives with many skills who are capable of working on their own without supervision.

Substitute substances and products will require the same manipulative skills as those they are intended to replace, and are normally assumed to be the province of an existing skill-group. For example, plumbers originally worked only with lead sheets and pipes. Over the years, their skills have been extended to include working with other non-ferrous sheets, with iron, steel, copper and now plastic pipes, gutters and fittings, and with the components associated with them.

Operatives skilled in a trade may work on their own, or with the shared or entire support of less skilled operatives called **craft labourers**. **General operatives**, or **general labourers** will have acquired their skills mostly from experience.

A group of any of these, probably called a **gang** or **crew**, may work together to achieve a common purpose such as a continuous road surface, a concrete foundation, or the external walls of a building. It will usually include someone with authority over the group, e.g. **supervisor, foreman, chargehand**, who will be accountable to site management for the quality and progress of their work.

The person who has overall responsibility for the progress of the Works may be called either the **site manager**, the **general foreman**, the **site agent**, or the **superintendent**. This individual will be accountable to a **contracts manager** whose base is likely to be an office elsewhere.

All processes involve the direct use of implements or plant. Craftsmen will usually own their hand tools. Employers will provide hand held power saws, paint sprayers, and other mechanical tools, and excavators, dumper trucks and other mobile mechanical plant. They will also provide implements for non-craftsmen.

OPERATIVE HEALTH AND SAFETY

Under the CDM Regulations, a competent **principal contractor** is appointed by the owner to develop and implement the construction stage of the health and safety plan. Their duties include disseminating information in the plan, co-ordinating the work of other organizations on the site, training, and monitoring compliance.

OPERATIVE TASKS

During their training and work-experience, operatives learn to carry out simple sequences of actions, or **tasks**, using those kinds of materials regarded as the province of their group. For example: 'lay and finish concrete ground slab'; 'construct brick wall in English bond'; 'set out for, cut and fix roof timbers'.

In practice, most sequences of actions are likely to be repeated many times, and such tasks are usually described as being 'cyclic'. Where cycles of tasks can be co-ordinated, the process begins to look like a 'flow'. Interruptions in the smooth cycling of tasks may be caused by the need to:

- move to another workplace;
- prepare materials;
- clean equipment;
- move scaffolding or other temporary workplaces; or
- move items of mechanical plant.

Operatives will apply their skills on any project, either as a result of an instruction from a supervisor, or because the work obviously needs to be done.

OPERATIVE EMPLOYMENT ANALYSIS

Stevens (1987) reports 'for house construction that only about half the time on site is spent directly "making the building grow"'. Construction processes involve a great deal of hard physical work, often in awkward, uncomfortable or distasteful conditions. If operatives are to maintain their efforts and concentration, work places must be safe, and periods of relaxation are essential.

Time will also be required for meals and refreshment, attending to personal needs, and for social contacts with colleagues. Particularly when working on large, high buildings or on dispersed sites, much time can be taken up in moving between the site entrance, the site accommodation and the workplaces.

At least once every day, operatives will also spend time receiving instructions, making ready for work, and clearing up afterwards. These last two, sometimes called **preparatory and concluding tasks**, can be further analysed as follows.

- **Making ready** includes preparing tools and plant, finding materials, and moving to the workplace. Protective coverings must be either installed or removed, e.g. putting covers over floors and furnishings before decorating ceilings; removing frost protective coverings from the tops of walls.
- **Clearing away** includes cleaning tools and plant, e.g. paint brushes, concrete mixer; putting away tools and plant; putting away surplus materials; removing rubbish; removing or fixing protective coverings; moving from the workplace; seeking further instructions.

These preparatory and concluding tasks have sometimes been described as being **indirectly productive**, in contrast to those which are **directly**

productive. As they are essential to doing useful work, they can both be included in the category of **effective employment**.

However, there will be times when operatives are forced to wait, either for instructions, assistance or materials, or because some other task on which their work depends has not been completed. They are then regarded as being **ineffectively employed**. This will also be the case when a process has been completed but there is insufficient time left in the working day for other effective employment. When either there is no work for operatives to do, or they are not working because of inadequate supervision, they can be described as being **unemployed**. These work-states may also be described as either **active**, **inactive** (can't get on) and **idle** (won't get on).

PLANT EMPLOYMENT ANALYSIS

The above analysis of effective and ineffective employment can also be applied to mobile plant which is on the site for specific processes such as excavating. Making ready will include periodic maintenance, refuelling, changing buckets or other equipment, and moving to and from the workplace. Clearing away will include cleaning and moving from the workplace.

The period during which the plant is waiting for or receiving repairs or maintenance (called **down time**) can be regarded as ineffective employment. Plant is often kept unemployed on the site to ensure it will be available when needed. Although this will suit the site manager, those responsible for minimizing the site costs may not be so pleased.

11.3 BUILDING PROCESSES

KINDS OF BUILDING PROCESS

Any distinct series of actions or events can be called a **process**. In the BS EN ISO 9000 series of standards relating to Quality Assurance which has superseded BS 5750, a process is

- a value-adding transformation involving people and other resources; or
- a set of interrelated resources and activities which transform inputs to outputs.

We need to understand the various kinds of process which can take place on a building site before we can plan and organize the work to be done and estimate its likely cost. We shall start with some broad distinctions.

The initial work of clearing the site and changing its surface configuration by excavating or filling so that it will accommodate the Works we will call **site preparation processes**. These are considered later in Chapters 12, 15 and in Appendix B. Setting up, maintaining and removing the site accommodation, cranes, and other facilities which will enable the construction processes to take place we will call **enabling processes**. Those actions associated directly with the construction of the Works we shall call **construction processes**. The totality of those actions on the site which result in the completed building we shall call the **total building process**.

ENABLING PROCESSES

A variety of mechanical and fixed plant, temporary constructions, services and individuals will provide essential support to the men and plant engaged directly in making the building grow. Cranes, hoists, mixers, pumps, etc, scaffolding, hoardings, screens, storage compounds and various kinds of accommodation all have to be brought to the site, unloaded, moved, installed and maintained. Temporary bases for these, together with temporary roads, water, telephone and electric power supplies will also be required. Then, at the end, they must all be taken down and removed, and the site reinstated.

Transport for operatives will be required to and from the site, and for the movement of men, materials and plant on the site. While some members of the site management team may be on the site for most or all of the time, others will only visit it occasionally. These resources may all be called **indirect items**, or **Preliminaries**, although, when thinking about their costs, we shall refer to them as the **Works overhead**. They are considered in more detail in Chapters 12, 16, 18 and 19.

CONSTRUCTION PROCESSES

Although sometimes called **assembly processes**, this by no means describes the complexity of construction processes. Most materials have to be stored first and then moved to where the operatives are working. Formless products have to be prepared and mixed, and formed products often have to be made to fit by reducing their size and shape before being secured in their final position. Also, while some processes directly 'make the building grow', other supporting processes might be going on at the same time so that the growing processes can take place, e.g. bringing materials, moving scaffolding, clearing rubbish.

The kinds of products to be processed and their quantities, the location of the work to be done, and the time available, will indicate which operative skills to employ on construction processes. They will also indicate the kinds of plant to use (if any), and the extent to which similar processes can be carried out simultaneously. Decisions on these matters constitute the **construction method** and are considered in Chapter 12. Obviously, the choice is limited by the resources available and the benefits forgone by not using them elsewhere.

- **Main product and quantity** The name of the main product will indicate the kinds of operative skills required to process it. The amount to be processed will indicate a suitable combination of operatives and plant.

  For example, a small quantity of tarmacadam for repairing potholes in an existing road will indicate that the work should be done by hand. A very large quantity for surfacing a new road will indicate the need for a paving gang with a mechanical paver, road rollers etc. and a suitable number of vehicles to maintain the supply of materials.
- **Location** If it is not practicable to use a machine, e.g. when excavating small pits, or a trench within an existing building, the work will have to be done by hand. When constructing the foundations, floors and walls of a small extension, there might be working space for only a single bricklayer and labourer. Where a safe workplace is not available, it will be necessary to construct one, e.g. an external scaffold from which an existing building can be repaired or redecorated.
- **Time available** Where time is a consideration, and suitable workplaces are available, more than one tradesman or gang may be employed on the same process, e.g. carpenters constructing a roof; joiners hanging doors; painters.

The majority of these processes involve constructing a part of whatever size is required, i.e. they are **dimensional parts**. Each part has a name that is different from that of its main product, e.g. a wall is built with bricks, a floor is made with concrete. Complementary processes may be involved, e.g. providing formwork and positioning reinforcement for a concrete wall.

In contrast, the process of making an **integral part** by fixing a component, e.g. a door, a washbasin, will be much the same wherever the work is done, and the name of the part will be the same as that of the component. Note that, when estimating the quantities of resources required and their costs, dimensional parts have to be measured, but integral parts are merely counted. Construction processes are further considered in Chapter 15 and in Appendix B.

QUALITY CONTROL

When constructing, it is vital to 'get it right' first time. Yet the Works will be unique, the drawings may be inadequate, and the operatives may not have worked together before. Thus, continuous and vigilant supervision and forethought by the construction team is essential, although not always the case.

Normally, design team members will only visit the Works at intervals to inspect and approve or condemn what has been done. On larger Works, though, a resident clerk of works (COW) or resident engineer may be employed (and paid by the owner) to see that the contract is being observed and to report on progress. Some authority may be delegated to the COW to give instructions to the constructor.

When some work is not good enough and has to be removed and replaced, this will result in unnecessary delays. It is also likely that the removal process will cause damage to other work, which will, itself, have to be **reworked**. Consider, for instance, the consequences of discovering that the reinforcement in a reinforced concrete structure has been positioned incorrectly. Should it be demolished? Can it be demolished? Can it even be tested to see if it might still be strong enough for its purpose?

Opportunities exist for organizations to demonstrate to potential customers what quality of products and services they can expect. One way is to obtain recognition by an accredited authority that the organization is meeting published standards. An internationally recognized standard is BS EN ISO 9000 – Quality Assurance, which has been developed from BS 5750. Another way is to show a commitment to continuing improvements in the management of the organization and its products. This may be called Total Quality Management (TQM).

Within the industry, as within others, individuals are beginning to recognize that they owe a duty to anyone who might be affected by the quality of their work, not just to the other parties to their contract. Members of this larger group with an interest in the project are sometimes called stakeholders. Other quality issues, including whole life costing, relations between individuals, and value engineering, are considered in Chapters 17 and 20.

OPERATIVE AND MANAGING SYSTEMS

All the operatives, management personnel, plant and other resources engaged on the Works in site preparation, enabling processes, and construction processes constitute what could be called the **operative system**, whereby inputs of materials are transformed into an output of building works.

The operative system is maintained by regulating and controlling the movement of operatives, plant and construction products across the boundary between the site and its environment (see Figure 10.10). This is the responsibility of members of the **managing system**, and is considered in the next chapter.

11.4 COMMENTARY

The application of an acquired skill in the production of an object is sometimes called an **art**, although, nowadays, that word is more likely to refer just to sculpture, painting, and other **fine art**.

The construction and maintenance of buildings depends on the work of experienced operatives with practical skills in the selection and manipulation of simple materials, i.e. on their arts. Such craftsmen used to be called **art**isans or **art**ificers. Objects made by hand are often referred to as **art**efacts (or artifacts).

Much of our interest in old buildings has to do with our appreciation of the arts that were demonstrated in their construction. Maintaining such buildings keeps these skills alive. However, as the process of building becomes more industrialized, so the importance of these arts diminishes. The skills of the designers of buildings can also be described as an art.

11.5 SUMMARY

The structure of the building industry is placed in its historical background. Many features of the total building process are the opposite of those of manufacturing. There, an input of materials is transformed by static human and plant resources to suitable outputs which are sold elsewhere. In construction, resources have to be temporarily supplied to the site to process inputs of materials and construct the Works on the customer's land. Other plant and human resources enable the building process to take place.

Site preparation processes and construction processes for making the building grow consist of series of repetitive tasks. Each process will be carried out by one or more specialist operatives with an exclusive range of skills in the preparation, manipulation and incorporation of a particular set of products.

While on site, they may be effectively or ineffectively employed, or unemployed. Effective employment includes the time taken in making ready and clearing away. Some of the factors that affect their productivity were considered. The responsibilities of the principal contractor for health and safety, and the importance of producing work of an assured quality were introduced.

The managing system that controls the movement of operatives and plant to and from the site was distinguished from the operative system which consists of all those who are employed on the site.

REFERENCES

BS EN ISO 9000 – Quality Assurance, The British Standards Institution, London.

Chartered Institute of Building (1983) *Code of Estimating Practice*, The Chartered Institute of Building, Ascot.

Latham, Sir Michael (1994) *Constructing the Team*, Department of the Environment, HMSO, London.

Nelson, J. L'a. (1969) *Instructions to Operatives*, Building Research Establishment, Current Paper 10/69, HMSO, London.

Stevens, A. J. (1987) Housebuilding productivity in the UK – results of case studies, in *Managing Construction World-wide*, Vol. II, p. 707, E & FN Spon, London.

Tressell, Robert (1965) *The Ragged Trousered Philanthopists*, Panther, London.

FURTHER READING

Andrews, Francis B. (1974) *The Mediaeval Builder and His Methods*, EP Publishing Ltd, Wakefield, Yorkshire.

Ball, Michael (1988) *Rebuilding Construction*, Routledge, London.

Briscoe, Geoffrey (1988) *The Economics of the Construction Industry*, B.T. Batsford Ltd, London.

BS 8000. Workmanship on Building Sites, British Standards Institution, London.

Department of the Environment. *Quarterly Housing and Construction Statistics*, HMSO, London.

Forster, George (1989) *Construction Site Studies, Production, Administration and Personnel*, Longman Scientific and Technical, Harlow.

Harvey, Roger C. and Ashworth, Alan (1993) *The Construction Industry of Great Britain*, Newnes, Oxford.

Illingworth, J. R. (1993) *Construction Methods and Planning*, E & FN Spon, London.

Manser, J. E. (1994) *Economics – a Foundation Course for the Built Environment*, E & FN Spon, London.

Patchett, J. M. (1983) *Construction Site Personnel*, Butterworths, London.

Powell, C. G. (1982) *An economic history of the British building industry 1815 – 1979*, Methuen & Co. Ltd, London.

Tavistock Institute of Human Relations (1966) *Interdependence and Uncertainty, a Study of the Building Industry*, Tavistock Publications, London.

Managing projects 12

... only the future can be altered, the past merely provides experience. (BS 6046)

12.1 MANAGING COMPLEXITY

INTRODUCING PROJECT PLANNING

Owners, professional firms and main contractors all have one feature in common. In each case, a number of individuals with different skills will work on a variety of interrelated tasks over an extended period of time to achieve a common goal by a predetermined date.

To do this, the managers of each organization must create a **plan**, or representation, of what is to happen and when. They must also arrange for the necessary production organizations, operatives, money and materials to be available when needed, and must monitor progress against the plan.

They are quite likely to use the same project management methods as those preparing to send a space capsule to Mars, build a ship, or organize a pop festival, and will call their basic planning and control unit an **activity**. Those involved are likely to refer to the totality of it all as the **project**. Even so, while the owners may regard the construction of the **Works** as just the final stage in their project, in the eyes of all those who work for the construction organization, the Works **is** their project.

A construction activity will:

1. require the same operative skills and other resources throughout;
2. involve the same kinds of main products and processes;
3. have a distinct start and finish; and
4. result in the construction of one or more clearly defined parts of an element, or the completion of some other process.

Activities are not carried out in isolation. Instead, they form sequences where being able to start any one activity will depend on certain of the others having been finished. Thus, each activity will have some **dependence relations** with preceding activities. Their completion/start dates are called **events** and are usually numbered.

Once the activities have been identified and described, including calculating the time they will take, i.e. their **durations**, they must be arranged in sequence to suit their dependence relations. The process continues with the planner going forward and backward through the plan, while adjusting the start and finish dates of the activities until they all fit together as efficiently as possible. Planning a project is another kind of iterative design process.

Construction differs from other industries in that the same kind of process may be needed at a number of different places on the site, on different floors and, once they have been enclosed, in different rooms etc. To meet cost and time targets, efforts must be made to bring operatives and plant to the site when needed and to keep them employed at the various **locations**.

While planners will be concerned mainly with human resources, excavators, cranes, concrete mixers and other mechanical plant also have limited outputs, and can only be in one place at a time. Money is also a scarce resource, and construction managers will seek to use the owner's money rather than their own working capital wherever possible.

But the abilities and dedication of the planners must be matched by the efforts of general and site managers and those in charge of subcontractors' operatives if the approach is to work and the risks implicit in the project are to be minimized. Planning is a team effort.

PROJECT MANAGEMENT COMPUTER SOFTWARE

A graphical display showing the dependence relations within a group of activities is called a **logic diagram**, a **precedence diagram**, or a **network**. The task of creating and modifying such networks, adding dates and optimizing the use of the available resources so as to achieve time and cost targets is called **planning**.

The purpose of a network diagram is to support members of the managing system when they are planning, organizing, co-ordinating, monitoring or controlling the time spent on activities. The approach enables them to investigate alternatives by doing 'what ifs' and, given sufficient data, computers are capable of working out the best use of the available resources.

Even so, a project which members of the constructor's managing system might regard as, perhaps, 50 work-packages of related activities, might be treated as several hundred quite separate activities by those managing the operating system on the site. At any one time, a large construction organization might be engaged in a number of projects involving, in all, very many hundreds of activities. Managing them will involve many individuals and the only practical way of dealing with such complexity is to use computers and project management computer software.

A number of (mostly) general purpose software systems are now available, each capable of producing a variety of graphical and other outputs from the same basic data. However, not all project management software can cope with the complexity of the building process.

As different individuals will be responsible for different groups of activities, e.g. the different projects or parts of projects, the chosen software system must be able to match this management structure. To manage all the resources available to the organization, managers must be able to treat any group of activities as both a project on its own and as a subproject of the organization as a whole. (In project planning, materials are also regarded as a resource.)

PROJECT COMPLEXITY

A single-storey, single space enclosure can be constructed in just a few activities. Complications arise from the following:

1. having an increasing number of spaces with different purposes;
2. having internal vertical divisions to create these spaces;
3. having more than one storey (story) to provide more spaces; this may call for temporary supports to loadbearing floors, installing staircases or hoists for operatives and cranes for lifting materials, erecting and dismantling temporary workplaces, and delays associated with these;
4. incorporating a structural frame;
5. having higher standards of thermal, sound, security and other attributes and the additional parts these demand;
6. providing services and installations, furnishings and equipment;
7. having more than one technical solution for, say, the external walls, or the roof, and, in consequence, having to employ more than one kind of skilled operative to construct that element.

In the main, the total building process consists of the following sequence of activities, although these will overlap:

1. equip the site with plant and other facilities;
2. site preparation;
3. structure (including providing temporary workplaces), drainage and external works;
4. completion of weathershield, including windows and external doorways;
5. non-loadbearing walls;
6. services and installations;
7. internal doorways;

8. finishes and decoration;
9. furnishings and fittings;
10. remove plant etc.

12.2 ANALYSING THE TOTAL BUILDING PROCESS

ENABLING PROCESSES

Enabling processes (see Chapter 11) will include bringing to the site and erecting the cranes, huts, compounds, scaffolding and other plant items, and providing a water supply, electric power and other temporary services. Each will constitute a separate activity as different resources will be involved, and some will depend on others having been finished, e.g. heavy vehicles will be able to deliver large items of plant only after a temporary road has been made on the site. When the time comes that they are no longer required, these facilities must be removed and the site reinstated. These processes can also be treated as activities.

SITE PREPARATION PROCESSES

The site is likely to be prepared in the following stages.

1. Demolish any unwanted buildings, cut down unwanted trees and bushes, collect rubbish, remove tree roots and other wood likely to encourage fungal attack and termites (not in the UK) and clear the site. Relocate any overhead cables and underground pipelines.
2. Construct the temporary benchmark at the level called for by the design.
3. Set out the outlines of the buildings and other works.
4. Strip the turf and remove the vegetable soil over the area to be occupied by buildings, pavings and other works or to allow the surface levels to be modified.
5. Establish the level of the bottom of the lowest floor construction or pavings, i.e. the **formation level** and, on a partially excavated sloping site, the line of the edge of the excavation, i.e. the cut and fill line (see Chapter 13).
6. Excavate the ground above the formation level, a process normally referred to as **excavation to reduce levels**.
7. Set out for and excavate the variously shaped voids needed to accommodate the construction, including basement voids, pits for column bases, and trenches for foundations, services and drainage pipelines.

Each of the above will constitute a separate activity, or more than one if it cannot be completed without being interrupted by others.

CONSTRUCTION PROCESSES

Identifying the activities and processes that are called for by the design is mainly a matter of identifying:

1. the main products to be used for each of the parts;
2. the locations of these parts;
3. the amount of each that can be constructed before interruption by those who must begin work on other activities.

These main products and their quantities will indicate the skills the operatives must have, what their work will involve, and any necessary complementary processes. (Calculating these quantities and predicting process durations are considered in Chapters 13, 14 and 15.)

Except for the initial ones, any one activity can be started only when certain others have been completed. Putting this another way, when an activity is complete, this **event** may allow another to start. If more than one can be started, either one of them must start first, or they can be run in **parallel**, i.e. **concurrently**. Thus, all construction planning involves looking for relations of **precedence**, or **dependence** between activities.

However, the sequence of these construction processes is determined as much by what cannot be done as by what can be. The most important constraint is, of course, the force of gravity. A process cannot be carried out if a strong, stable and safe workplace is not already in position. This can be either a permanent construction or a temporary scaffold. Neither can any products be placed in position unless there are other products already there to support them. Other constraints might be:

1. other processes already being carried out at that workplace;
2. suitable resources or materials not available;
3. weathershield not yet complete;
4. materials cannot be positioned;
5. continuity with other elements or parts cannot be achieved;
6. necessary prior work not completed.

LOCATIONS OF ACTIVITIES

Every activity is likely to involve working at a number of locations, and this is where construction differs from most other kinds of project. Even the simplest building is likely to contain a number of rooms and require some external works as well. Many projects involve a number of different kinds of buildings and their external works.

It is possible that an entire part of each of these might be built without interruption. But, for example, the external walls of a multi-storey building might be constructed one storey-height at a time in readiness for the next structural floor. Also, the building may be sufficiently large for it to be regarded as consisting of a number of different locations, or **zones**. When planning the work to internal finishes, heating and other services, and to windows and doorways, each room might be regarded as a separate location.

The work at each location might then be called an activity, a sub-activity, or, perhaps, an operation. As an aside, readers are warned that others may use the terms 'process', 'operation', and 'activity' without distinction in everyday speech.

Continuity of employment can be maintained and completion time minimized if the work of a particular gang is concentrated at one location at a time. Then, with careful management, the gangs of operatives working on dependent activities can follow each other from location to location. This is called **parallel working**, and is essential for achieving time and cost targets. However, it is not always easy to get operatives to finish completely before moving on. Outlines of some processes and the constraints that affect productivity are given in Chapter 15 and Appendix B.

12.3 ACTIVITIES

ACTIVITY ANALYSIS

During the planning process, the **method statement** on which the bid was based (see Chapter 18) is extended and, possibly, reconsidered.

For each activity, answers must be found to the following questions and written down.

1. **What** is the purpose of the activity and its quantity?
2. **Where** is it located?
3. **How** will it be identified? If events are numbered, activities can be labelled by giving the numbers of their start and finish events (sometimes referred to as their *i* and *j* events) although, obviously, this must wait until the network is drawn. Alternatively, the activities themselves can be numbered.
4. **What** are those immediately preceding activities that must be finished before this one can start?
5. **What** immediately following activities can start when this and possibly some other activities as well are finished (see Figures 12.1 and 12.2)?
6. **How** will it be carried out? The sequence of processes to be performed should be described.
7. **Who** will carry them out, and what will they use, i.e. **what** labour and plant resources will be employed?
8. **What** operatives and plant performance rates are likely?

9. **How long** are they likely to take, i.e. what will be the estimated total working time, or duration?
10. **What notice** will be required before the resources will be needed?
11. **When** should the procedure of obtaining, i.e. **procuring** these resources be started (see Figure 12.4)?

Although all projects are unique, they are all likely to include many of the same little activity sequences relating to particular technical solutions. These, like the technical solutions themselves, can be identified and learned, and then applied to a variety of different jobs, e.g. for any reinforced concrete construction, the sequence on the site is:

1. make or repair and assemble formwork,
2. prepare and fix reinforcement,
3. place and consolidate concrete,
4. delay while concrete hardens,
5. strike formwork,
6. finish exposed surfaces.

SPLIT ACTIVITIES AND DELAYS

Some of the work of a particular group of operatives may have to be carried out both before and after the internal finishes. These are often referred to as **first and second fixings**. For instance, it may be necessary to interrupt the work of those installing the plumbing and electrical services and the joinery to allow the walls to be plastered. Wherever possible, operatives should be given continuity of employment on that site, as once they have moved to another site, it may be difficult to get them back.

Unavoidable delays Where a part is constructed with a Category A substance, the next activity must be delayed until the substance has developed its properties, including sufficient strength. Examples will include delays occurring between pouring concrete foundations and starting to build masonry walls and between plastering walls and applying decorations.

The same considerations apply to composite constructions. For instance, there will be a delay after pouring a reinforced concrete floor before (1) its supporting formwork can be removed, (2) it becomes strong enough to act as workplace, and (3) it will support formwork for the next floor. Masonry walls are another example, as there will be a limit to the number of courses that can be raised while the mortar of the bottom course remains unset. Any construction containing free water may have to be allowed to dry out before finishes or decoration can be applied. This can take months.

12.4 PROJECT MANAGEMENT GRAPHICS

USING COMPUTER SOFTWARE

Once the data on each activity has been entered, the computer will perform the following functions.

1. Calculate the earliest and latest dates for each event. Adding the durations from left to right gives the earliest finish date for each event and for the project. Working backwards from that finish date, and subtracting the durations gives the latest dates. The earliest finish date of the last activity will be the earliest completion date for the project.
2. Identify the longest continuous path through those activities, called the **critical path**. This will connect events whose earliest and latest dates are the same. Here, a delay or speed up in any activity will also delay or speed up the project completion date, although a speed up may cause the critical path to be re-routed through other activities. Activities not on the critical path will have time to spare, called **float**.
3. Aggregate the demand for resources, day by day.
4. Express the results in a suitable graphical form.

The planners can then consider how best to deal with any imbalance in the demand for each kind of resource, and what overtime, weekend working, or additional resources (if available) might be employed to help with peaks in demand, bearing in mind their cost. Depending on the method employed, this process may be called **resource smoothing** or **resource levelling**. Finally, tasks are assigned to individuals.

Like any other design procedure, this process involves iteration, and is called **scheduling**. Assigning the resources available to the organization as a whole is called **multi-project scheduling**.

BAR OR GANTT CHARTS

Figure 12.1 shows an array of rows and columns called a **bar** or **Gantt chart** (named after its originator) where:

1. each row represents an activity or a resource;
2. each column represents a period of time, e.g. a day, week or month; and
3. Each activity is represented by a horizontal line drawn between its planned start and finish dates. Later, a second bar can be extended to show when work is done and what percentage has been completed.

For simplicity, the chart assumes that every activity finishes at the end of a time period. A bar chart provides an easily grasped summary of the data on the network and can be drawn manually where there are only a few

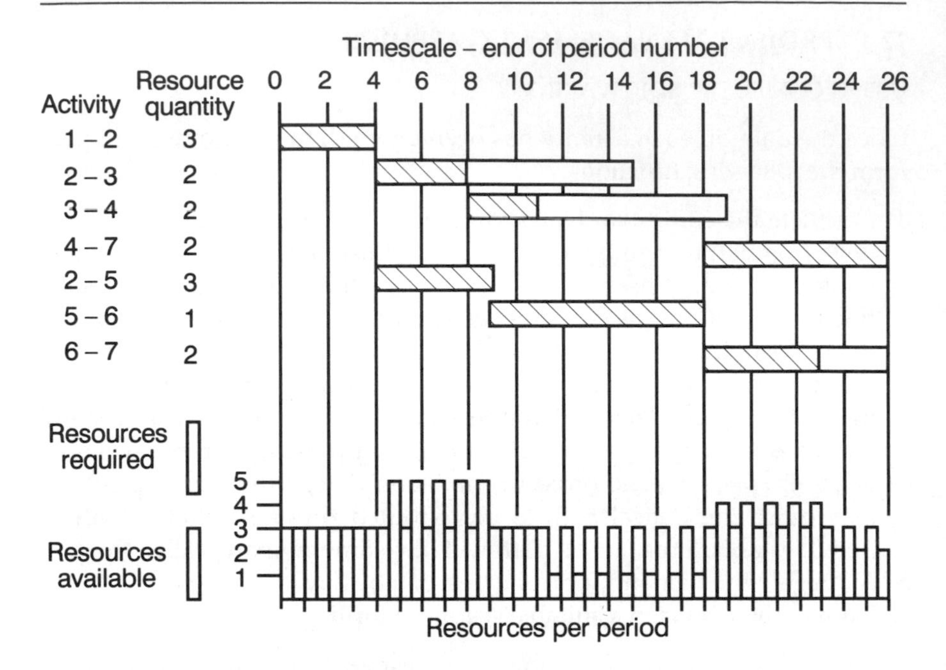

Fig. 12.1 An example of a bar chart and resources histogram.

activities. The chart shown in Figure 12.1 is based on the data displayed on the network in Figure 12.2. The two should be read together as being different expressions of the same data.

A bar chart can also be helpful when working out how best to achieve a complex and repetitive sequence of processes within an activity, such as constructing the loadbearing brick walls to an upper floor and hoisting and fixing precast floor beams and staircases. Arrow headed lines are sometimes used to indicate the dependence relations between the start of one activity and the finish of another.

HISTOGRAMS

For each period, which could be a day, week or month, throughout a project, the supply of and demand for a particular scarce resource can be displayed as a pair of columns, either side by side or superimposed. Such a diagram is called a **histogram**. They are invaluable for showing the excess demand for or under-utilization of each resource, and will indicate where smoothing or levelling should be applied.

Figure 12.1 shows the demand for resources to match the bar chart. Activities (2–3), (3–4) and (6–7) are shown in their 'earliest start' position, the bars being extended to their 'latest finish' times. The demand for resources can be smoothed by utilizing this extension, or float. For example, (2–3) could be moved to periods 11–14, and (3–4) could move somewhere within periods 14–18. To finish on time, one additional resource will be required for five periods between 18 and 26.

NETWORK ANALYSIS

Network diagrams take two main forms. In **critical path analysis (CPA)**, the activities are denoted by arrows between circles, or **nodes**. These nodes represent events when one or more activities are finished and others can start. **Milestone** events can be labelled so that achieving them can be reported to the manager responsible. This approach is called **activity-on-arrow**. Where more than one event must occur before an activity can start, the ones finishing later are linked to the other one with broken lines denoting **dummy** activities (see Figure 12.2).

The alternative is an **activity-on-node** network or **precedence diagram**. These can take a variety of different forms, one being called **programme evaluation and review technique**, or PERT. They are quite different from CPA as the details of an activity are given in a box (see Figure 12.3). Arrowheaded lines from preceding boxes will identify those activities whose finish will allow it to start. Similarly, lines are drawn from it to boxes detailing activities that can start when it is finished.

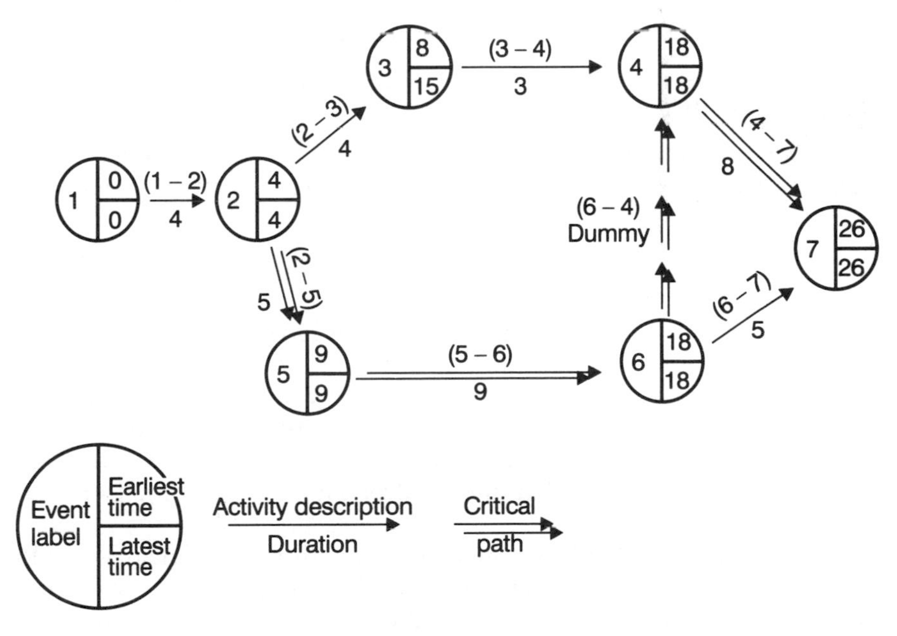

Fig. 12.2 A schema of an **activity-on-arrow** network.

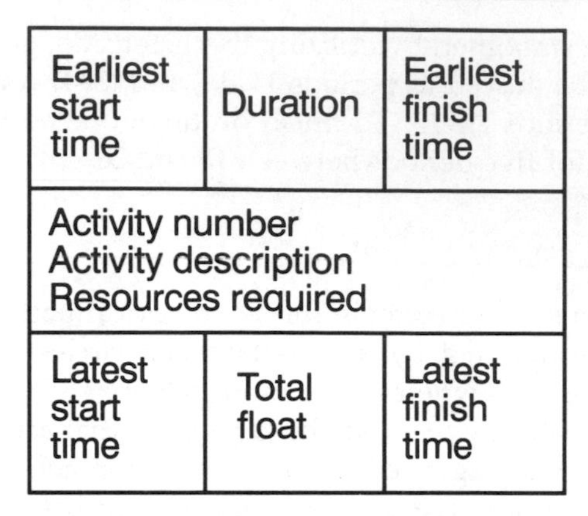

Fig. 12.3 Data to be found in the activity boxes of an **activity-on-node** network.

OTHER GRAPHICS

Frequently, a sequence of activities is to be repeated at different locations, e.g. constructing a reinforced concrete multi-storey frame; refurbishing a street of terraced (town) houses, building an estate of houses. In such cases, the demand for the various resources and their movement from place to place can be planned and harmonized using a graphic technique called **line of balance**.

Flowcharts are helpful when analysing or planning a procedure. An example is given in Figure 18.4 at the end of Chapter 18.

12.5 RESOURCE PROCUREMENT

Nowadays, an important attribute of a contracting organization will be their managing system's ability to organize their operative systems by supplying the following when they are needed:

1. **Products** In the UK, the materials supply industry is able to respond speedily to demands for the delivery of most products. However, before this can happen, management must have obtained quotations and placed orders. Not only does this take time, but the delivery of some products may be delayed, perhaps because they have to be specially made, or because their manufacturers have a backlog of orders.
2. **Their own operatives** To keep their own operatives in work, managers must be able to plan and control the activities on all of their sites.
3. **Subcontractors' operatives** Subcontracts with busy specialists should be agreed well in advance as they too will have work to do on a number of sites. In such cases, of course, the main contractor's management will not be in direct control, but must work through the subcontractor's

management staff. Even so, some managers tend to delay letting sub-contracts until the operatives are needed, and then to bargain with specialists who are looking for work.

4. **Plant** Mechanical plant with their skilled operators will be required for activities such as groundworks. Plant will also be required for the enabling processes. All items of plant and, where appropriate, their operators, are available from plant hire specialists. Where an organization decides to own and operate plant for itself, their plant department will compete with these specialists, and, where successful, will charge the various Works for their services.
5. **Supervisory and other staff** Even when preparing the bid (see Chapter 18), the management may have taken into account who would be available to take charge of the site. While it is desirable to keep the same site staff throughout, less effort is required from them once the structure is complete. The services of surveying and engineering staff may be required from time to time.

The dates for these actions can be calculated by working backwards from the earliest start dates for the activities concerned. They can be recorded on a bar chart or calendar.

Figure 12.4 illustrates the stages in the procurement of materials, subcontractors' operatives, and other resources so that they will be available when required. Such resource procurement programmes would be prepared by the planners in collaboration with the buyers, quantity surveyors and other members of the managing system.

12.6 COST CONTROL DURING CONSTRUCTION ACTIVITIES

This section uses costing ideas that are more fully considered in Chapter 16. Controlling a process implies measuring the difference, or **variance**, between its planned and actual outcome as it is happening, and taking remedial action if necessary. Measuring the variance after the process has been completed gives no opportunity for remedial action, and is merely reporting. In seeking to control a contractor's costs, the duration of a Works contract is as important as the total of its site costs, i.e. its **prime cost**.

Control will be effective only if information can flow freely between individuals in the various departments of the organization and on each of the sites, and if management reports can draw freely from all the data sources.

CONTRACT DURATIONS

The staff of an organization can manage only a limited number of contracts at any one time. The income from each contract will include a contribution to the costs of running the business which, when added to

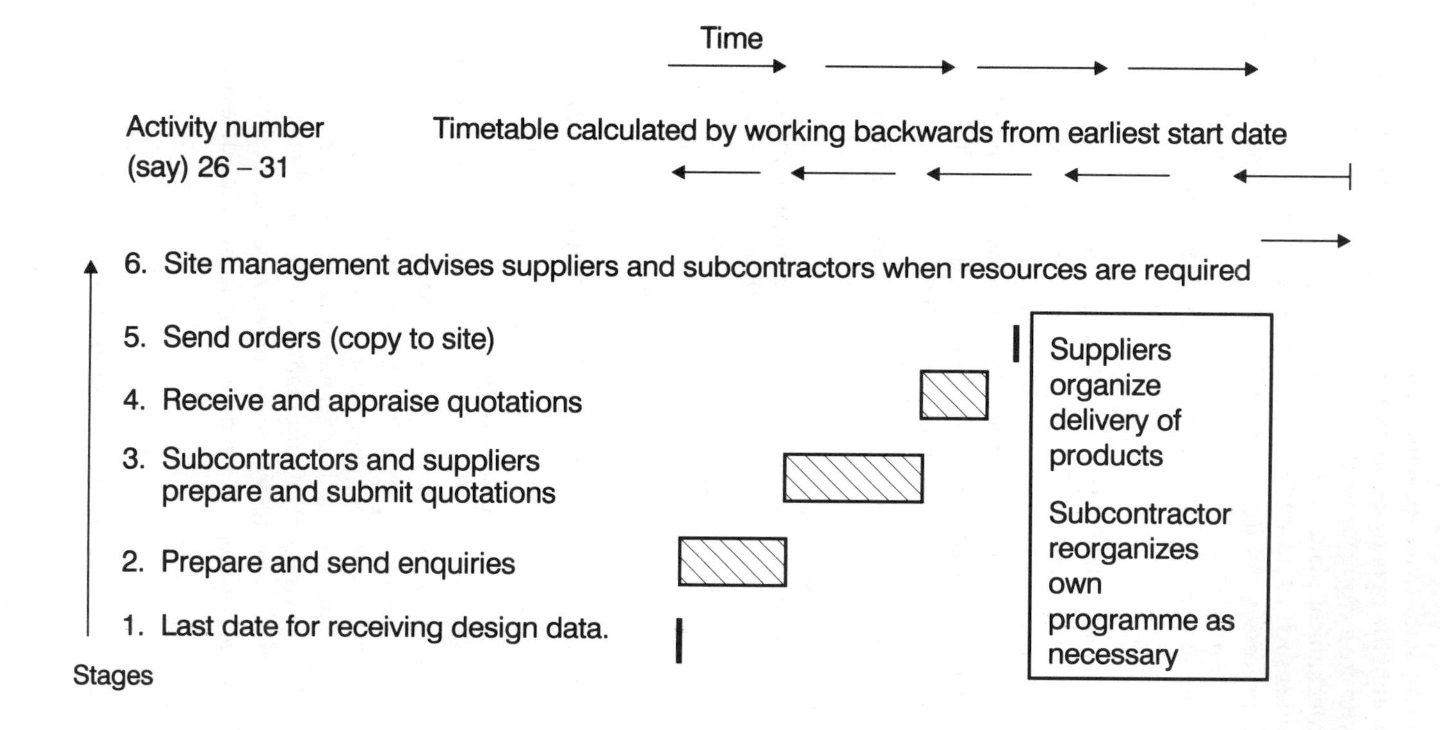

Fig. 12.4 Relating the start date of an activity, the dates for the stages in procuring its resources, and the last date for receiving the relevant operative information from the designers.

contributions from other contracts, is intended to enable the organization to at least survive financially (see Figure 16.1). A bid for a contract will assume a certain elapsed time for the Works.

If contracts take longer than planned, the rate of flow of the contributions lessens, the actual amount received each year is less, and the return on capital and the surplus available for distribution to the owners is less. Also, the costs of time-related enabling processes will be higher. Hence the importance of having a construction plan and striving to keep to it.

Planning the progress of the Works has already been considered. **Monitoring** requires a flow of progress reports such that trends can be established in time to take any necessary remedial action. One approach is to make the effort and take the time to measure the partially completed work of every current activity and compare it with the amount that should have been done by that time. Another is to first analyse each major activity into a number of stages, each with its own duration. Then, each actual completion date can be reported as it occurs and compared with that in the plan.

Controlling progress involves:

1. modifying the estimated quantities and costs of the activities and their start and finish dates to allow for changes in the design;
2. regularly collecting data on the extent that activities have been completed;
3. predicting their likely finish dates;
4. calculating the variances between (3) and their planned end dates;
5. considering these trends and taking remedial action if necessary;
6. modifying the plan and forecasting the revised completion date for the Works.

THE DIRECT COSTS

The financial objective of a Works contract will be for the total cost directly attributable to the Works to be less than the estimated cost; that is, for any variance to be positive. A framework for financial control can be created by:

1. choosing the cost centres to be used for cost control and assigning them to the individuals who will be making financial decisions; each should be based on a **work package** of activities;
2. allocating the constituents of the estimate to the work packages as cost targets;
3. estimating and plotting the planned cumulation of cost at intervals throughout the contract period; Figure 12.5 shows the upper and lower limits to a cost envelope within which the actual cumulative expenditure should lie;

4. plotting the total expenditure to date against this plan; Figure 12.5 also shows the alternative inferences to be drawn when the cumulative expenditure is outside this envelope.

Actual expenditures can then be compared with cost targets, and their variances calculated and reported to management.

TACTICS

Delays Where no formal project planning is practised, the tactic of **delays** can still be used to maintain a continuity of work for oneself or for employees. Having a backlog of orders enables a favourable starting date to be chosen and, once started, the workloads of individuals can be kept in balance by delaying progress on one site while they work on other sites. Main contractors can apply the tactic themselves and will be its victims when it is employed by subcontractors.

Although inconvenient and irritating to building owners, such delays may not affect them financially. However, a delay in completing the Works may cause the owner actual financial loss; for instance, the loss of Christmas sales due to a shop not being ready. A condition in the building contract may provide for the repayment to the owner of a predetermined sum for each day or week that completion is delayed, and a clause in their subcontract is likely to pass on this liability to each specialist subcontractor.

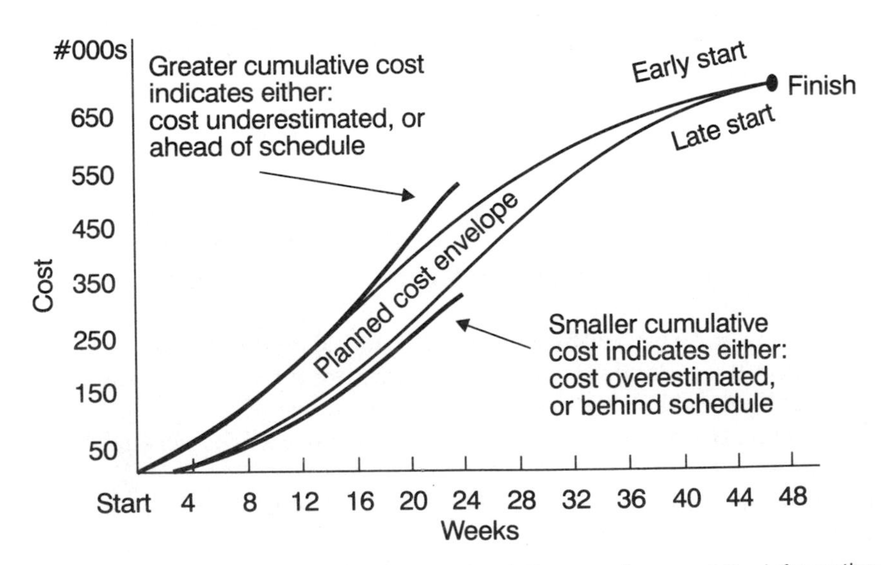

Fig. 12.5 An example of a planned cost cumulation envelope and the information that can be obtained by plotting the actual cumulative cost.

Contractor's own work Site managers should control at least the phased delivery of significant products such as ready-mixed concrete, bricks and blocks, but their authority should be restricted to the quantities in the estimate. A tidy site will usually indicate a concern for site safety and minimal materials wastage – and vice versa. The costs of each of the fixed and time-variable enabling processes can be controlled separately.

Subcontractors Much of the estimate is likely to have been based on the lowest quotations received from specialists. However, with hard bargaining, even lower prices may be negotiated nearer the time when the work is to be done. The later this is left, though, the greater the risk that the Works might be delayed.

Cash flow management A contract will usually require the owner to make retrospective interim payments to the main contractor based on the value of the completed work minus a small retention. Suppliers and subcontractors may offer discounts for prompt payment.

The margin between expenditure and income will be financed from the contractor's working capital. This will have its own interest charges. Graphs of the various planned and actual cash flows in and out can be plotted to show the extent of this margin from week to week.

Value/cost comparisons Many organizations will, having prepared a claim for an interim payment on a contract, compare the income earned thus far with the costs incurred. However, this can hardly be regarded as cost control as it fails to identify operations with excessive costs in time for remedial action to be taken.

Contract variations and claims Changes in the quantity and quality of the Works will increase the uncertainty of cost control unless their financial effects are agreed before they are implemented. Delays caused by the design team or by the owner will upset the smooth progress of the Works and extend time-variable costs. Information from the planning graphics discussed in this chapter may support claims for the reimbursement of extra costs arising from such delays.

12.7 SUMMARY

The chapter was concerned mainly with the information required for planning and controlling a Works project. This consists of a complex sequence of self-contained activities, each requiring a particular set of resources, and resulting in the completion of an identifiable portion of work or some related service. The information required on each activity was considered, including its dependence relations with other activities. Some practical problems that only apply to construction works were identified and discussed.

The main features of bar charts and networks were illustrated and their use in managing projects was discussed. Other planning graphics were introduced, including one for the timely direction to the site of resources controlled by other organizations. It was argued that controlling the contractor's construction costs involves adhering to the durations in the construction plan as well as keeping within the estimated prime cost.

FURTHER READING

BS 5964, Building Setting Out and Measurement, British Standards Institution, London.

BS 6046, Use of Network Techniques in Project Management, British Standards Institution, London.

Illingworth, J. R. (1993) *Construction Methods and Planning*, E & FN Spon, London.

Lock, Dennis (1988) *Project Management*, Gower Publishing Company, Ltd, Aldershot.

Reiss, G. (1992) *Project Management Demystified*, E & FN Spon, London.

Dimensional relations 13

Chapter 8 explored the regular relations between the S-dimensions of a building and the quantities of its enclosing elements. The object was to generate design efficiency indexes early in the outline or schematic design stage of the design process.

Throughout the remainder of the design and construction process, measurements will be made for various purposes, every building and its site being unique. In our joint Report to the RICS on the Rationalization of Measurement (1967) Douglas Ferry commented 'during the progress of a project ... up to fourteen different sets of measurement ... may take place'. All will be based on the relations between S-dimensions, the widths and/or thicknesses and other dimensions of the parts, and their quantities. This chapter considers these relations, and develops both the approach and the examples in Chapter 8.

13.1 DIMENSIONS

Two of the meanings of **dimension** given in the Oxford English Dictionary are:

1. 'Measurable or spatial extent of any kind, as length, breadth, thickness ...'
2. 'A mode of linear measurement, magnitude, or extension, in a particular direction; usually as co-existing with similar measurements or extensions in other directions.'

Thus, a dimension can be said to have both direction and magnitude, rather like a vector and its scalars. We shall refer to the extent of a dimension as its extension or measurement. The process of finding out what this is we shall call **measuring**.

DIRECTION OF DIMENSIONS

The following terms are used to indicate the directions of dimensions, although not always consistently by everyone.

- **Length** The greatest of the three dimensions of a space or an object; the full extent of something in a particular direction.
- **Width** A measurement from side to side of a space or a loose object. Usually the width is the second dimension in order of magnitude of anything.
- **Breadth** An alternative to width that we shall only apply to spaces.
- **Height** The measurement from the base upwards.
- **Depth** The measurement from some datum downwards. Cupboard manufacturers sometimes use depth to mean the horizontal dimension from front to back.
- **Thickness** The third dimension of a solid in order of magnitude, at right-angles to the length and the width.
- **Girth** A continuous measurement round the inside or the outside of an object; the sum of the lengths of adjoining objects.
- **Size** The magnitude of the dimensions of anything.

Terms that are used to describe a loose object may change when the object is incorporated into a building. For example, both the width of a skirting and the length of an architrave to the side of a doorway may, when fixed, be called their heights.

SCALES AND UNITS

An ordered set of numbers where each successive number is one unit greater than the previous one is called a scale. The most reliable is the **absolute scale** which is used to count **how many** things there are, e.g. cans of paint, doors, operatives. The scale starts at zero, consists only of whole numbers, or **integers**, and has no unit. Conventionally, we state the number of objects only where there are two or more.

The **ratio scale** is used to **measure how much** of any one thing there is, i.e. its extent. All ratio measurements of extension start from zero, can include fractional values, and can be negative. Arithmetical operations can be performed on both the quantity of a thing and the number of similar things. For instance, where some walls are identical in size, their length, height and number can be multiplied together to give their total area. Measuring tapes and scale rules are physical examples of ratio scales.

The unit will be the one in common use in the country concerned, e.g. the metre, the imperial foot (305 mm). Most local units have now been replaced by the SI (Système International d'Unités) and the derived SI units of the metric system. Even so, we shall sometimes refer to imperial (or British or English) units as well as to metric units for the benefit of readers in North America. There, the transition to metric units is making slow progress.

Measurements in one scale can be transformed to those in another by applying a **constant of proportionality**, e.g. a length in metres can be transformed to its equivalent in feet by multiplying by 3.28, and vice versa. Sub-units, e.g. millimetres, centimetres, inches, may be used instead of fractions of the unit. Multiple units, e.g. hectare, yard, can be used for large quantities. The centimetre is not used in the UK construction industry.

The dimensions on a drawing need not be labelled with their unit or sub-unit symbol unless both units and sub-units are used, e.g. both metres (m) and millimetres (mm). As dimensions are seldom given to less than one mm, using just mm has the advantage of avoiding fractions. In any case, most people find whole numbers easier to deal with than fractions.

On drawings, to be consistent, dimensions in metres should be given to three decimal places, including trailing zeros. Where a measurement is less than one metre, give the leading zero. Conventionally, measurements for bills of quantities, materials ordering etc. are taken to two decimal places, and we shall do the same in our examples.

SPATIAL DIMENSIONS

In Chapter 8 on design efficiency, the floor areas and wall girths of orthogonal, i.e. right-angled, buildings were calculated from the lengths and breadths of the rooms, etc, and of the space within the building envelope. These dimensions will have been decided during the **scheme design** stage and will be given against dimension lines on **general location drawings**. We used the notation *'i'*, *'j'* and *'k'* to indicate lengths, breadths and heights.

We learned that a dimension line on a drawing was really just a representative of any number of identical lines that could have been drawn between pairs of opposite walls, or between the floor and the ceiling. We imagined each space or constituent space as being filled with identical, parallel arrowed dimension lines indicating the length or the breadth or the height. Thus, we can measure them wherever we happen to be within a space. These **spatial** or **S-dimensions** also applied to the constructions surrounding the spaces, e.g. the lengths and widths of ceilings and floors, and the lengths and heights of walls (see Figure 13.1). Conventionally, S-dimensions are measured between the faces of the structure and ignore the thicknesses of any applied finishes.

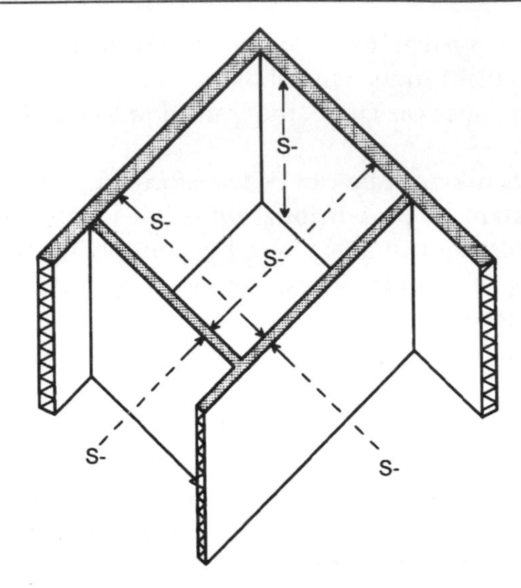

Fig. 13.1 The spatial, or S-dimensions of a corner room.

TECHNICAL DIMENSIONS

The physical constructions can also be regarded as being full of arrows in one direction or another. Some will be based on the S-dimensions of the spaces, e.g. the lengths and heights of walls mentioned above. The others, such as the thicknesses of floors and walls, and the widths and thicknesses of strip foundations, are aspects of the **technical solutions** for the parts. These we have been calling **technical dimensions**, or **T-dimensions**. Of course, all parts are three-dimensional, and strip foundations, copings, skirtings etc. with one S-dimension will have two T-dimensions, while walls, ground slabs etc. with two S-dimensions can have only one T-dimension.

These are decided during the **detail design** or **working drawings**, or **design development** stage of the design process. They may be stated in the specification or shown on an assembly or component drawing. Most will be based on public knowledge, i.e. on the widths and thicknesses given in the Building Regulations and other Building Codes, in national or international Standards, and in building construction textbooks.

The study of building construction consists in the main of a study of the parts of buildings and their T-dimensions, particularly at their junctions with other parts (see Figure 13.2). Although a width or thickness can be measured where it is exposed to view, it is often difficult to get at the T-dimensions of the parts of an existing building.

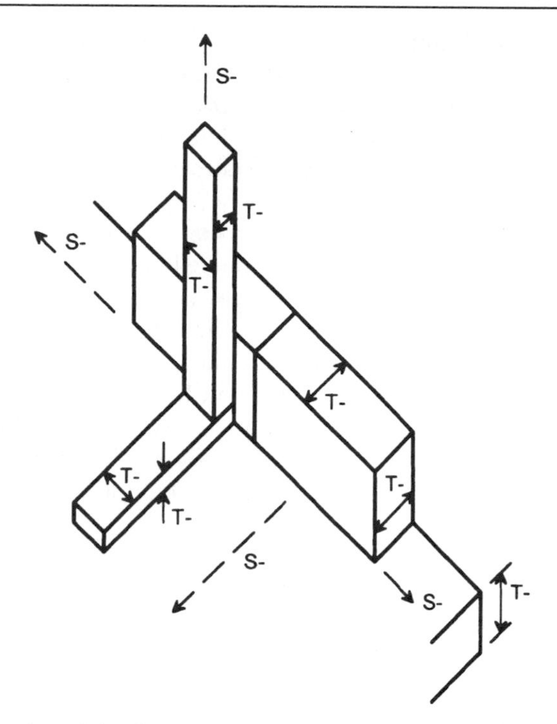

Fig. 13.2 Examples of the S- and T-dimensions of open frameworks and continuous constructions.

QUANTITIES AND QUALITIES

S-dimensions are essentially different from T-dimensions in that they will vary with the size and shape of the spaces. For instance, when the walls of a room are to be plastered, the plasterers will start at one corner, and work their way round, extending their work both horizontally and vertically. S-dimensions can be combined using the rules of arithmetic.

In contrast, T-dimensions are **intensive**, and apply invariably wherever that technical solution is employed, say 13 mm thick wall plaster. An intensive measurement of an object, such as its thickness, is one of its **qualities**. Increasing or decreasing any intensive measurement results in a different quality of object, and adding such measurements together would be meaningless.

For instance, the total area of the (say) 13 mm thick finish to the wall surfaces of a simple rectangular room is the sum of the individual wall areas calculated from their S-dimensions. Alternatively, it is the sum of their lengths multiplied by their common height. However, even if all four wall surfaces were the same size, you cannot say that there is one wall with a 52 mm thick finish.

S-dimensions are **independent variables**, and originate elsewhere, i.e. in the design brief. In contrast, T-dimensions are **dependent variables** although, once chosen, they become **invariant**. They depend on the size of the S-dimensions and on technical details such as the loads to be resisted, and the technology of the products being used. This was discussed in Chapter 9.

The difference between S- and T-dimensions is not widely recognized in the industry. This is, perhaps, understandable because both kinds are measured in the same units and with the same scale rule or measuring tape.

Some dimensions will change their nature, depending on the purpose of the measurement. This applies particularly to constructing with class A products. For example, when calculating how much wet-mix concrete will be required for a foundation, it is proper to regard all three dimensions as extensive. But when designing or describing a strip foundation, both the width and depth are qualitative T-dimensions, and only its length will be based on S-dimensions.

This idea that the quantity of a part should be based solely on the S-dimensions will be called the **quantification rule**. The process of carrying out arithmetical operations on the dimensions of a part to obtain its total quantity we shall call **quantifying**.

13.2 AN INTRODUCTION TO DIMENSIONAL RELATIONS

In this section we will be looking at the relations between the relatively few room dimensions decided by the designers and shown on the drawings etc. and the much larger number of different dimensions and quantities that can be derived from them for construction management purposes.

Initially, we shall assume that each room has only one kind of ceiling finish, wall finish and floor finish etc. and that the envelope has only one kind of roof, external walls etc. These we shall call the **main technical solutions**. Measuring **alternative technical solutions,** including windows and doorways, and adjusting the main solutions, will be considered in a later section.

Our example will be based on a one-room orthogonal building whose plan is the same as that in Figure 8.9 and which is repeated below in Figure 13.3. The relations and procedures can be applied to the measurement of most buildings.

Obviously, being a one-room building, the same S-dimensions will apply to both the room and the envelope.

ASSEMBLING THE DATA

We shall continue to use a table for the data and calculations. On Figure 13.4, which is based on Figures 8.11 and 8.14, a column for T-dimensions

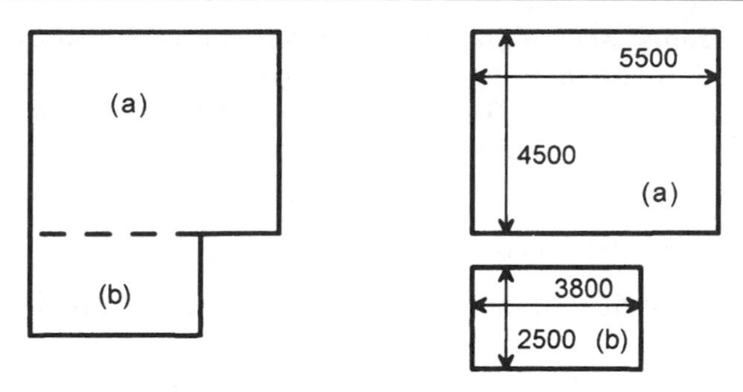

Fig. 13.3 Floor plan of a one-room building with two constituent spaces used in the example of the 'spaces' approach to measurement.

has been inserted at the side of each column of S-dimensions. The S-dimensions have been entered as before.

Note that the row of data on the compensating line is now given between the rows of data on the constituent spaces. This line is really the plan view of the surface between the two constituent spaces and will be called a compensating plane where appropriate.

DIMENSIONAL RELATIONS WITHIN A ROOM

In Figure 13.4, three quantities are calculated from the data in each row and totalled. The formulas provide for adding the T-dimensions to the S-dimensions although, in this example, some are zero. Each quantity is given the appropriate sign by including the number of spaces in the calculation.

1. The **horizontal girth** on each row is calculated by the formula $(i + j) * 2 * no$, e.g. in row 3, the girth in cell I3 is ((B3 + C3) + (D3 + E3)) * 2 * H3.
2. Ignoring doors and windows for the time being, the formula for the wall surface area is **total girth** * k. E.g. in row 3, the formula for the area is I3 * (F3 + G3). The effect of G3 will be to reduce the height (k) by 150 mm, the assumed height of the skirting.

	A	B	C	D	E	F	G	H	I	J	K
1		(i)		(j)		(k)					
2	Ref	S-	T-	S-	T-	S-	T-	nr	Wall girth	Wall area	Horiz. area
3	(a)	5.50		4.50		2.60	−0.15	1.00	20.00	49.00	24.75
4	(a-b)	3.80		0.00		2.60	−0.15	−1.00	−7.60	−18.62	0.00
5	(b)	3.80		2.50		2.60	−0.15	1.00	12.60	30.87	9.50
6	Totals	5.50		7.00					25.00	61.25	34.25

Fig. 13.4 Three-dimensional data model of room dimensions and quantities.

3. The **horizontal area** of the ceiling or floor in each constituent space is calculated by the formula $i * j *$ no in column K. E.g. for space (a) in row 3, the formula is (B3 + C3) * (D3 + E3) * H3. The compensating line has no horizontal area.

THE ALTERNATIVE ENCLOSING RECTANGLE APPROACH

In the above, all the S-dimensions were confined, or **closed,** to those of the spaces being measured, and we can refer to this as the **spaces approach**. An alternative approach is first to obtain the dimensions of the rectangle that will enclose the plan of the room. Any area outside the room is then deducted as shown in Figure 13.5.

Measured data can be derived as follows:

- The overall i and j dimensions of the **enclosing rectangle** can be calculated by aggregating the i and j columns in Figure 13.4 after multiplying the dimensions by their numbers, e.g. the formula for the total i dimension of 5.50 given in cell B6 of Figure 13.4 is ((B3 + C3) * H3) + ((B4 + C4) * H4) + ((B5 + C5) * H5). Of course, in such a simple example, the overall dimensions are obvious.
- In any rectangular shape such as the small area outside the room, any two adjacent sides will have the same lengths as those of the other two sides. In our case, the two sides against the enclosing rectangle will have the same lengths as those against the room and so the **girth of the room** will be the same as that of the enclosing rectangle. Thus, twice $(i + j)$ will be seen to equal the **wall girth** of the room, i.e. 25.00, and this can be multiplied by the height to give the **wall area**.
- The **horizontal area** of the room can be calculated by the formula **area of enclosing rectangle minus areas outside the room.** In our case we need to deduct the crossed area in the bottom right-hand corner of

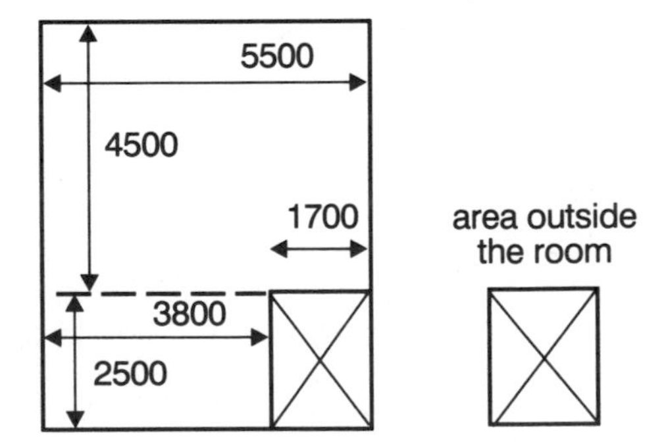

Fig. 13.5 Floor plan of a one-room building showing the **enclosing rectangle approach** to calculating horizontal areas.

Figure 13.5. The dimensions of the area outside the room can be obtained from the data on Figure 13.4, i.e. 1.70 and 2.50.

THE EFFECT OF ADDING T-DIMENSIONS

As we are dealing with a one-roomed building, the dimensions of the room and those of the internal surfaces of the envelope will be the same. We will now consider the effects of including horizontal technical T-dimensions in our calculations.

Suppose the roof has the same kind of projecting eaves all round and we wish to calculate its area on plan. The projection from the inner face of the external wall might consist of the thicknesses of the inner wall leaf, the cavity, the outer wall leaf, the width of the eaves soffit, the thickness of the fascia, and a projection of 60 mm over the gutter.

The total of these T-dimensions (say 100 + 50 + 100 + 200 + 20 + 60 mm, or 530 mm) must be doubled for both ends, and entered in columns C and E as 1.06 m. The details can best be seen by looking out of a top-floor window.

Their effect is shown in Figures 13.6 and 13.7, the latter being much exaggerated. In all calculations, each pair of T- and S-dimensions is first added together to give its overall dimension.

The *i* and *j* columns have also been aggregated to give the dimensions of the enclosing rectangle as above. Note that:

- the enclosing rectangle method would give the same result as that of the spaces approach;
- the dimensions of the crossed unwanted area outside the eaves perimeter are the same as those calculated earlier for the area outside the room;
- the compensating plane has been given width and has become a kind of space, but with a negative sign.

	A	B	C	D	E	F	G	H	I	J
1		(i)		(j)		(k)				
2	Ref	S-	T-	S-	T-	S-	T-	nr	Horiz. girth	Horiz. area
3	(a)	5.50	1.06	4.50	1.06			1.00	24.24	36.47
4	(a-b)	3.80	1.06	0.00	1.06			–1.00	–11.84	–5.15
5	(b)	3.80	1.06	2.50	1.06			1.00	16.84	17.30
6	Totals								29.24	48.62
7	Enclosing lines		6.56		8.06					

Fig. 13.6 Calculating the horizontal girth and area after adding T-dimensions.

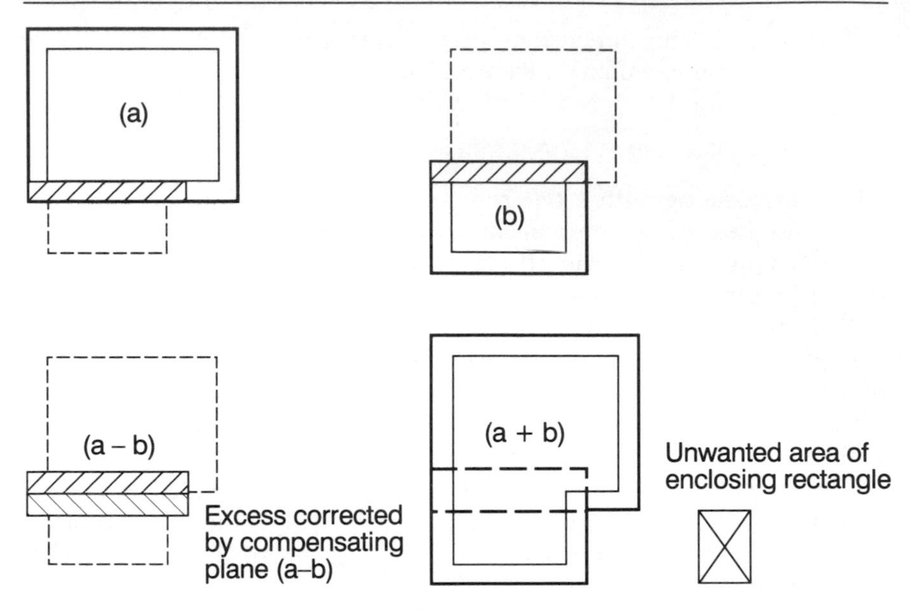

Fig. 13.7 Schema showing the effect on the horizontal girth and area of adding T-dimensions.

THE EFFECT OF DEDUCTING T-DIMENSIONS

Figures 13.8 and 13.9 show T-dimensions of twice 300 mm being deducted from the S-dimensions. The example is just to show that the approach works both ways, and their effect is exaggerated for clarity.

Note that:

- the compensating plane is given a reduced length of 3.20 and a width of –0.60 but when these are multiplied by the number –1.00, the length becomes negative and the width becomes positive;

	A	B	C	D	E	F	G	H	I	J
1		(i)		(j)		(k)				
2	Ref	S-	T-	S-	T-	S-	T-	nr	Horiz. girth	Horiz. area
3	(a)	5.50	–0.60	4.50	–0.60			1.00	17.60	19.11
4	(a-b)	3.80	–0.60	0.00	–0.60			–1.00	–5.84	1.92
5	(b)	3.80	–0.60	2.50	–0.60			1.00	10.20	6.08
6	Totals								22.60	27.11
7	Enclosing lines		4.90		6.40					

Fig. 13.8 Calculating the horizontal girth and area after deducting T-dimensions.

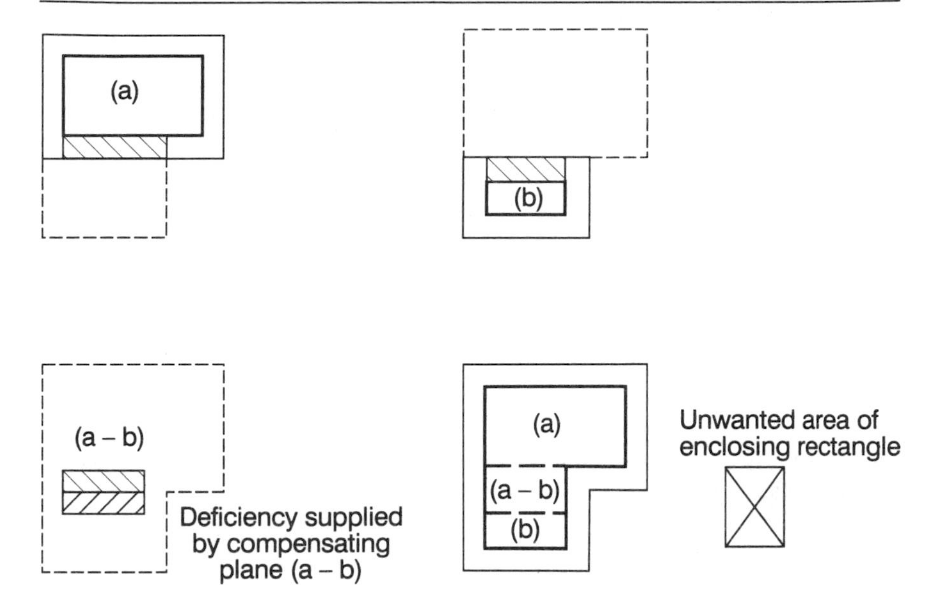

Fig. 13.9 Schema showing the effect on the horizontal girth and area of deducting T-dimensions.

- similarly, the area quantity is changed from negative to positive and so fills the gap between (a) and (b);
- The dimensions of the area to be deducted from that of the enclosing rectangle remain the same as before.

13.3 SOME APPLICATIONS

THE EFFECTS OF CORNERS ON THE MEASUREMENT OF WALLS

When building the corner of a masonry wall, the first unit to be laid will form the actual corner and the start of the wall in one direction. The next unit will be laid at right angles to it as a start to the other wall, as shown on Figure 13.10. Note that the combined lengths of the two units is the same as their centreline length when laid.

We shall adopt the rule given by William Salmon (1748) in *The London and Country Builder's* VADE MECUM. He wrote 'when you measure two Walls that constitute an Angle, the Length of one must be taken to the Outside, and the other to the Inside'. We shall take the '*i*' dimensions to the outside.

Figure 13.11 shows the calculation of the total length of the external walls to our one-roomed building. These are assumed to be 150 mm thick. Such centreline girths will also apply to other continuous constructions

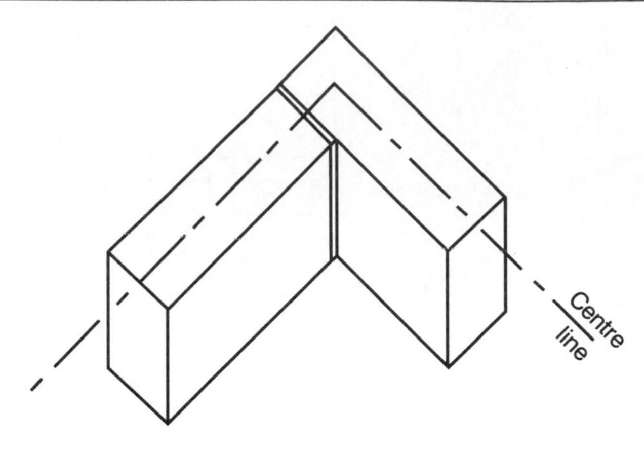

Fig. 13.10 A corner of a masonry wall showing that one wall should be measured to the outside and the other to the inside.

that might show on a vertical section through the wall, such as strip trenches and foundations, foundation walls, foundation dampproof course, parapet wall, dampproof course, and coping.

Figure 13.12 gives an exaggerated illustration of this, and of how the compensating plane, when given thickness, removes the surpluses that are in the neighbouring spaces.

Figure 13.13 shows the corner of a cavity wall consisting of a 150 mm inner leaf, a 50 mm cavity and a 100 mm outer leaf. In this case, the T-dimensions data for the outer leaf would be:

(*i*) (0.15 + 0.05 + 0.10)* 2 and
(*j*) (0.15 + 0.05) * 2.

	A	B	C	D	E	F	G	H	I
1		(i)		(j)		(k)			
2	Ref	S-	T-	S-	T-	S-	T-	nr	Girth
3	(a)	5.50	0.30	4.50	0.00			1.00	20.60
4	(a-b)	3.80	0.30	0.00	0.00			–1.00	–8.20
5	(b)	3.80	0.30	2.50	0.00			1.00	13.20
6	Total								25.60
7	Overall lengths	5.80		7.00					

Fig. 13.11 Using S- and T-dimensions to calculate wall lengths and centreline girths.

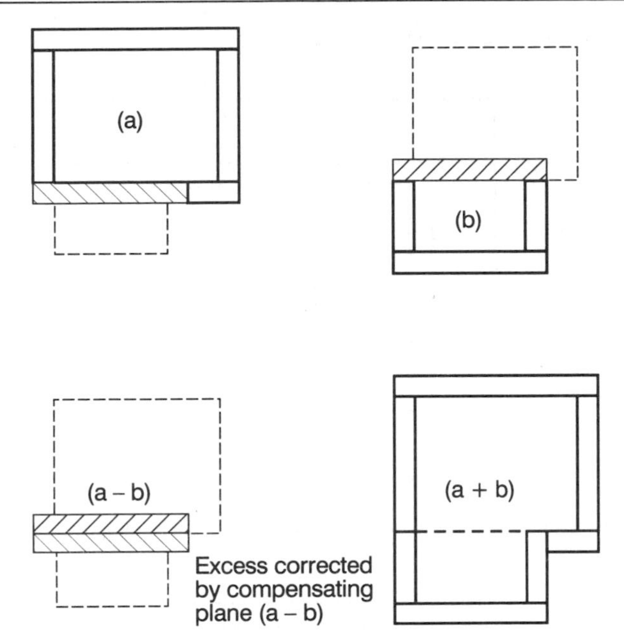

Fig. 13.12 Schema showing the generation of the horizontal lengths of external walls using the spaces approach.

CONVENTIONAL MEASUREMENT OF WALLS

Conventional measurement is likely to be based on the overall S-dimensions of enclosing rectangles plus or minus the T-dimensions at the corners. Whatever the shape of the envelope, provided it is orthogonal, there will always be four more external corners than there are internal ones.

Thus, there will always be a difference of **overall wall thickness * 2 * 4** between the girth on the internal face and that on the external face. Also, the difference between the internal or external girth and the centreline girth of the wall will always be half this, i.e. **overall wall thickness * 4.**

When the *'k'* dimension applies throughout, the centreline girth of an enclosing wall is obtained by applying Salmon's rule, i.e. doubling the sum of the overall outside dimension in one direction and the overall inside dimension in the other. Alternatively, the thickness of four corners can be added to the total girth on the inner face of the wall or deducted from that on the outer face.

Figure 13.14 shows that for each wall that returns upon itself, or is **re-entrant,** twice its length must be added to the girth of the enclosing rectangle. Note that where there is either an external or an internal corner at both ends of a length of external wall, the difference between the length of the internal face and that measured on the outside is equal to two external wall

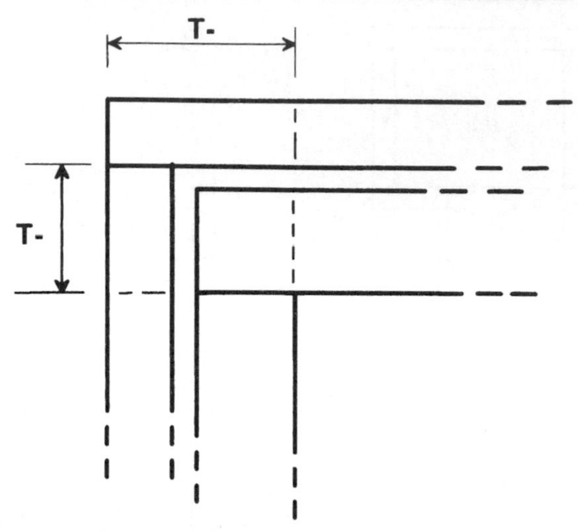

Fig. 13.13 Plan of corner of cavity wall showing the T-dimensions to be added for the outer leaf.

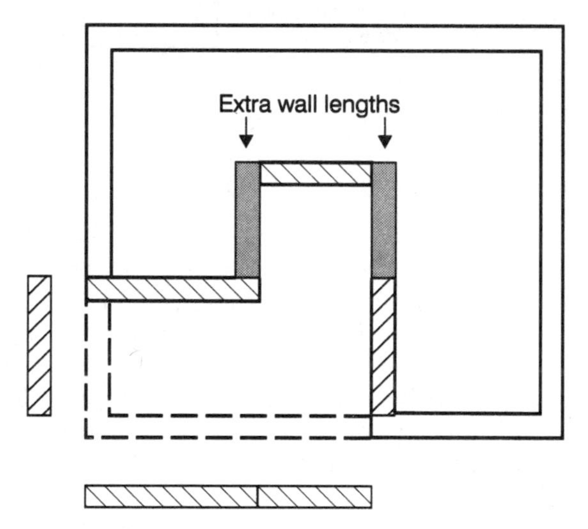

Fig. 13.14 Floor plan showing re-entrant external walls and their effect on corners and wall lengths.

thicknesses. Where there is an internal angle at one end and an external one at the other, the internal and external lengths are the same.

PITCHED ROOFS

A pitched roof is best seen from above. Any sloping length such as that of a rafter or verge can be calculated from its length on plan by either divid-

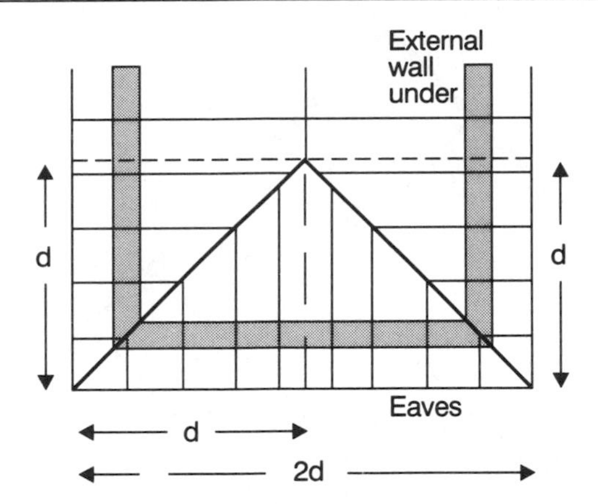

Fig. 13.15 Plan of a hipped end to a pitched roof structure referred to in the calculations for the length of a hip or valley.

ing this by the cosine of the angle of slope, or multiplying it by the secant. This angle, sometimes called the **pitch**, is measured from the horizontal.

Similarly, the area on plan of the roof covering will be transformed to the sloping area by dividing/multiplying it by the cosine/secant of the angle of slope. This sloping area will be the same whether the roof has hipped or gabled ends.

Figure 13.15 is a plan showing hips as the 45° diagonals of squares with sides of horizontal length *d* formed by the eaves, the ridge, and lines parallel to the rafters. Valleys are similar. Because hips and valleys are longer than rafters, their true angle of slope is less than the pitch. If θ° is the roof pitch, then the formula that will calculate the hip length can be expressed formally as:

$$\sqrt{d^2\left(\sec\theta^2 + 1\right)}$$

or operationally as:

$$(d \wedge 2 * ((1 / \cos\theta°) \wedge 2 + 1)) \wedge 0.5 = \text{hip length}$$

THE ENVELOPE OF A MULTI-ROOM BUILDING

If the envelope is a simple rectangle on plan, then each horizontal S-dimension within the external walls will be the sum of the room S-dimensions in the same direction plus the thicknesses (T-dimensions) of the walls in between (see Figure 13.16).

The '*K*' dimensions will be the sum of the floor-to-floor heights of each storey including the thicknesses of the structural floors and their finishes in between. This may best be visualized when standing in a doorway. 'L'

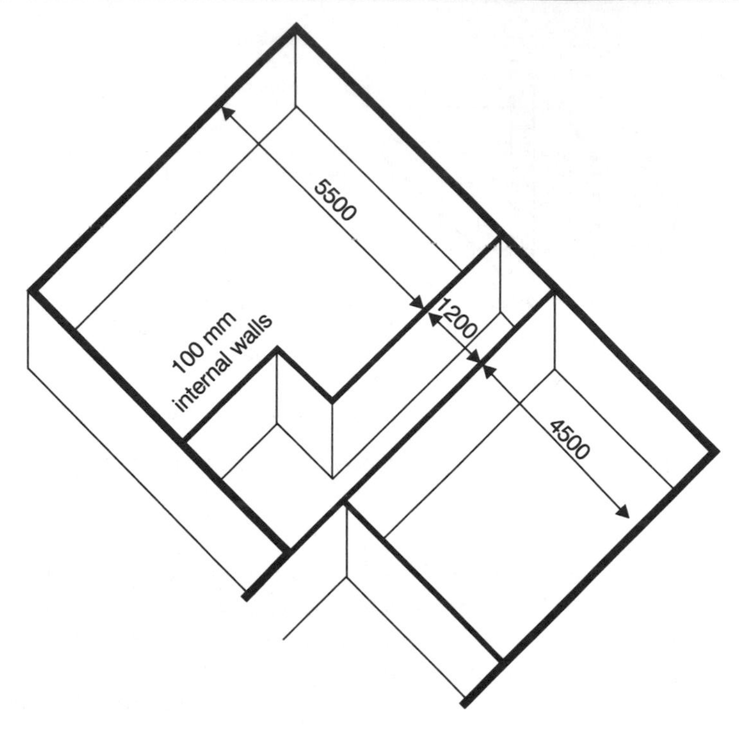

Fig. 13.16 Planometric view of an envelope showing the build-up of an S-dimension for the building space.

shaped or other complex envelope shapes must first be resolved into constituent spaces, with their compensating planes in between, as described earlier.

GIRTHS OF INTERNAL WALLS

In Chapter 8, the total girth of the internal walls on a particular floor was derived from those of the spaces and the envelope. That is, the sum of the length and breadth (i.e. half the girth) of the space inside the building envelope (notation in capitals) was deducted from the sum of the lengths and breadths of the rooms etc. This approximation assumed that the internal walls have no thickness, and was expressed as follows:

For any one floor,

$$\Sigma(i + j) - \Sigma(I + J) = \Sigma \text{ lengths of internal walls on that floor.}$$

However, the actual I and J dimensions will include the thicknesses of abutting internal walls. An apparently reliable correction for this over-deduction is to add $+ (n - 1) * t$, where n = the number of rooms and t is the thickness of the internal walls main technical solution. $(n - 1)$ is used in the formula because an internal wall is only required if there are two or more rooms.

CHECK ON DIMENSIONS

The horizontal area within the envelope should equal the sum of the horizontal areas of the rooms, stairwells etc. plus that taken up by the internal walls. If not, there must be an error somewhere.

BEAMS AND COLUMNS

The length of a beam or the height of a column is likely to be related to one of the S-dimensions of the room. The other dimensions required for concrete or other casings, fabric wrapping, painting etc. will be based on their T-dimensions.

HIGHER OR DEEPER CONSTRUCTIONS

Where a building is complex, it should, if possible, be resolved into portions with the same storey heights, numbers of storeys and technical solutions. Some portions will have complete envelopes and can be measured independently. Where lower portions butt up to them, the shared walls will become internal and their functions and technical solutions will change. Also, the remaining weathershield above the shared wall must be made continuous with the lower roof.

SERVICE AND INSTALLATION PIPEWORK AND CABLING

The dimensions of conductors of all kinds have the character of S-dimensions. Inside buildings, they consist in the main of straight lengths at right-angles to each other, going round and up or down the surfaces of the rooms and other spaces. The individual lengths of parts of drainage pipelines will be the overall lengths between chambers etc., less the appropriate T-dimensions at their ends.

NON-ORTHOGONAL SPACES

The dimensions and quantities of constructions around orthogonal spaces can be transformed to those around non-orthogonal spaces by applying suitable mensuration formulas.

13.4 MORE THAN ONE TECHNICAL SOLUTION

In Chapter 8 on design efficiency, we were only concerned with the relative quantities of those elements that could be calculated at the outline stage of the design process. Thus far in this chapter, we have looked at the relations between the S-and T-dimensions which enable us to calculate the dimensions and total quantities of the parts of the various elements, assuming that each is constructed throughout with the same **main** technical solution. This is usually the case with the ceiling and floor finishes to each room etc. and with floors and roofs.

It is never the case with external and internal walls or their finishes, as they will always have windows and doorways. We must now consider the dimensional relations of some of these **alternative** technical solutions to the enclosing elements. Where an alternative solution will apply around one or more sides of one or more component spaces, the formula for calculating the area or girth can be modified so that the actual quantities of both the main and the alternative solutions are obtained.

Otherwise, the quantity of each alternative solution must be accompanied by an **equivalence deduction** from the quantity of the main solution. The dimensions of alternative solutions will all be decided during the detail design stage (see Chapter 9).

INTERNAL WALLS

Where there is more than one technical solution for the internal walls, one of them must be regarded as the main solution. Measure each of the others individually and make an equivalence deduction from the girth of the main solution. Alternatively, directly measure each kind of wall and use the formula as a check. In the formula, t will be the average thickness of the abutting walls.

DOORWAYS

The dimensional relations of a doorway are similar to those of windows, particularly where a doorset, consisting of the door, its frame or lining, and possibly its glazing and hardware, is used. However, the parts of internal doorways are more likely to be selected individually.

The door itself will be chosen first. Decisions concerning its frame or lining and other trims will depend on the kind of wall the opening is in and its thickness, and on the wall finish and skirting on either side. Please study existing doorways while considering the following.

When moving furniture through an open doorway, and assuming the door will open back, the important dimensions are those of the **clear opening** between the faces of the door lining or frame. These will be the width and height of the door, less the depth of the rebates. Note that a door will only open fully into a room if it will clear the skirting, architrave and its own frame or lining.

Add a T-dimension to each end of the clear opening dimensions to give the dimensions of the other parts and the lengths and heights of equivalence deductions. Note the overall girth of the decoration on the faces and edges of the architraves and the lining and how that on a door edge matches the decoration of the room from which it is visible. If the details of any finished constructions are unclear, remember that all wood members were originally rectangular sections.

WINDOWS

The co-ordinating length of the window unit will provide the dimensional basis for the extra finish and decoration to the underside (or **soffit**) of the opening, and the extra dependent products:

1. in the wall, e.g. the lintel;
2. for the finish, e.g. metal beads; and
3. for the window, e.g. sill and window board.

Suitable T-dimensions must be added where ends will be built in.

Its co-ordinating height will apply to the vertical finish and decoration at the **jambs**, and to the extra dependent products, e.g. damp-proof course, insulation or walling to close the cavity, stone dressings, jamb linings. Their widths will depend on the position of the window unit within the thickness of the wall.

Equivalence deductions from the structural wall and its finishes will be based on the co-ordinating width and height of the window unit, although the actual work size of these units is likely to be 5 mm or so less than the co-ordinating size (see the next chapter).

The measurements of glazing for standard windows will be published by the window manufacturers, but those for other windows should be taken from the actual units. Before this, glazing sizes can only be estimated by assuming the finished thicknesses of the framing, the widths of casements, and the sizes of rebates.

STAIRCASES

The dimensions of a staircase will be based on the floor-to-floor storey height, the angle of inclination, or **rake**, of the stairs, and the size and number of steps. Equivalence deductions for the **stairwell** will be required to the upper floor and to its floor and ceiling finishes. Finishes will be required to the exposed edges of this floor and to the staircase itself.

13.5 SITE PREPARATION

Excavation, and its consequent disposal, is always measured as the volume it had occupied before it was broken out. This is usually referred to as **solid measure,** or **in bank**.

AREAS OF SURFACES

The area of an irregular-shaped surface with straight-line or **rectilinear** boundaries can be ascertained by:

1. dividing it into simple triangles or other component shapes whose mensuration formulas are known;
2. measuring the boundaries of these component shapes;
3. applying the formula for each shape; and
4. totalling the results.

To find the area of an irregular plane surface with two parallel ends, first, divide it into an even number of strips with the same width, and measure the lengths of their boundaries. Counting the first, there will obviously be an odd number of these.

Then, add the end lengths to twice the other odd numbered lengths and four times the even numbered lengths and multiply the total by one third of the strip width. This is often called Simpson's Rule.

Curvilinear boundaries can often be transformed to equivalent rectilinear ones by substituting straight **give and take** or **average** lines which cross and re-cross the actual boundary line. Such lines are positioned visually on the plan so that the area cut off on one side appears equal to the extra area on the other side.

USING CONTOURS

A line on the ground whose points are all at the same height above sea level, or a representation on a plan of such a line, is called a **contour line**. A series of such lines with the same height interval between them will show the configuration of the surface of the site. They are useful also when seeking to identify the boundary line between excavating and filling.

Where the ground is sloping and only a portion of the site of a building is above the formation level (see Figure 13.17), the excavation will be wedge-shaped in section. Over the rest of the area, a wedge of filling material will be required. The boundary between excavation and filling is called the **cut and fill line**. If vegetable soil of depth d_v is removed, all surface levels will be correspondingly lower.

Thus, the cut and fill line will be that surface contour whose level is (**formation level** + d_v). This can be marked on the plan to indicate the edge of the area of excavation. Both the average depth of the excavation to reduce levels, and that of the extra filling below the formation level will be the difference between their average surface level and the level of the cut and fill surface contour.

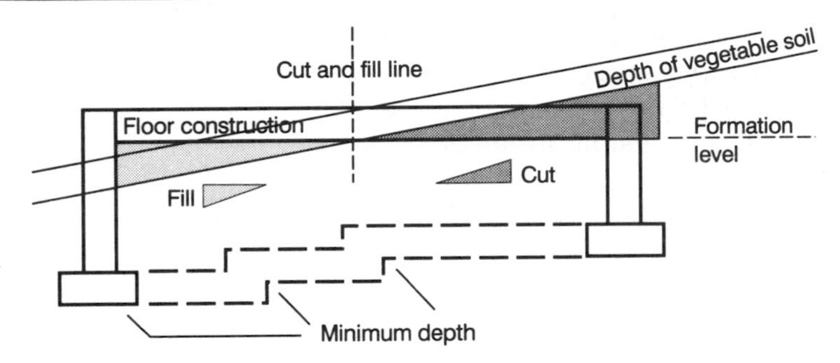

Fig. 13.17 Schema of a section through a substructure construction on a sloping site showing the cut and fill line, the triangular sections of excavation and filling, and stepped foundations.

CALCULATING THE AVERAGE SURFACE LEVEL

Normally, a series of regularly spaced 'spot' levels is taken at the intersections of a uniform square grid that has been set out on the site. The more undulating the surface, the more frequent the levels should be. If the site is fairly flat, it may be sufficient to average the values of the levels at the corners of the void in question.

Otherwise, the grid should be imagined as being further subdivided into quarter-squares with the level at the centre of each set of four representing the average over the set. Each level must be 'weighted' by multiplying it by the number of quarter-squares that are actually over the void (some judgement must be exercised here). Their sum is then divided by the total number of quarter-squares in the calculation.

Of course, where part of the site will be excavated, and the rest filled, the average surface level on each side of the cut and fill line must be separately calculated.

Computer software linked with electronic distance measuring (EDM) equipment will transform readings at the various points to their co-ordinates and levels. It will also treat the lines joining sets of three adjacent points as constituting triangles, and will calculate the total of their surface areas. When given the reduced level to be achieved, the software will calculate the depth at each point, average them for each triangle, and calculate their volumes. This approach is more accurate than averaging the depths at the corners of squares.

VOIDS FOR BUILDING SPACES BELOW GROUND

The horizontal dimensions of voids for basements, ducts, manholes etc. will be based on the S-dimensions of the spaces and the T-dimensions of walls and foundations. However, these dimensions may need to be extended by further T-dimensions of as much as 600 mm all round, to provide a **working space** for operatives. Their tasks might include applying an external damp-proof membrane, building a protective wall, and erecting and dismantling formwork. The extra excavation will then be backfilled.

The depth of a void will depend on whether the surface level is:

1. wholly at the formation level;
2. partly at and partly below the formation level; or
3. wholly below the formation level.

A basement void with battered sides will resemble the frustum of a pyramid. Its volume can be calculated by the Prizmoidal Formula. This is: **multiply the sum of the area of the top, the area of the bottom, and four times the intermediate area by the depth and divide by six.**

VOIDS FOR FOUNDATIONS

The major horizontal dimension of a foundation trench will be based on the S-dimensions of the building envelope, e.g. the centreline girth of the external walls. The other horizontal dimension will be one of the T-dimensions of the construction to be accommodated, e.g. the width of a strip foundation, possibly increased to allow for working space.

On a sloping site, as shown on Figure 13.17, a resistant and economical depth is maintained by stepping the horizontal bottoms of the foundation trenches. Their average depth can be calculated from their top and bottom levels taken at frequent and regular intervals. The volume of a trench with battered sides is the product of its cross-section and the girth of its centroid.

The horizontal i and j dimensions of voids for isolated column bases will depend on their T-dimensions. Their depths will be the difference between their designed bottom levels and either the formation or the natural level after stripping the topsoil.

VOIDS FOR DRAINAGE PIPELINES

The location drawings or the site plan should show the levels of pipelines at the bottoms of manholes or inspection chambers, called the **invert levels**. The mean depth of any one length of pipeline trench will be the average of the invert depths at the ends plus any thickness of pipe and bed. The levels at the tops of the voids will depend on whether the natural ground levels have been reduced by other excavation work.

The mean depth of a system of trench voids can be derived from their longitudinal sections. Multiply each trench length by its average depth, add these longitudinal sectional areas together, and divide by the total length. The gross volume of a system of trenches with the same width will be its longitudinal area multiplied by its width.

13.6 COMMENTARY

The relations between the S-dimensions of spaces, as modified by T-dimensions, and the quantities of the technical solutions, have been illustrated by entering the data on a table with lettered columns and numbered rows. This is similar to the screen display of a computer spreadsheet.

One advantage of using an actual spreadsheet is that once entered, the various sets of S-dimensions, T-dimensions, formulas for relating them, and labels for the quantities, can be copied and linked as required. Furthermore, any change in a dimension will immediately affect all related data. Where another building has the same technical solutions, its S-dimensions can be entered on a copy of the spreadsheet.

The method is possible only when measurements are given using the same ratio scale, e.g. metres or feet. In the UK, we have been liberated from multiple or dual units, such as:

- measurements in feet and inches;
- quantities measured in feet and reduced to yards;
- areas in rods (272 sq ft) and square feet;
- areas in squares (100 sq ft) and square feet;
- volumes of timber in board feet (1 sq ft 1 inch thick, and sometimes called 'feet super as inch'), calculated by multiplying the length and width in feet by the thickness in inches;
- weights in tons, hundredweights, quarters and pounds.

13.7 SUMMARY

Some of the terms used in connection with dimensions were considered. Technical (T-) dimensions which provide qualities were distinguished from the S-dimensions of spaces which determine quantities.

The tabular approach used in Chapter 8 was extended to demonstrate the relations between S- and T-dimensions and the quantities of the various parts of orthogonal buildings. Both the spaces approach to the measurement of constructions surrounding complex spaces and the conventional method of deducting from enclosing rectangles were illustrated.

Generally, each extensive dimension of a part of an enclosing element will consist of:

1. a central S-dimension, with
2. a set of either negative or positive T-dimensions at either end, although not always the same set, and sometimes zero.

The results of adding or subtracting T-dimensions were illustrated. This included the effects on constituent spaces and the action of compensating planes between them in maintaining the correctness of the calculations.

These dimensional relations were applied to the measurement of various parts of elements and of site preparation works. Initially, it was assumed that there was only one technical solution for each part of an element.

REFERENCE

Ferry, D. J. O. and Holes, L. G. (1967) *Rationalisation of Measurement*, Royal Institution of Chartered Surveyors, London.

Products quantities 14

In this chapter, we continue to assume that we are preparing to build, and that all the drawings and other design information is available. The previous chapter concentrated on the extensive dimensions of buildings and other constructions, and on the quantities of the various parts that could be generated from them.

We shall now consider the numerical relations which will transform these dimensions or their quantities to the net quantities of the products required to build them. As a rule, each type of class A formless product is ordered by giving its volume or mass, and this can be derived from the measured quantity of the part. The number of each size of class B formed product is best calculated from the individual dimensions of the parts. The quantities of their fixing and/or jointing products will also depend on this number.

However, products may only be supplied in multiple units and sizes decided by their manufacturers or suppliers, e.g. paint container sizes, cement bags (sacks), brick and block packages, board and sheet sizes. They may be paid for in different units. Thus, when ordering, each net calculated quantity of a formless product must be transformed to the next whole number of units of supply, and each size of formed product may have to be increased to the next larger available size.

Inevitably, the excess will be wasted, as it is not normally economical to remove and store these surpluses until a use can be found for them. In any case, where reductions are made in the sizes of boards, sheets, timbers, and other formed products to fit the dimensions of the construction, the excess is of no use. All such wastage is **unavoidable**, but still has to be paid for. **Avoidable waste** is discussed in Chapter 18.

In spite of the trend towards harmonization, ordering and buying units may vary from place to place. For instance, walling blocks may be sold by number, or per square metre of walling. (We adopt the usual abbreviations of m, m^2, and m^3 for linear, square and cubic metres.) Up-to-date publications from the manufacturers are essential sources of technical information.

We shall separately consider each class of products identified in Chapter 5, the dimensions, quantities and units involved in the transformations being given at the start. Many transformations involve multiplication or division by constants of proportionality which are usually called **conversion factors**.

14.1 CALCULATING QUANTITIES OF FORMLESS MATERIALS

Formless materials consisting of loose particles are usually measured by giving their weight (or, more correctly, their **mass**) in tonnes. To obtain this quantity from knowledge of the measured volume of a part of a building, we need to know the **bulk density**, or mass in kg of one m^3 of the loose material. This will depend on the **mass density** in kg/m^3 of the solid material and on the size and shape (or **grading**) of the particles, i.e. on the proportion that is solid.

(A1) CONCRETES

First obtain the volume of the construction. This will be the product of its constituent S-dimensions, or their quantity, and any T-dimensions. Examples include: the area of a floor and its thickness; the centreline girth of a strip foundation and its width and thickness; the number in a set of identical column bases and their three dimensions.

For any construction, an equal volume of wet-mix concrete will be required. Depending on the specification, which may refer to BS 5328, the supplier will determine the amounts of constituent materials.

Where formwork for concrete constructions is to be made on the site from plywood and wood sections, first decide how often the formwork is likely to be reused. Then sketch the likely shapes of forms and supports (their dimensions will be based on those of the constructions) and calculate suitable numbers of plywood sheets and lengths of wood sections.

(A2) MATERIALS FOR THICK COATINGS (I.E. PLASTER TYPE)

There are two possible cases:

Case (1) where coating thicknesses reflect the nature of the materials or are specified by the manufacturer. The transformation is from the surface area to weight. Plasters are supplied dry in bags for mixing with water before use. Their coverages per buying unit are published by the manufacturer although site conditions may prevent these from being achieved. For example, besides providing the minimum thickness, it may be necessary to fill irregularities in the surface of the background. Also, controlling the thickness during application may be difficult.

Case (2) where the thickness is specified by the designer. The transformation is from a surface area of a given thickness to its volume and thence to its mass or those of its constituents. The substance may be a compound of plaster or cement and water, extended by being mixed with sand or other inert aggregate. (See class AB2 later for an introduction to calculating the constituent quantities of mortars.) The lengths of angle and edge beads etc. will be based on those of the S-dimensions of the rooms.

(A3) MATERIALS FOR THIN COATINGS (I.E. PAINT TYPE)

The transformation is from the surface area to liquid measure by dividing by the coverage factor published by the manufacturer. However, these will be affected by the porosity and irregularity of the surfaces being treated, the extent to which the materials are thinned, the ambient temperature and the skill of the operatives.

(A4) MATERIALS FOR FILLINGS (I.E. GRAVEL TYPE)

The lengths of pipeline beds or surrounds of known width and thickness, and the areas of surface beds of known thickness are transformed to their volume and thence to the mass of the loose material. The thickness of a bed may have two components:

1. the designed minimum thickness (less the minimum thickness of any blinding to provide a smooth surface); and
2. the average thickness of any extra filling required where the surface of the ground (after removing the vegetable soil) is below the formation level. Where (2) only applies to part of the bed, the extra volume must be calculated separately.

The depth of sand blinding will be the specified minimum depth plus the average depth of surface irregularities.

14.2 CALCULATING QUANTITIES OF FORMED PRODUCTS AND FIXINGS

'THING THINKING'

Formed products are **things**. We order them by stating how many we need, we fix them one at a time, and each has its own requirements for fixing materials. We shall be thinking of them as separate objects both in this section and in Chapter 15 which considers their processing.

All units are ordered by giving the number of each size. They may be supplied and/or paid for in multiples, e.g. tens; thousands; boxes of 50; bundles of 500; per 100 lin m.

Co-ordinating size and work size Where similar units are fixed together in a building to form a continuous construction, they may be:

- spaced apart, e.g. by mortar joints between bricks and blocks;
- placed edge to edge, e.g. carpet tiles; or
- overlapped, e.g. roofing slates and tiles, tongued and grooved flooring.

The dimension(s) of the visible, effective contribution made by each unit is called its **co-ordinating dimension(s)**. These may be either more or less than its actual, or **work size**. For example, a 440 × 215 mm (actual **work size**) block will contribute 450 × 225 mm when bedded in mortar. In contrast, the effective, or **face** width of a nominally 125 mm wide tongued and grooved floor board is only 108 mm. Where a co-ordinating dimension is less than the work size, the amount of the overlap is sometimes called the **margin.**

Calculating numbers in continuous constructions Each plane should be dealt with separately, and it is helpful to visualize units from classes B1 to B6 as being fixed in rows and/or columns as in Figure 14.1. The whole number of units in each row and in any related column should be calculated separately.

Obtain this number by:

1. counting how many times the unit co-ordinating dimension will go into the S-dimension; or by
2. dividing the one into the other.

Take any fraction as a whole number as it should not be assumed that cut-off portions can be used elsewhere. That is, give the answer to the next integer. Where a planar construction includes windows or other

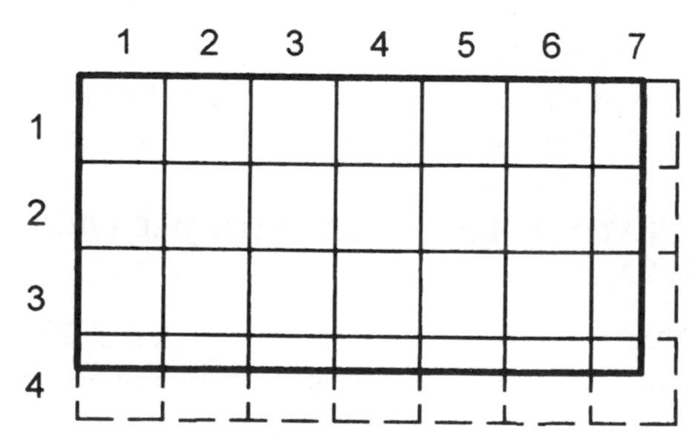

Fig. 14.1 Rows and columns of units in continuous constructions showing unavoidable wastage.

alternative technical solutions, either separately calculate the numbers of units between them or do an equivalence deduction of the number of whole units that are not required. A spacing or margin may be adjustable to some extent to suit site conditions and for economy.

Numbers of members in open frameworks Calculate the number of members by dividing their spacing into the S-dimension (see Figure 14.2). Regard a fraction as a whole number and add one for the member at the start. They may be set out to match the dimensions of sheet finishes and to provide support for their edges.

Jointing, fixing and supporting products The quantities of these will be the amounts required to secure each unit of main product multiplied by the number of units. Many class B1–B6 products are used in making coverings. Some require a continuous stable background to which they can be secured by adhesive or mortar bedding. Others are intended to be secured at intervals to either a solid or an open framework background by means of nails, clips, bolts, plaster dabs, or other unit fixings.

(B1) SMALL UNIFORM-SIZED RIGID FORMED UNITS, THEIR FITTINGS AND FIXINGS (I.E. TILE TYPE)

Obtain the number of units in the surface by counting or calculating the number of units in each row, and the number of rows. Overlapping units may have an adjustable margin. Some units at the edges and ends may be supplanted by **fittings**, e.g. single-lap verge roofing tiles. The number of units will determine the number of fixings, and the lengths and numbers of rows will determine the total length of any supporting open framework, e.g. tiling batten.

(B2) LARGE UNITS AND THEIR FIXINGS (I.E. TRUSSED RAFTER TYPE)

These will usually be shown on drawings and can be counted. If not, the number in each row can be calculated as above. Each unit will require its own fixings.

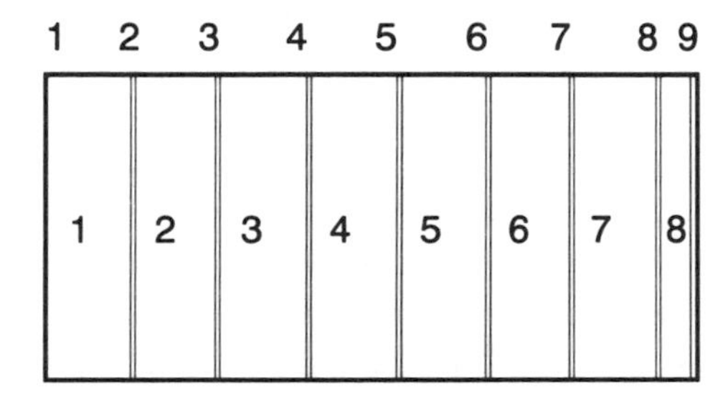

Fig. 14.2 An open framework construction showing the number of structural units .

(B3) RIGID FLAT SLABS AND SHEETS AND THEIR FIXINGS (I.E. PLYWOOD TYPE)

Case (1) Multiple sheets The transformation is from the dimensions of the surface area to the number of sheets of the appropriate size. Normally, flat sheets are butted at the edges.

Depending on their rigidity, they will be fixed at suitable intervals both along the edges and at intermediate positions to an open framework. Either the sheets will be cut to fit the locations of the supporting structure, or the framework members will be set out to suit the sheets. Manufacturers publish advice on fixings, supports and spacings.

Case (2) individual panels will be cut from whole sheets of, say, plywood or glass. Usually, panels are fixed at their edges.

(B4) PROFILED SHEETS, THEIR FITTINGS AND FIXINGS (I.E. CORRUGATED ROOFING TYPE)

Profiled sheets consist of thin materials formed into deep sections. Their longitudinal resistance to bending enables the sheets to span across widely spaced supports. Each type will be supplied in a variety of lengths of the same width and will be fixed with their lengths either vertical or sloping and with adjoining edges overlapped.

Where a single row of the longest sheets will not fully cover the surface, two or more rows of sheets of suitable lengths will be used instead, each lapped over the one in the row below. This margin can usually be varied to suit sheet lengths and the spacing of supports. Calculate their number as shown in Figure 14.1. The total length of the supports and the frequency of fixings will indicate the number of fixings.

The numbers of verge, ridge, corner and other sheet fittings and their fixings can also be calculated from the dimensions of the surface.

(B5) FLEXIBLE SHEETS AND QUILTS (I.E. FELT TYPE)

The transformation will be from one or two spatial S-dimensions to a number of rolls. There are three cases.

Case (1) Rolls of flexible damp-proof course (dpc), brickwork reinforcement etc. of widths to match the different thicknesses of the walls. Add laps to the net length and divide by the roll length.

Case (2) Products placed side by side in rows, possibly overlapping on both edges, e.g. roofing felt, insulating quilt, vapour control membrane. Count or calculate the number of rows. Add the lengths of all the rows and divide by the effective length of one roll. Sheets will be nailed at intervals wherever support is available.

Case (3) Products placed side by side in rows or columns, in unbroken lengths with no end jointing, and supplied in rolls, e.g. wallpaper, carpet. The quantity of any adhesive will depend on the area covered. To obtain the number of rolls,

1. calculate the number of rows or columns as previously described;
2. count the number of full lengths (which might be just one) that can be obtained from one roll;
3. divide (2) into (1). Regard any fraction of a roll as one.

(B6) MALLEABLE SHEETS (I.E. SHEET LEAD TYPE)

The transformations are similar to B4 above. These products are mainly used in the weathershield of a building. To minimize the effect of their relatively high thermal movement, units will be rather small and independently fixed. Fixing pieces are often made from otherwise surplus material, and nailed to the background.

Some metals may be described by their mass per unit of area, and may be purchased per unit of mass. Because of their high price, surplus materials may be preserved for use elsewhere. They may have a scrap value.

Case (1) Various thicknesses of non-ferrous metal flashings, damp-proof courses, gutters etc. at junctions between elements. They are likely to be cut from rolls of strip metal of the required width.

Case (2) Metal sheets cut from standard sized sheets or rolls to cover surfaces. The sheets are joined at their edges by complex folding (see Figure 2.4) and will be turned up or down at the perimeter of the surface. Add the effect of these to the S-dimensions in each direction to obtain the gross area of metal.

(B7) SOLID SECTIONS AND FIXINGS (I.E. TIMBERS TYPE)

Sections can be regarded as being either supporting (i.e. structural) or supported. Each supporting unit, e.g. timber joist, steel beam, must be in one piece. Its length is likely to be dependent on one of the S-dimensions, and it will be fixed at its ends. Count or calculate the number of units as described earlier.

In the UK, the sectional sizes of rectangular and profiled timbers will probably be 'nominal', i.e. as sawn, and their finished size can be expected to be much less due to drying and planing. In North America, the quoted sizes are likely to be the actual ones at a stated moisture content. Designers and suppliers should state which size convention is being used. Standard lengths are at 300 mm intervals between 1200 and 4800 mm.

(B8) HOLLOW SECTIONS, THEIR FITTINGS AND FIXINGS (I.E. PIPE TYPE)

The transformation will be from each length dimension to one or more units of the standard length in which they are supplied. However, the surplus cut from simple sections without sockets, e.g. plastic drain pipe, copper tubing, may be used elsewhere. Fittings will be required at joints in the running length, junctions, ends, and changes in direction. Sections may be supported at intervals, e.g. brackets for wastes and service pipes, or continuously, e.g. beds for drainage pipelines.

(C1) COMPONENTS AND FITTINGS, AND THEIR FIXINGS (I.E. DOOR TYPE)

The numbers of each are counted. No transformations are involved except, perhaps, for fixings such as screws, which will be ordered in multiples and rolls of mastic glazing ribbon.

14.3 CALCULATING QUANTITIES OF MATERIALS FOR COMPOSITE CONSTRUCTIONS

Composite constructions are made from a mixture of formless and formed materials.

(AB1) REINFORCED CONCRETE

See (A1) for concretes. Reinforcing bars (also called rods) may be purchased and supplied to the site cut to length and bent ready for fixing. If not, list the numbers and lengths of the different diameters of bars on the reinforcement schedule and match them with a sufficient number of standard bar lengths. Provide tying wire, clips and other accessories.

The number of sheets of fabric reinforcement is obtained by dividing each S-dimension by the equivalent co-ordinating dimension of a sheet.

(AB2) SMALL STRUCTURAL UNITS IN MORTAR AND ASSOCIATED MATERIALS (I.E. BLOCK TYPE)

The number of blocks or bricks in a one-unit thick wall can be calculated as described earlier for class B1 products. As brick walls can be built to any thickness, the total number will be that in a half-brick thickness multiplied by the thickness in half-bricks.

The brick bonding will indicate the proportions of facings and common bricks, or other mixtures of brick types, in a wall. For example, in a

215 mm thick wall faced on one side only and where each course consists of alternately headers and stretchers (i.e. Flemish bond), two out of every three bricks will be facings. In English bond faced on one side, the header rows will be all facings, and the stretcher rows will be half facings; thus three bricks in every four will be facings.

The quantity of mortar for a wall can be calculated by deducting the volume of the units from the volume of the wall. A 'rule of thumb' for calculating mortar constituents is to assume that every m^3 of mortar-mix will require one m^3 of dry or saturated sand. Multiplying this volume by its bulk density will give the required mass. Although coating the sand particles with cement paste will increase the bulk of the mix, this is likely to be offset by its increased workability and improved compaction.

Calculate the volume of the cementitious materials by multiplying the required volume of mortar-mix by their specified proportion (e.g. $1/8$). The bulk density of ordinary Portland cement can vary from 1300 to 1500 kg/m^3, but is usually taken to be about 1.44 tonnes, or 29 50 kg bags, per m^3.

Walls will have dependent products built into them to help maintain their attributes. These products, although classified in AB2, will have characteristics of other classes, e.g. ties (B1), lintels (B2), reinforcement (B5) or damp-proof courses (B5).

Use the spacing of cavity wall ties (e.g. 450 × 450 mm) to calculate their number. Each tie can be visualized as being in the centre of a rectangle of those dimensions. More will be required at the jambs of openings.

14.4 COMMENTARY

In this chapter we have assumed that all the dimensional information on the design is available. However, at the bidding stage, contractors often work with information from bills of quantities provided by the design team (see Chapter 19). As these give the total quantities of parts rather than their dimensions, many products quantities cannot be calculated precisely, and estimators use approximate methods instead.

Readers will be well aware of the relative simplicity of manual multiplication compared with the complexity of manual division. Perhaps this is why the industry has, in the past, tended to work with data from which fresh data could be derived by multiplication. Calculators and computers have changed this, as it is now just as easy to divide as it is to multiply. Thus, when calculating numbers or quantities of formed products, instead of referring to tables calculated by someone else, we can derive them for ourselves by dividing by some of their co-ordinating dimensions.

Note that the US gallon is only five-sixths the volume of the UK gallon. The US gallon originated as a measure of wine; that of the UK as a measure of ale!

14.5 SUMMARY

The chapter considers the different kinds of numerical relations between the dimensions of parts and the quantities and units of the different classes of main products used in their construction. Readers were encouraged to calculate these quantities for themselves rather than to rely on tables produced by others. Each of the product classes from Chapter 5 was considered in turn.

A concrete part requires its own volume of wet-mix. Coverage data is used to transform the surface areas of (A2) plaster type products to weight, and areas of (A3) paint type products to liquid measure.

Formed main products are all bought by number. In the case of a planar part, the number of formed units in each direction can be calculated by dividing by the equivalent co-ordinating dimension of the formed unit, and giving the result to the next whole number. Each member of an open framework must be in one piece, and their number will depend on their spacing, not forgetting the one at the start. Product quantities calculations for reinforced concrete and masonry were introduced.

Process durations 15

Controlling the overall time and cost of the Works involves predicting the durations of the individual activities. Incorporating them into a construction plan was discussed in Chapter 12. Anticipating them when preparing a bid is discussed later in Chapters 18 and 19.

In Chapters 11 and 12 we suggested that construction can be seen as a succession of processes, each consisting of a series of tasks. A group of processes with a distinct start and finish and an identifiable result we call an activity, although a self-contained portion of that work may be called a sub-activity, or operation.

Site managers will be responsible for the planning and control of sub-activities. Managers with wider responsibilities are more likely to think of the Works as a set of work packages, each consisting of a group of activities. This chapter introduces the factors that influence the durations of these activities. For a more detailed discussion, see Appendix B.

Although every Works is unique, many of the processes and materials employed in their construction will be similar to those used in other Works. Only their extents will differ. Is it possible then, to obtain performance data from current jobs for future use?

15.1 PERFORMANCE DATA

COLLECTING DATA

Obviously, the rules for collecting data on existing jobs must match the ways in which the data will be used in estimating etc. Thus, we shall need to record all those times for which operatives are paid. Where, during any one day, an operative or gang works on more than one activity, the appropriate number of paid hours must be assigned to each.

Reliable data is difficult to obtain. If a member of a gang is absent, the performance of the remainder may be disproportionately affected; they could even be prevented from working altogether. Also, not all gang members may be working on the same activity. For example, a bricklayer's labourer may be repositioning scaffolding in readiness for the next activity. Balancing the workloads of operatives may involve changing the composition of gangs.

This suggests that the times to be recorded should be those of individual operatives rather than that of gangs. To do this, you must be able to recognize them. It must also be possible to identify and measure the amount of work constructed.

Whoever is given the task of recording and collating the data may either prefer or be forced by circumstance to give priority to more immediate matters. Thus, it is far from easy to collect performance data on site, and it is seldom done. However, this is not as serious as it might seem. As each subcontractor will be concerned with a limited variety of work, perhaps even just one type of process, durations can be taken from the operatives' time sheets. Even so, the overall duration of an activity offers no indication of how the time was spent.

PROCESS PERFORMANCE

The rate at which operatives can process products to the required quality standard is called their **performance rate**. The rate for processing formless materials is likely to be expressed as a quantity per hour or per day; that for formed units as the number of units per hour or per day, or the time per unit.

The hours or days between the start and end of an activity is called the **elapsed time**, but the hours or days actually worked is called its **duration**. Note that we are not concerned in this chapter with who the operatives are working for, just with the factors that affect their performance.

FACTORS AFFECTING PERFORMANCE

- **Capability of operatives** The capability of operatives to perform construction tasks will depend on their natural physical and mental abilities, their training and experience, and on opportunities for learning how best to carry out repetitive work cycles. An increase in performance rate with experience is sometimes described as a 'learning curve'. Thus, the performance rates of specialists is likely to far exceed those of generalists.
- **Motivation of operatives** Operatives choose to work, and so will only achieve their optimum output if they are suitably motivated. Pride in achievement, competition with their equals, and financial rewards all play their part.

Management can encourage this motivation by treating workpeople as human beings, by maintaining their supply of materials, by providing uninterrupted opportunities for work, by keeping the work-loads of individuals properly balanced, by paying them fairly and by maintaining a disciplined site.

- **Constraints on performance** Most operatives will have their own idea of what is 'a fair day's work for a fair day's pay'. A gang of operatives will tend to develop a similar notion, sometimes called a **group norm**. In some parts of the world, local norms or their equivalent money values may be enforced by a trade union.

  Other constraints might be the site conditions, e.g. muddy, or cluttered with materials, the distance from materials sources, the lighting level, and having to co-ordinate work with that of others, e.g. plumbers and electricians. Also, operatives might be fatigued from too much travelling, overtime, noise or socializing, or not enough food and rest.
- **Beginning and end effects** The need to make ready so that work can proceed, and to clear away at the end will apply to the Works as a whole as well as to individual construction processes. At the start, performance rates are usually low. They then build up as operatives become familiar with their tasks. Near the end, output will slow as minor jobs that had been passed by are finished off.

  Directly employed operatives who are facing an uncertain future may lose their motivation to work. **Labour-only** subcontractors can be unwilling to complete their work if they have already been paid almost in full for the bulk of it. Not only will this cause disruption and delay; it will also cost proportionately more to employ another subcontractor to finish the work. Once the site has been handed back to the owner, there will be a general unwillingness by all concerned to return to complete the Works or to deal with defects.
- **Technical constraints** Each kind of product will have its own technical characteristics in use, and these are discussed in Appendix B.

PERFORMANCE RATES

The data that arises from an activity will include:

1. its duration;
2. the quantities of materials used; and
3. the quantity of work completed.

Note that (1) and (2) are **inputs**, and are independent variables; (3) is the resulting **output** and is a dependent variable.

A rate is the ratio, or **constant of proportionality** prevailing between the quantities of two variables such as the cost and quantity of some products, e.g. the price rate for each thousand bricks. However, so-called performance rates are not like this although, historically, they have often been called **constants**.

On the contrary, there is no such constant proportionality between the duration of a process and the quantity of materials used or the quantity of work produced. A **performance rate** is just the ratio between the totals reached by the end of an activity or other particular set of circumstances. It would not have been achieved constantly throughout the period. Neither is it likely to be repeated exactly ever again.

Performance rates are obtained by dividing (1) above by either (2) or (3), or the other way round. To do this, all three variables must be single quantities. Apart from fixing components, only one kind of main product is normally used in an activity, and in this chapter we shall express performance, or **productivity**, as the rate at which the main product is processed. That is, we shall be using the independent variables (1) and (2). Obviously, performance rates will be influenced by the quantities of dependent products that are processed at the same time, but it is seldom possible to distinguish the times for fixing these.

In contrast, and for the historical reasons mentioned later in Chapter 17, the UK construction industry tends to express its estimating data using (1) and (3) above. This is considered in Chapter 19. Readers may wonder where the industry obtains the data it uses for estimating. Much of it is simply personal opinion. Data is also available in 'price books', although seldom with any reference to its sources.

However, we should treat this problem sympathetically. Would there be any point in collecting a mass of durations and quantities if the ineffective time could not be separated from the effective? And even if this were done, could the factors likely to affect future construction processes be predicted and quantified?

COMMITMENT TO A SINGLE RATE

When estimating the duration of an activity, the uncertainties and differing factors might be expressed by giving a range of likely durations and your opinion of their percentage chances of being correct. This approach is sometimes called **probabilistic**, and might be helpful when planning.

When preparing a bid, however, a construction enterprise has to commit itself to a series of single figures, and this can only be done using the likeliest performance rates. Using single rates in this way is sometimes described (disparagingly) as **deterministic**, but what is the alternative? In any case, some experienced practitioners have a quite remarkable intuitive ability to predict future outcomes.

In an organization where estimated performance rates are regarded as constants of proportionality rather than 'best guesses', individuals may be blamed when the job takes much longer and costs that much more. Equally, they may be congratulated for guessing correctly. However, in view of the unpredictable nature of the factors discussed above, it is likely to be just a coincidence if the actual duration turns out to be very close to the estimate.

The risks associated with these uncertainties can be minimized or shared when construction enterprises decide either to specialize in the construction of one particular occupancy type such as housing or fast food restaurants, or to subcontract the actual production to specialists.

In this chapter and also in Appendix B, we shall concern ourselves solely with the technical factors that might influence the performance of operatives during effective employment, and with possible ranges of performance rates. Our object is to offer an approach to analysing and predicting times for activities based on the quantities of main products processed.

15.2 FACTORS AFFECTING PERFORMANCE RATES

GENERAL CONSTRAINTS

In Chapter 11, we analysed working hours into effective employment, ineffective employment and unemployment. We also considered some of the general factors that affect performance during effective employment. Significantly lower than normal performance can occur during effective employment because:

- the work at each location is small, or 'bitty', and operatives are unable to develop a rhythmic cycle of working;
- the tasks being performed are unfamiliar, and so have to be learned, e.g. using a new substitute product;
- workplaces are being shared with those working on other processes because of pressure to complete.

Ineffective, or **non-productive** employment will also occur when operatives are prevented from getting on with their tasks and have to wait. Causes of this waiting, which will vary from job to job, and may depend on the time of year, can include the following:

- unsuitable weather conditions;
- not all members of a gang being kept fully occupied because the gang is unbalanced, e.g. too many or too few labourers;
- waiting because other operatives are still occupying the workplace;
- waiting for materials to be delivered, e.g. by tower crane;
- waiting for formless materials to set, e.g. a concrete foundation, a floor screed, a coat of paint;

- interference with one process by those working on another, e.g. drainage pipeline backfilling delayed because the pipeline is incomplete due to the late delivery of some drainage fittings;
- inadequate attendance by other trades. Services subcontractors in particular expect help with the unloading and distribution of their materials. They will expect holes and chases to be cut, and other work to the structure to be carried out to accommodate their pipelines, cable runs, etc, whenever and wherever they are required.

Much of such ineffective time reflects the **buildability** (**constructability**) of the Works. Removing and re-executing unsatisfactory work is also ineffective employment.

ESTIMATING TIMES FOR ACTIVITIES

Appendix B contains commentaries on performance rates for the classes of construction products given in Chapter 5. Each subsection begins with a list of possible tasks, and the classes of operatives who might have these skills. A process where the highest performance rate is likely to be achieved is then described, and this is followed by a discussion on factors that might reduce this rate.

Ranges of performance rates that might be achieved with some of the main products are shown on tables. These rates will include for the incorporation of dependent products as well and are for one tradesman with suitable assistance unless the membership of a gang is stated. The scales are approximately logarithmic, and indicate the performance that might be achieved in an eight-hour day and the average amount per hour.

The daily output has the advantage of being more easily derived from progress on the site although this will include for both effective and ineffective employment. It is also a more convenient unit to use when comparing the estimated with the actual progress on the site. The tables are intended as a guide and the data is not based on any formal site research.

When estimating the duration of an activity, the stages will be:

1. calculate the total quantity of the main product to be processed (see Chapters 13 and 14);
2. consider the constraints on the construction process;
3. select a point in the range of performance rates that seems most likely to represent the hourly or daily output for one operative or gang;
4. multiply (3) by the number of craftsmen, gangs of operatives, or excavating or other equipment to be employed, to give the overall rate of output;
5. divide (1) by (4) to give the duration of the activity.

As an introduction to Appendix B, the factors that constrain productivity when working with the different classes of products will now be summarized. A major general factor will be the amount of work to be carried out.

TECHNICAL FACTORS AFFECTING PERFORMANCE WITH FORMLESS COATINGS

1. **Nature of the base for the coating,** e.g. smooth, narrow, sloping, porous, rough.
2. **Physical characteristics of the product**, e.g. coarsely textured material, viscosity affected by temperature.
3. **Surface finish,** e.g. trowelling.
4. **Accessibility,** e.g. working low down, working from scaffolding.
5. **Incorporating dependent products**, e.g. plaster angle and edge beads.
6. **High viscosity of the coating material,** e.g. a bitumen paint.

TECHNICAL FACTORS AFFECTING PERFORMANCE WITH FORMED PRODUCTS

1. **Size, shape and weight of units** will determine the nature of the assembly process, e.g. blocklaying, pipefitting. Assembling the units may be a one- or two-handed task, or may require two or more operatives, possibly using lifting equipment. Obviously, the heavier the units, the more tiring they are to handle.
2. **Substance of units** will indicate if they can be bent, cut or drilled, their method of fixing or jointing, and their strength.
3. **Type of fixing or jointing products.** Often, most of the operatives' time will be spent on these products, e.g. nailing, pipe jointing, laying mortar beds.
4. **Surface finish** and **nature of the work generally** will depend on the kind of main product.
5. **Unusual arrangement of units** and **incorporating dependent products** will interrupt and slow down the operatives' routines.
6. **Interdependent tasks**, i.e. those which have to be done to suit the convenience of others and which will interrupt and delay normal working.

TECHNICAL FACTORS AFFECTING PERFORMANCE WITH REINFORCED CONCRETE

1. **making and erecting** formwork;
2. **bending and securing** reinforcement;
3. **time for mixing cycle** and **volume of wet-mix** produced;

4. **time for one cycle** of moving, placing and compacting wet-mix;
5. **volume of wet-mix** placed in each cycle;
6. **workability** of wet-mix concrete and its need for compaction;
7. **constraints on compaction**, e.g. small T-dimensions of constructions, intensity of reinforcement;
8. **releasing and dismantling** formwork;
9. **surface finishing.**

TECHNICAL FACTORS AFFECTING PERFORMANCE WITH SOILS

1. **the nature of the soil or rock,** e.g. cohesive or non-cohesive soil; wet, moist or dry; the presence of tree roots;
2. **machine or hand digging;**
3. **work done in each cycle,** e.g. working capacity of bucket;
4. **ease of movement on the site;**
5. **preserving underground services;**
6. **supporting the sides of voids;**
7. **removal of surplus spoil.**

15.3 THE 'ESTIMATING FUNCTION'

OBTAINING DATA

A 'snapshot' of the performance rate being achieved during a cyclical process can be obtained by timing the duration of one or more cycles and identifying the amount of main product processed. Given that

A = **amount processed,**
D = **duration of process,** and
R = the performance rate **quantity of product processed in one unit of time** then

$$A/D = R \tag{15.1}$$

Thus, if the contents of a 5 m^3 ready-mixed concrete vehicle are delivered, moved and deposited in 50 minutes by a gang of 4 operatives, their performance rate is:

$$5/(50/60) = 6 \text{ m}^3 \text{ per hour.}$$

When applying this to excavation, the product quantity will be the volume of soil before it was dug. Let us assume that its coefficient of bulking is 33%. Thus, if an excavator has a 0.4 m^3 bucket, it can effectively dig 0.3 m^3 per cycle. With an average cycle time of three minutes, its rate of output is:

$$0.3/(3/60) = 6 \text{ m}^3 \text{ per hour.}$$

The performance rate being achieved when working with formed main products will be revealed by counting the number of units processed in a particular time, e.g. if 18 bricks are laid during a 15 minute period, the performance rate is 18 / (15 / 60) = 72 per hour.

However, all these outputs will have been achieved during effective employment. To allow for ineffective employment and for unemployment, the average cycle time must be increased and the performance rate reduced. Of course, if the same kind of process continues for a whole day, the daily output will reflect all those factors.

APPLYING DATA

Another version of the above function will predict the duration (*D*) of an activity. If (*A*) is the quantity of the work to be constructed or the quantity of the main product to be used and (*R*) is a suitable performance rate, then

$$A / R = D \qquad (15.2)$$

This can be called the **estimating function**. Obtaining A was discussed in Chapters 13 and 14. Deciding a suitable performance rate R is considered in Appendix B. For instance, given a performance rate of laying 400 bricks per day, the time to lay 600 bricks will be:

$$600 / 400 = 1.5 \text{ days.}$$

If only the area of the wall is known, e.g. 10 m^2 of half-brick thick wall, it can be transformed to the equivalent number of bricks by dividing by their co-ordinating size, i.e.

$$10 / (0.225 * 0.075) / 400 = 1.5 \text{ days.}$$

15.4 SUMMARY

The construction process can be thought of as a series of activities, each based on the work of a single skill-group, using one class of main product and its dependent products and perhaps some components, and having a clear start and end and an identifiable result. Planning and controlling the progress of the Works and cost estimating is based on estimates of the durations of these activities using performance rate data obtained from experience.

The chapter began with a consideration of the uncertainties affecting these durations, and the problems of data collection. Then, some of the factors that might influence performance rates with the various classes of products were outlined. Performance rate data is being expressed by giving upper and lower limits to the quantities of main product that might be processed in one day, and the hourly equivalents. However, estimating data in the UK is more often expressed as the average time in hours to construct one unit of quantity of finished in-place work.

The rate selected for use in estimating will depend to some extent on the amounts of dependent products fixed at the same time. A consideration of the factors that constrain productivity and ranges of performance data for some important main products are given in Appendix B.

FURTHER READING

BS 8000. Workmanship on Building Sites, British Standards Institution, London.

PART FOUR
The Business Environment

INTRODUCTION

The government of every country empowers its citizens to make and enforce financial agreements with each other. Those who are members of the professional team or who manage businesses must have a working knowledge of the law as it affects, e.g. contracts for professional services, building work and the supply of products.

They must also have some knowledge of what the law says concerning employment, taxation, health and safety, trading standards, the ownership of land, and local government, including planning and building control. It is not our purpose to consider these or any other aspects of the Law in any detail.

Previous parts have been almost entirely concerned with the technical side of building. This final part is concerned with:

1. analysing the expenditures of a business so that they can be understood and recovered through its trading activities;
2. the financial factors that can influence decisions regarding both new and existing buildings;
3. the stipulations of the more usual types of building contract;
4. the two main tendering or bidding procedures used by contractors.

CONTENTS

1. Chapter 16 **Cost analysis and cost equivalents** introduces some of the ideas that can help us to analyse and control expenditures on building work. Buildings are expensive and are likely to be financed by loans from others. They take a long time to design and erect, and are intended to have a long, but indefinite life. Thus, when making business decisions, we must be able to compare expenditures at different times in the future. This chapter introduces the appropriate arithmetic,

although the formulas will be found on calculators and in computer spreadsheets. Examples are given in Appendix C.

2. Chapter 17 **Contracts** The choice of legal agreement between an owner and a building organization, and between a main contractor and subcontractors, will depend on the urgency, the stage reached in the design process, and those other circumstances discussed in Chapter 6. A number of different standard forms of contract have been developed over the years, and the important features of some of these are introduced.
3. Chapter 18 **Estimating and tendering** Before agreeing to start, the owner will usually need to know what the Works will cost and whether that price represents value for money. One way of finding out is to invite offers from a number of suitable contractors, thus creating a market for the Works. This chapter describes the procedure that might be followed in a contractor's office when preparing a bid based on the operative information discussed in Part Two. It involves preparing some of the directive information discussed in Part Three.
4. Chapter 19 **Pricing bills of quantities** In this alternative way of preparing a bid, the design team supplies each competing contractor with a list of the descriptions and quantities of the various parts of the Works and other related items. Measurement for these quantities was introduced in Chapter 13. Unit price rates are applied to the quantities of these items and the total amount of all the items will provide the basis of the bid. Any bid should be sufficient to pay the costs of all the materials, plant and operatives' time etc., incurred on the site and still leave a surplus that, when added to the surpluses from other contracts, will pay for head office costs and some profit for the owners.
5. Chapter 20 **Managing costs and value** The cost of a new building or the price of an existing one should be based on its future usefulness to its occupiers. This chapter is concerned mainly with the relations between an occupier's costs, the owner's costs, the capital cost of a project, and its scheme design. It also discusses the risks associated with a project and the significance of time and cost targets. It ends by suggesting a procedure for checking the appropriateness of decisions made during the design process.

Cost analysis and cost equivalents 16

This final part considers some of the financial, commercial and economic aspects of building projects and existing buildings. We begin with some ideas that can help us to analyse and think about expenditures, particularly those on construction.

Costing usually refers to the process of accumulating detailed records of expenditures as they occur. It is also used loosely to describe the process of predicting future costs. By **price** we mean the money value at which something is bought and sold.

Much of this part is concerned with factors such as interest rates and market prices which will influence a business decision and its outcome. While information on such factors might be accurate at the time, their values are likely to be changeable and there will be **uncertainty** regarding their accuracy, both initially and later. All too often, assumptions made regarding these factors will turn out to have been wrong. Thus, every business or other decision involves **risk.** This is the likelihood that the actual outcome will be unexpected and have an adverse rather than a beneficial effect on the fortunes of the organization.

No prudent would-be owner of a building project will make a commitment to build without knowing what the price will be. This transfers the risk of financial uncertainties to the constructor(s). In an emergency, though, getting the work done will be more important than its final cost, and the financial risks will be accepted by the owner. Of course, uncertainty implies that costs may be greater or less than expected. The features of the various types of legal agreement for construction works, including how the financial risks are shared between the owner, the designers and the constructors will be considered in Chapter 17.

A constructor may be an individual, a partnership, a small business, or a legal entity incorporating many owners, such as a public limited company. Alternatively, it may be a building works department within a larger organization. As mentioned at the start of Part Three, we shall refer to them all as **organizations**.

16.1 COST CONCEPTS AND COST BEHAVIOUR

AN APPROACH TO COST ANALYSIS

Cost analysis ideas can be applied to any historical or future expenditures, although, in the main, we shall be considering those of a general building organization. Effective cost control depends on being able to measure the differences, i.e. the **variances** between estimated and actual costs. To do this, the same system of cost analysis must be used for both tender preparation and site costing.

Expenditures by constructors are of many kinds. For example, some will be for activities at head office, while others will be in connection with specific Works contracts. Operatives are paid for their past services, usually weekly. Buying an excavator will provide a continuing benefit over many years and is, in effect, an advance payment for the future use of the asset. Even so, its use in the future could hardly be regarded as costing nothing even if no cash payments are involved.

Annual financial accounts are compiled for the benefit of owners or shareholders, the Inland Revenue, and others not directly involved in the organization. They are concerned with the past, and give a 'snapshot' of the financial condition of the organization at the end of its financial year.

A quite different set of ideas can be used to classify costs for economic management (see Figure 16.1). Forecasting costs, setting budgets and controlling expenditure involves distinguishing the various Works locations, individuals and items of equipment that incur costs. These are often called **cost centres**. The Works locations can be called production cost centres.

Among the ideas used to analyse expenditures are the following:

- A cost that can be identified and assigned, i.e. **allocated** in full to a cost centre is called a **direct cost**.
- The sum of a group of direct costs, particularly those relating to one production cost centre, is often called the **prime cost**.

Costs may be treated as being:

EITHER	OR
direct	indirect
variable	fixed or period costs
part of prime cost	part of overhead costs
allocated to production cost centres	only apportioned to production cost centres

Fig. 16.1 Costing terms.

- **Variable** production costs include those of materials and products, and operatives and plant. Their quantities will, to some extent at least, vary directly with the quantities of work produced. Variable costs will normally form part of the prime cost.
- A cost that cannot be allocated in full to any one production cost centre is classified as an **indirect cost**. These may be shared out, i.e. **apportioned** amongst production cost centres on some basis or other.
- The sum of a group of indirect costs is called an **overhead** (sometimes, the **on-costs**). The costs of items too small to be accounted for individually can be added to the overhead.
- Where, on any one construction site, there are a number of production cost centres, costs that cannot be allocated in full to any one of these constitute the **Works overhead** (or **Project overhead** or **Preliminaries costs**). Of course, if the Works is regarded as a single production cost centre, all costs incurred on the site can be allocated to it and so become direct costs.
- a cost that cannot be allocated to a construction site becomes part of the **general overhead,** i.e. part of the cost of being in business.
- **Fixed costs** are those which, in the short term, will be incurred even though no production takes place. Where they depend on the passage of time, they may be called **period costs**. They will include the costs of the enabling processes on sites and all the costs incurred at the office of a construction organization including owning and operating plant and buildings. Thus, fixed costs are associated with the Works overhead or the general overhead.
- Expenditures are often classed as being on either labour, material, or expenses. These are called the **factors of production**.
- when calculating charges to customers, general overhead costs may be shared out, i.e. **apportioned** to the various production cost centres. This process is sometimes referred to as **absorption**. When a stable organization decides in advance what proportion of the prime cost this is likely to be, it may be called a **predetermined overhead absorption rate.** Old text books on estimating for general builders may include phrases such as 'add 15% for establishment charges and profit' to net prime costs, but this assumes that the organizations are stable.
- In the long term, all overhead costs must be recovered through the selling prices.

16.2 RECOVERING COSTS

MARGINAL COSTING

Marginal costing assumes that, in the short term, the head office staff will be able to undertake extra work at no extra cost, i.e. at no increase of general overhead. Thus, in the short term, the lowest price for which a contract could be undertaken without loss will equal the total of the direct costs, i.e. the prime cost expected to be incurred on the site.

This may at least partially explain the wide differences between tender prices (bids), particularly when work is in short supply. The prime cost of doing one extra contract can be called its **marginal cost**.

Generally, though, each Works contract will be expected to make a **contribution** towards the running of the organization, i.e. to the general overhead, as indicated on Figure 16.2. It is misleading to call this surplus a 'profit'. One contract is shown to require a subsidy. For the organization to remain financially viable over time, these contributions, less subsidies, must exceed the general overhead costs, and this overall surplus will be the **gross profit** from the business. Some of it may be retained, some may be invested in new plant, some will be paid in tax, and the remainder, the **net profit**, will be available for distribution to the owners.

On Figure 16.2, this surplus is shown as the difference between the area of the contribution above the broken balancing line and that of the under-recovery spaces below it. Although the actual contribution from a contract will not be known until the final account is agreed, which may be years later, an uncertain proportion will be included in every payment from the owner. Their irregularity and uncertainty is indicated by the fluctuating upper line in the figure.

A **marginal costing** approach is appropriate when an organization undertakes a limited number of projects of widely different types and sizes. When preparing a tender (bid), the process of calculating the likely

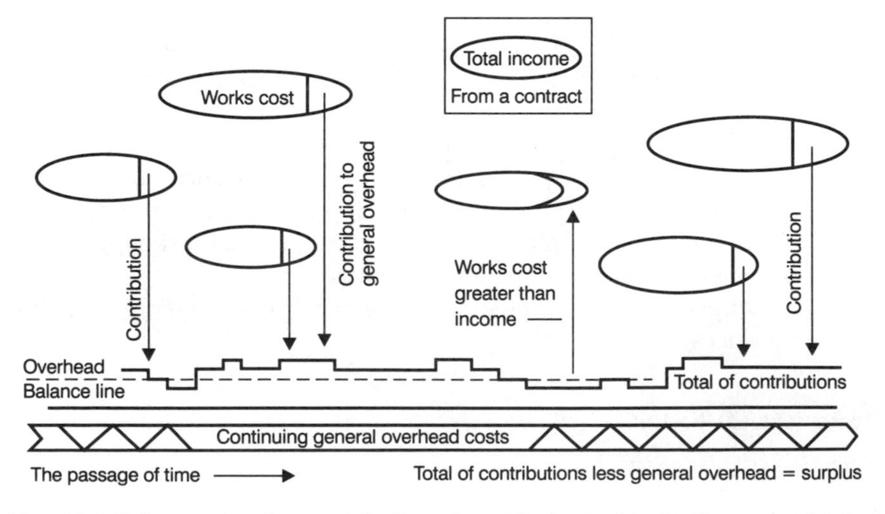

Fig. 16.2 Schema showing contributions from Works contracts, the general overhead costs, and the gross profit of a general contracting organization.

prime cost is called **estimating** and deciding what contribution the contract should make to the organization is called **adjudication**. The procedure is considered in Chapter 18.

THE TREND TOWARDS INDUSTRIALIZATION

It would seem that the process of cost analysis is essentially that of identifying fixed costs within what would otherwise have been regarded as variable ones. The proportion of fixed costs increases as construction becomes more industrialized.

In medieval times, all building process costs would have been variable, both for materials and for operatives and direct plant. There were few kinds of materials, and surpluses could be transferred to other sites relatively easily. Operatives were paid by the day and would be discharged when there was no work for them. There were no costs in connection with their employment, e.g. insurance, sick pay. This situation can still apply in less industrialized countries.

As the variety of materials and plant increases, operatives must become more specialized, and more supervisors, technicians and managers are needed. Society may require that employees are given more security of employment, health care and other benefits. Organizations will need to insure against their liabilities as employers. When work is lacking, personnel may be retained so that they will be available when needed.

The transfer of processes from site to factory involves investing capital in buildings and fixed plant and substituting managers and machine operators for construction operatives and site plant. An implication of industrialization is that an increasing proportion of the materials and operatives' and plant time, i.e. the factors of production, cannot readily be employed elsewhere. Thus, in the short term at least, their costs have to be treated as being indirect, and so become part of the overhead of the undertaking. This trend is unlikely to be reversed.

STANDARD COSTING

Where a limited range of work is carried out for many customers, e.g. by a specialist subcontractor, a **standard costing** approach is more appropriate than marginal costing.

A standard cost is an estimated cost for one unit of an item of work. This will be based on a technical specification and identified factors of production plus a share of the relevant overhead costs. It will relate to a specified period of time and to a specified level of production. The latter is important, as if the expected level of production is not achieved, the overhead costs will be under-recovered, and vice versa. Standard costs will form the basis of **standard prices**.

RECOVERING THE COSTS OF A STABLE ORGANIZATION

Many professional firms, specialist subcontractors and other small businesses are stable organizations, each with an established labour force through whose efforts it will earn its income. The total number of productive hours the workforce is likely to work in a year can be predicted, as can the annual costs of all the other resources employed in the organization. By 'resources' we shall mean those agencies by which the construction processes are carried out, and which continue to be available to the organization. Expenditure will be incurred on the following:

- employees' wages, including those of the office staff and the owner;
- employer's expenses in connection with employees, e.g. National Insurance contributions, employer's liabilities insurance, sickness and holiday pay, transport, costs of guaranteed weekly minimum earnings, training;
- plant, equipment, tools and vehicles;
- office costs, including the rent for accommodation, furniture, heating, lighting, telephones, computers, copiers, printers, fax machines and other equipment, stationery, postage, insurance, accounting and other services, audit and legal fees, cleaning, security;
- promotion, including advertising and hospitality;
- annual cost of working capital employed. When the owner provides this, the opportunity to earn interest through some other, possibly safer, form of investment is lost, and the cost of a bank loan is avoided. It should, therefore, still be regarded as a loan. Its annual cost will be at least equal to what it might otherwise have earned, or what a bank loan would have cost.

This annual expenditure on the resources of the organization can be divided by the hours which directly productive employees are expected to work for customers, and for which payment will be received, i.e. allowing for **bad debts**. The resulting **all-in** hourly rate can be used when preparing quotations and invoices. (Readers who are shocked by the hourly rates charged for vehicle servicing and repairs might ponder over the above list.) During any one period, the difference between the expected and the actual hours gainfully worked will indicate whether costs are being over- or under-recovered. Of course, in that period, once the above costs have been recovered, any further income will add to the gross profit.

Those involved with project planning (see Chapter 12) regard construction products as resources, presumably as they enable a project to take place. But when considering the organization as a whole, products are just the **throughput** worked on during the construction processes. Its resources, i.e. its means of support, are people, plant and money.

REPORTING VARIANCES

The deviation between a budgeted or planned expenditure and the actual amount spent is sometimes called a **variance**. Depending on whether the actual amount is greater or less than the planned one, so the difference is called either an unfavourable or a favourable variance. But simply calculating the difference between the planned expenditure, or **budget** for a construction item and the actual expenditure on it may not identify the manager responsible.

This is because a cost is the result of multiplying a quantity by a price rate. Normally, site managers will be responsible for quantity variances, while general managers will be responsible for price variances.

16.3 THE EFFECTS OF TIME AND INTEREST RATES ON THE VALUE OF MONEY

Lending, borrowing, investing, receiving and spending money all imply benefits to both parties to the transactions. Financial decisions should also take into account future costs and benefits, although the effect of these will depend on their timing.

Individuals and organizations will pay rent for the privilege of using other people's money. Thus, a computer that has been bought with #1000 borrowed from a bank at 10% interest will, after three years, have cost an additional #300. If the intention is to buy it in a year's time, investing #950 now may provide the required amount. If bought from savings, the opportunity of earning interest on that amount will have been lost for ever, although the computer may be obsolete after four years.

This section is about how we can transform money values at different times to their equivalents at other times. One way of looking at the arithmetic is to say that there is a known value, and an arithmetical function that will transform this given value to the value we need to know. We shall call the whole expression a **formula**.

Among the groups who have developed applications for these formulas are actuaries, valuation surveyors (realtors), economists and financial specialists. Although they have given the functions a confusing variety of different names, they mostly use the same symbols for the formulas, many of which will be found in computer spreadsheet manuals. The following terms are associated with the formulas:

- the amount of a transaction is called the **principal**, usually designated *P*;
- the equivalent at the present time of some future value of *P* is called its present value, or **PV**;

- the equivalent in the future of the current value of P is called its future worth, or **FW**, or, possibly, its future value, or **FV**;
- the amount per period equivalent to either a PV or a FW is called a **payment** or **Pmt**;
- **interest** is the rent paid by the borrower to the lender of a sum of money;
- the rate of interest i, or r, is usually expressed as the percentage of the loan that would be paid by the borrower at the end of each year, were the calculation to be made at that time, it has a powerful influence on business decisions;
- **simple interest** is reckoned on the principal only, and paid at fixed intervals;
- **compound interest** is reckoned on the principal and any accumulated interest;
- when interest is being compounded, the number of periods included in the reckoning is usually represented by n.

For calculation periods of less than a year, say half-yearly, or per month, the related fraction of the annual interest is used in calculations. Because this interest is paid early, it can, itself, earn interest before the end of the year, thus effectively increasing the annual rate. For example, when the calculations are made monthly, an annual rate of 10% becomes effectively almost 10.5%. This larger rate is called the **annual percentage rate**, or APR.

THE COMPOUND INTEREST FUNCTION

The basic function is the one for the effect of compound interest and can be designated A.

$$(1 + i)\hat{\ } n = A$$

This is similar to the **exponential function**. It will calculate the **compound amount of 1 after *n*** periods, and its reciprocal, **1 / A**, will calculate the **present value of a future worth of 1**.

Its operation is illustrated in Figure 16.3, which shows an investment of #100 at a compound interest of 10% per annum. At the end of the first year, #10 will have been earned. This is added to the principal to give a compounded amount of #110. By the end of the second year, this compounded sum will have earned interest of #11. This is also compounded, and during the third year the compounded sum will earn interest of #12.1, and so on.

Each of the following three subsections is concerned with a particular formula and its reciprocal. All are based on the compound interest function, which, for simplicity, we are calling A.

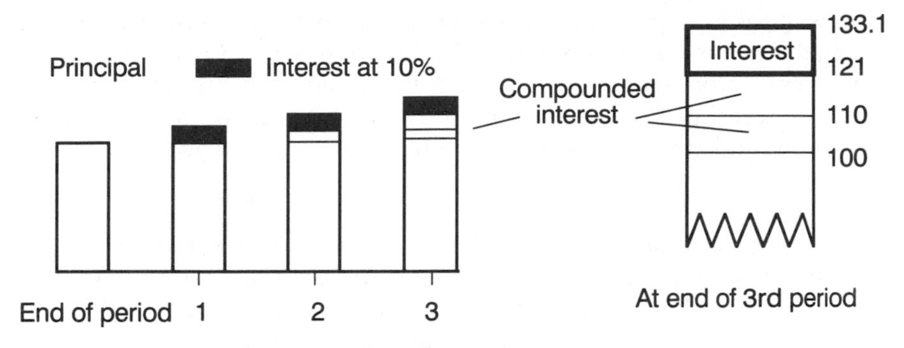

Fig. 16.3 The growing effect of compound interest on a principal sum.

PRESENT VALUE (PV) AND FUTURE WORTH (FW)

Formula (16.a) expresses the equivalence between the PV of, say, one's savings and their FW when lent to a building society, as illustrated in Figure 16.3.

$$\text{PV} * A = \text{FW or, more fully}$$
$$\text{PV} * (1 + i)\hat{\ } n = \text{FW} \qquad (16.a)$$

It is assumed that the savings are invested at the beginning of a period, and the FW is valued at the end.

The reciprocal formula is:

$$\text{FW} / A = \text{PV or, more fully}$$
$$\text{FW} / (1 + n)\hat{\ } n = \text{PV} \qquad (16.1/a)$$

This is more useful in decision-making, as it will give the PVs of future single expenditures. For instance, if a roof covering is to be renewed in five years' time at an FW of #5000, and $i = 8\%$, (1/a) will calculate its PV.

$$\#5000 / (1 + 0.08)\hat{\ } 5 = \text{PV}$$
$$= \#3403$$

This process is sometimes called **discounting**. The total PV of a number of payments or receipts at different times in the future is sometimes called the **discounted cash flow**.

PERIODIC FUTURE PAYMENTS AND THEIR ACCUMULATING FUTURE WORTH

The function $(A - 1) / i$ will calculate the **compound amount of 1 per period.** It is incorporated into the following formula (16.b) which will calculate the growth of n regular future payments, each being made at the end of the period.

$$\text{Pmt} * (A-1) / i = \text{FW or, more fully}$$
$$\text{Pmt} * ((1 + i)\hat{\ } n-1) / i = \text{FW} \qquad (16.b)$$

This will, for example, calculate the growth of regular savings or the loss of regular income, say, from rents while a building is being repaired after a fire. Thus, when saving #1000 each year, and investing at an annual rate of interest of 10%, when the fifth annual payment has been made the compound amount will have grown to:

$$\#1000 * ((1 + 0.10)\hat{}5 - 1) / 0.10 = \#6105$$

The reciprocal function $i / (A - 1)$ will calculate the **uniform series of periodic payments that will amount to 1 by the end of *n* periods**. It is sometimes called the **sinking fund function** and designated S. The following formula will calculate the amount to be put aside at the end of each period at a particular rate of interest so that, when the final payment is made, the required future sum will have been accumulated.

$$\text{FW} * S = \text{Pmt or, more fully}$$
$$\text{FW} * i / ((1 + i)\hat{}n - 1) = \text{Pmt} \qquad (16.1/b)$$

Applications include the annual amounts to be invested out of income for the replacement of plant, vehicles and parts of, or whole buildings. Such **sinking funds** must be invested securely, and the interest rate will be correspondingly low. Their pattern of growth is shown in Figure 16.4.

Thus, to accumulate enough money (say #4000) to take you round the world in three years' time, you must invest (at 7% after, or **net of**, tax):

#4000 * 0.07 / ((1 + 0.07)^3 – 1) = #1244 at the end of each of the three years. Payments of #100 per month would give the same results.

A SERIES OF FUTURE PAYMENTS AND THEIR PRESENT VALUE

Elderly people sometimes buy themselves an **annuity**, i.e. an annual income for life, based on their statistical life expectancy. Similarly, a lender of funds can, in effect, buy an income from the borrower for a

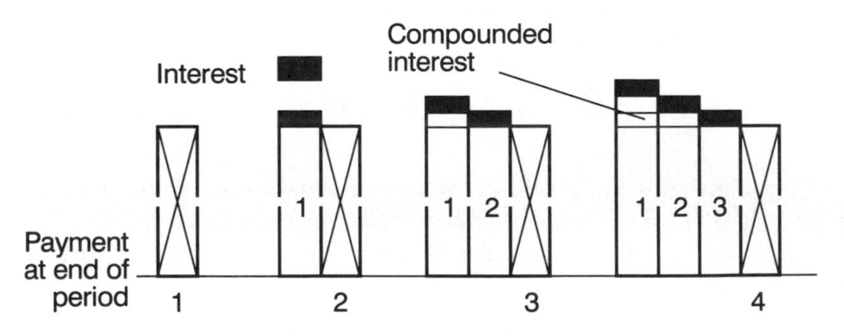

Fig. 16.4 The effect of periodic payments on the growth of a sinking fund.

number of periods. Where a building is used as security for this loan, the agreement will usually take the form of a mortgage. This will give the lender (the mortgagee) a legal interest in the property until the mortgagor has repaid the debt. To calculate these periodic payments:

1. add simple interest on the principal borrowed to
2. the annual contribution to a sinking fund that will accumulate to the value of the principal when the final payment is due, and which will repay the debt. As this sinking fund contribution is paid to the mortgagee, it has the effect of gradually paying off the debt (see later).

The function (16.c), called the **annual equivalent**, or the **annuity function**, will calculate the **uniform series that 1 will buy** and can be expressed as $i + S$.

$$\mathrm{PV} * (i + S) = \mathrm{Pmt} \text{ or, more fully}$$
$$\mathrm{PV} * (i + (i / ((1 + i) \hat{\ } n - 1))) = \mathrm{Pmt} \qquad (16.c)$$

Thus, the annual repayments of a mortgage for #40,000 over 30 years at 8% can be calculated as follows:

$$\#40{,}000 * (0.08 + (0.08 / ((1 + 0.08) \hat{\ } 30 - 1))) = \#3553 \text{ per year.}$$

Doubtless, readers will turn with relief to a computer spreadsheet containing these functions, or to published tables variously called 'actuarial tables', 'valuation tables', 'tables of compound interest factors', or 'annuity tables'. Some spreadsheet generated examples are included in Appendix C, including a table showing the rate at which the amount of the loan declines as regular repayments are made. This is important only when a mortgage is to be repaid early.

The less the outstanding loan, the less the interest content of the regular payment, and the more the amount available for repaying the loan. Figure 16.5, which is based on the table at the end of Appendix C, shows this as a graph. Note that only 5% has been repaid after five years, 15% after 11 years, and 50% after 22 years. Thus, if a property is bought with a 100% mortgage and house prices then fall by 5%, the owner will, for the next five years, owe more to the mortgagee than the property would fetch if sold.

In one sense, the payment of simple interest on the loan is for the benefit of the lender, and the contributions to paying off the debt is for the benefit of the borrower. As an alternative, the borrower can enter into a contract of insurance, the contributions to which are intended to both pay off the loan and pay for insurance against risks such as becoming unemployed. Borrowers should always investigate the additional expense of premature repayment, say, when moving house, and the risk that the sinking fund might not grow as expected.

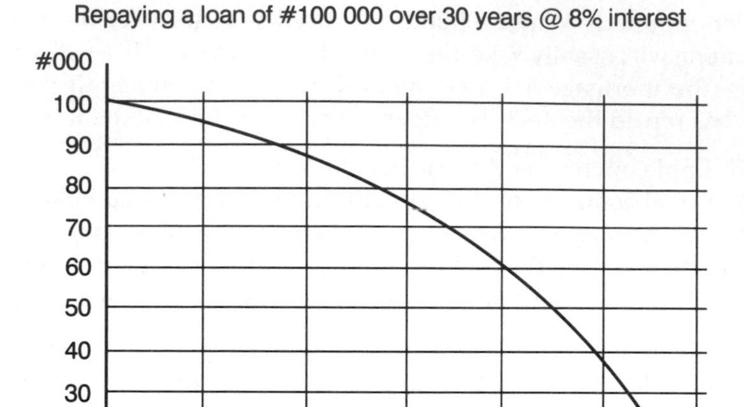

Fig. 16.5 Graph showing the reducing amount of a loan as regular payments of capital and interest are made.

The reciprocal of the annuity function is $1 / (i + S)$. It will calculate the **present value of 1 per period**, and is incorporated into the following formula which will calculate the present value of regular future payments or income.

$$\text{Pmt} * 1 / (i + S) = \text{PV or, more fully}$$
$$\text{Pmt} / (i + (i / ((1 + i) \hat{} n - 1))) = \text{PV} \qquad (16.1/c)$$

For instance, perhaps an investor is considering the purchase of a shop which is let at an annual rent, and has a life expectancy of 15 years. Suppose the net income from the property, after deducting taxes, management, maintenance and other owner's costs-in-use, is $9000, and an annual return of 10% is looked for from the investment. The present-day capital value of the property that is equivalent to this net income for 15 years can be calculated as follows:

$$\#9000 / (0.1 + (0.1 / ((1 + 0.1) \hat{} 15 - 1))) = \#68\,454$$

Of the investor's #9000 income, #6845 will be interest on the capital invested. The remainder (#2155) will be the annual contribution to a sinking-fund that pays 10% per year. Then, at the end of the 15 years, the investor will have received back the original investment.

Unfortunately, such a high rate of interest from a secure sinking-fund is highly unlikely, and it would be wiser to assume, say, 6%. If we recalculate the above using 10% for the first i and 6% for the other two (i.e. those calculating S), we get a capital value of about #63 000.

Valuers (realtors) and others sometimes refer to the ratio between capital value and income (in this case 7.0) as the **year's purchase**. This is just a way of saying that the current capital value is equal to seven years of the net annual income.

However, the above is a gross oversimplification of the full valuation process as, for instance, it takes no account of the residual value of the site when the building comes to the end of its life.

16.4 COMMENTARY

When dealing with plant costs we must distinguish between the viewpoint of the historical cost accountant and that of the manager. The accountant has a legal obligation to give a 'true and fair view' of the finances of the organization, including the value of its assets. This includes reducing, or 'depreciating' the initial purchase prices of plant and machinery to values closer to what they might now fetch in the open market.

Often, an item will be fully depreciated in the accounts well before the end of its working life. This can lead to an exaggerated charge for its use early on, and an undercharging and misuse later. Economic management requires a more realistic approach to the actual costs of an asset over its working life.

Costing ideas can be applied in different ways. However, once a cost analysis system has been established, it should be maintained so that comparisons with the past can be made. The form of the system is of more lasting importance than its contents at any one time.

16.5 SUMMARY

Some of the ideas used by business managers to analyse their expenditures were introduced. Marginal costing, which distinguishes the prime cost incurred on a site from the contribution that a contract will make to the general overhead of the organization, seems more suitable for organizations with large but fluctuating workloads.

Standard costing, which assumes a stable labour force and a certain volume of production of a limited variety of services is more appropriate for specialist businesses. As industrialization increases, costs change from being direct and variable to being indirect and fixed.

Money can be borrowed or lent, the charge or rent for each period of time being expressed as a percentage of the principal sum, called the rate of interest. At the end of each period, the interest is normally added to, i.e. compounded with the capital sum. In the next period, interest is payable on the compound amount.

We are concerned with the equivalence relations between any two of the following three values,

1. the present valuc **PV** of a sum of money P,
2. its future worth **FW**, and
3. one or more payments **Pmt**.

The relations are expressed by three arithmetical functions and their reciprocals, based on the rate of compound interest i, and the number of periods n.

In each of the six formulas shown in Figure 16.6, the known value is shown in the domain. Given values for i and n, the equivalent value in the co-domain can be calculated by multiplying the known value by the appropriate function. Applications of these formulas were discussed and illustrated.

Ref	*Domain*	*Function*	*Co-domain*	*Function will calculate the following:*
a	PV	$(1 + i)^{\wedge n}$ or A	FW	Compound amount of 1
1/a	FW	$1 / A$	PV	Present value of 1
b	Pmt	$(A - 1) / i$	FW	Compound amount of 1 per period
1/b	FW	$i / (A - 1)$ or S	Pmt	Uniform series that amounts to 1
c	PV	$i + S$	Pmt	Uniform series that 1 will buy
1/c	Pmt	$1 / (i + S)$	PV	Present value of 1 per period

Fig. 16.6 Money value equivalence formulas based on compound interest functions.

FURTHER READING

Scarrett, Douglas (1991) *Property Valuations*, E & FN Spon, London.

Contracts 17

We shall now take an overview of the more important types of contract which, in effect, stipulate the obligations and entitlements of the owner, members of the design team, and contracting organizations. We shall also consider some of the features of subcontracts.

17.1 CONTRACTING FOR THE WORKS

A contract is a legally binding agreement between two parties to exchange items regarded as of equivalent value. Almost invariably, one of the items exchanged is a money payment. In the USA, such a payment is known as **compensation**. The usual way of creating a contract is for the party needing a service to accept an offer (or **tender** or **bid)** from the other party to provide it in exchange for a stipulated amount of money called the **consideration**.

For work that can be described easily, and carried out quickly, possibly by just one or two operatives, e.g. redecorating the outside of a house, or installing a new bathroom, such an offer and acceptance, preferably in writing, will be sufficient. But for work that is complex and will require the efforts of many different skilled operatives over a lengthy period of time, the contract will normally be based on one of the published standard forms.

These set out the obligations, duties and rights of the parties and the procedures to be followed in various situations likely to be met during the construction of the Works. They will have been agreed between the representatives of contracting and professional organizations and, in some cases, representatives of building owners and public authorities. The parties will be familiar with them, which is an advantage. Indeed, if unauthorized modifications are made to a standard form, or if a special contract has been prepared, tenderers will wonder why, and whether it would be more prudent to withdraw.

Normally, the party providing the service is expected to carry out their obligations before the other party becomes liable for theirs. While this would be acceptable for minor Works lasting a short time, arrangements for periodically valuing and paying for the work completed thus far will be included in the Conditions of most agreements for building and civil engineering work.

Obviously, the contract forms will reflect the national culture, so they will differ between countries and will, in any case, need revising from time to time. For instance, in the UK recently, those entering into contracts for professional services or for construction works have sometimes been presumed to have a duty to unknown persons who might have an interest in the building in the future. Such obligations are called **collateral warranties**.

17.2 STIPULATED OR LUMP SUM CONTRACTS BASED ON DRAWINGS AND A SPECIFICATION

We shall regard as the norm an agreement to pay a lump (stipulated) sum for buildings and/or other work to be constructed in accordance with the drawings and the specification, i.e. the operative information. These are sometimes called **fixed price** contracts as the price is fixed in advance. In such contracts, the financial risks are almost entirely passed to the contractor. They may or may not be **firm price** contracts, where the contract sum cannot be adjusted to allow for changes in the market prices of materials and labour. The contract agreement document will have two main sections, the Articles of Agreement, and the Conditions of Contract.

THE ARTICLES OF AGREEMENT AND THE CONDITIONS OF CONTRACT

The Articles contain the main points on which the parties agree. They are likely to begin with the names of the parties to the contract and a list of the drawings, specifications and other documents that describe what is to be provided. In the Agreement proper, the contractor agrees to the contract conditions and promises to complete the Works. In return, the owner promises to pay the contract sum. Certain clauses in the Conditions will stipulate when and how this sum might be modified during the course of the contract.

The name of the person who will act as the owner's representative and administer the contract on behalf of both parties will also be given. We shall call this person the **administrator**, although, by training and experience he or she may be an architect, surveyor, engineer, project manager or other professional, and can be self-employed or, perhaps, an employee of the owner. In many cases, the architect will also be the administrator. The administrator's duties will include exercising independent judgement on the quality and value of work done.

In contrast, the Conditions deal wholly with matters of practice. They will reflect the ways in which the industry operates in that locality and at that time, and the problems that are associated with construction works. Some of the more important Conditions are given below.

THE MAIN DUTIES AND RIGHTS OF THE OWNER

The duties of the owner will be limited to:

- handing over the site at the beginning;
- retaking possession when the Works are fit to be utilized; in the UK, this event is known as **practical completion**, and in the USA as **substantial completion**;
- paying the instalments of the contract sum to the builder as certified by the administrator and within the stipulated time limit;
- if necessary, nominating a successor to the administrator;
- insuring against specified liabilities.

The owner will be entitled to:

- receive the completed Works, including extra works, by the stipulated time as modified by the Conditions;
- appoint another contractor if the first is in default, and charge any extra expense to the first;
- have the Works maintained for a specified period after practical completion;
- retain a stipulated but small proportion of the value of all interim payments until the end of the maintenance period. This will provide some security against the costs of correcting non-conforming works or of arranging for another contractor to complete the Works;
- be compensated for any delay in completion. Usually, the amount will be predetermined and written into the contract, where it is called **liquidated and ascertained damages**.

THE MAIN DUTIES AND RIGHTS OF THE MAIN CONTRACTOR

The duties of the main contractor will be to:

- supply everything necessary to complete the Works to the required quality standards and by the agreed completion date;
- plan, organize, execute and control the Works;
- employ suitably skilled operatives and management staff, observe the laws relating to employment, and pay wages, etc. which are at least the equivalent of those generally paid in the locality;

- accept liability for, and insure against claims for injury to persons or property resulting from the Works;
- enter into subcontracts with specialists nominated by the Administrator where prime cost (PC) sums for their work have been included in the contract documents (see Chapter 9);
- renew Works damaged by fire or any other cause, and insure against the costs of this reinstatement and the owner's losses due to delay, up to the full value of the contract plus the costs of demolition and additional professional fees;

and not to:

- assign the whole of the Works to another organization;
- sublet any part of it to another organization without permission from the administrator.

The main contractor will be entitled to:

- receive payment of the amount certified by the administrator at regular intervals as being the value of the work done since the previous payment, less any retention; the basis for valuing such payments should be agreed in advance, e.g. by apportioning the contract sum to clearly recognizable stages in the building process;
- cancel his contractual obligations if the owner seeks to avoid making payments, or if the works are delayed for stipulated reasons; in this event, all losses must be recompensed, including the total amount the job was expected to contribute to their general overhead; this may be described as the loss of 'profit';
- be recompensed for extra expenditure due to the incompetence of the administrator or the designers, e.g. operative information incorrect or delayed.

THE DUTIES OF THE ADMINISTRATOR

The administrator's duties will be to:

- keep the contract documents;
- provide the drawings and other documents, and give all the instructions necessary for the Works to be constructed;
- visit the Works and ensure that they are being constructed correctly;
- employ and supervise a resident clerk of works, site engineer, or (US) project representative;
- ensure that the buildings are correctly set out on the site;
- when necessary, issue instructions modifying the design so that the Works will, on completion, meet the requirements of the owner; in the UK, such instructions may be called **variation orders**, or **architect's instructions** or **change orders**; in the USA, if all parties agree on a

change and on the consequential adjustment of the contract time and sum, it is called a **change order**; in the absence of total agreement, the owner and architect can issue a **construction change directive**;

- exercise independent judgement when certifying that particular stages have been reached, that obligations are owed or have been discharged, that moneys are due to the contractor, that the time for completion is extended, and that the Works are satisfactory or unacceptable;
- instruct the contractor to appoint certain subcontractors, called **nominated subcontractors**;
- adjust the amount of the contract sum to take into account the unexpected occurrences, or **contingencies**, of the construction process, and certify the value of the final payment. (A sum of money included in the contract sum to cover the likely extra costs of such occurrences is called the **contingencies sum**.)

DEALING WITH THE FINANCIAL EFFECTS OF CHANGES

The contract conditions may provide for the reimbursement of changes in the unit costs of materials and operatives, and in the quality or quantity of work, as follows.

- In financially unstable times, either the contractor or the owner may be reimbursed for increases or decreases in the market prices of stipulated materials, and in the wages and expenses of operatives. The unit costs of these items at the tender date will be recorded in the contract documents. As an alternative, official indexes showing the effects of such increases or decreases can be applied to the values of work constructed month by month by the different trades.
- As the nature of the subsoil is somewhat unpredictable, extra foundation works may be required. Also, changes in the design must be expected as most designs will be unique and untried. Human errors must also be expected.

A contract condition will stipulate the basis on which the financial effects of such changes are to be agreed. Ideally, a lump sum figure should be negotiated and accepted in advance. Alternatively, the extra or omitted work can be measured and valued at agreed rates, or, possibly, at rates recorded in the contract documents.

Failing that, extra work can be carried out as **daywork**. This involves identifying the costs that can be allocated to the extra works, i.e. the direct costs of the materials and the operatives' and plant times. Previously agreed percentages are added to these prime costs as contributions to the contractor's general business costs, or **general overhead** (see Chapter 16, and cost reimbursement contracts – later).

A contract may also include provisions for:

- payment of local or state taxes;
- a separate agreement, called a **performance bond** or **guarantee bond,** whereby, if stipulated circumstances arise involving some failure to perform, a third party agrees to make a predetermined payment to the sufferer;
- a bond whereby a third party, perhaps a bank, agrees to pay the owner a predetermined amount 'on demand' and without reasons; such bonds can also be included in subcontracts, and have been widely condemned;
- the resolution of disputes by a person with appropriate technical knowledge and suitable personal qualities, rather than by legal action. The process is private, speedy and relatively inexpensive, and encourages the parties to continue to work together. The three main alternatives are conciliation/mediation, adjudication, and arbitration. Mediators attempt to reconcile the parties and help them to settle their disputes between themselves. An adjudicator's duty is to make a judgement on any dispute arising directly or indirectly from a contract that cannot be resolved by the parties themselves, including those relating to subcontracts. Arbitration procedures are based on law and are the traditional way of solving disputes out of court. Unless ruled out by agreement, any dispute may still end in court where the financial risks can become even greater (see Commentary).

17.3 CONTRACTS BASED ON BILLS OF QUANTITIES

THE EVOLUTION OF BILLS OF QUANTITIES

Construction measurement and valuation customs were in existence before the Great Fire of London in 1666, and were applied to the 'after-measurement' of finished work for the purpose of calculating payment. Hence, the final document, which listed item descriptions and quantities, their prices, and the total cost, was called a 'bill'. As early as 1688, the Edinburgh Town Council appointed John Ogstoun, a wright, or carpenter, to be the town's measurer, and stipulated his fees.

In the 1700s, the task of measurement and valuation was undertaken by **custom surveyors**, often working in pairs, one acting for the tradesman, and the other for the architect, builder or owner who had employed him. Some of the customary rules of measurement were rather odd, to say the least. However, when the customary prices were applied to the measured quantities, it must be assumed that the overall results were acceptable to both parties, most of the time. Skyring (1816) called his book a 'correct list of Builders' Prices; calculated to do justice to the Employer, Master Builder, and their workmen'. As that is the date of the sixth edition, it had obviously been well received.

In the 1800s, general builders competing for a contract had to measure the Works before they could estimate its cost. One way was to work from the drawings, and apply the 'after-measurement' rules and customary prices with which they were familiar. Such a detailed approach was expensive, particularly as most of their bids would be unsuccessful.

The cost could be minimized if an independent **measurer** prepared an unpriced, or 'blank' bill of quantities (B/Q) and provided a copy to each tenderer willing to pay for it. When this procedure became general, such fees were shown in the B/Q, included in the bids, and paid by the successful contractor. Later, the architect and the owner grew to appreciate the benefits of the analytical approach of those they now called **quantity surveyors**, and the information on costs that could be obtained from priced bills. This led to their employing them directly, and to their issuing bills of quantities with the other production information.

A bill of quantities has been described as 'providing a complete measure of the quantities of material, labour and any other items required to carry out a project'. Early B/Q were entitled 'Bill of Quantities of Materials, etc.'. However, both statements are somewhat misleading. In the main, a B/Q consists of a number of statements, or 'items', each giving the description and quantity of either a part or one of its dependent products. For example, there would be an item for each thickness of brick wall, describing its main product and mortar and giving its area (but not the number of bricks). Other items would give the length of its dpc, the numbers of each size of lintel, etc. The intention is that a price should be attached to each item that will cover all the costs of the materials and the construction process (see Chapter 19).

Measurement rules are conventions which rely on the agreement or acquiescence of all concerned. Thus, the rules must be both specific and comprehensive and are, inevitably, lengthy. They evolved locally, many providing for the higher costs of complex works through over-measurements, or, as the saying was, 'making material pay for labour'. As Hawney wrote in 1721, 'It is customary, in most places, to allow double Measure for chimneys'.

Elsam, in his *Practical Builder's Perpetual Price-book* of 1825, described such customs as 'absurd', although, to some extent at least, they must have achieved their purpose. Such practices were still rife in 1906, according to Leaning's paper to The Surveyors' Institution. He compared the rules then in use in Glasgow, Edinburgh, Manchester and London and commented on the movement towards uniformity. This was achieved in 1922 with the publication of the first 'Standard Method of Measurement of Building Works' (SMM). Complete uniformity in billing units had to await the third (1948) edition, and the current edition is the seventh.

While having many different SMMs in the world might be lamented, they do provide the industry in any one place with a set of conventions that enables contractors to tender and enter into contracts with confidence, and owners to compare bids knowing they are for the same amounts of work. They also provide an equitable basis for valuing interim payments and changes in the quantity and quality of the works.

LUMP SUM CONTRACTS BASED ON BILLS OF QUANTITIES

The lump (stipulated) sum contract described earlier was one for the provision of the Works described on the drawings and in the specification. Defining the Works more precisely by issuing a B/Q is more appropriate for major projects, but only where contractors and subcontractors are familiar with their use. Their item descriptions and quantities will stipulate in detail what works are to be provided in return for the contract or subcontract sum. Where a specification also forms part of the contract, the B/Q item descriptions will be brief, and should refer to the specification rather than replace it.

An unpriced copy of the B/Q is supplied by the professional design team to each competing contractor. During their estimating and tendering process, a money rate is assigned to a unit of each item, and the whole is extended and totalled. After bids have been received and opened, a provisional decision is made on which one to accept. This bidder is then asked to submit their priced B/Q for checking.

When arithmetical errors have been corrected, unsatisfactory rates modified and adjustments made so that the total price is still the same as that of the bid, the priced B/Q is designated the **contract bill(s)**. While a contract based on a B/Q is still for a lump sum, it is also a contract for the execution of the quantities of work described in the individual items, and at the prices attached to them.

The quantities and their unit prices will be used when calculating the amounts of interim payments, and for all adjustments to the contract sum, including the correcting of errors in measurements. The procedures to be followed will be described in the Conditions, and the name of the quantity surveyor who is to carry out these functions will be written into the Agreement. Otherwise, the contract Agreement and Conditions will be much the same as in the one based on drawings and a specification.

Some standard forms of contract do not allow the B/Q to include items for enabling processes (see Chapter 11) or for the contribution to the general overhead of the organization (see Chapter 16). Instead, each measured item rate is expected to include a suitable proportion of these costs. More recently, however, the need to itemize and separately price the enabling processes has been recognized.

Contract arrangements based on a priced B/Q can be abused, as a list of guesses about what the design might eventually contain will look much the same as an accurate representation of a fully detailed design.

However, when it is important to get work started on the site as quickly as possible, bills of approximate quantities are a quite legitimate device. In such cases, both parties to the contract must be made aware that the quantities are provisional, that they will have to be remeasured, and that the extent of the work and the financial outcome is uncertain.

17.4 OTHER TYPES OF CONTRACTS

SCHEDULES OF RATES

Owners with many properties often need to arrange for their ongoing maintenance or refurbishment. In such cases, the kinds of work are predictable, but their extent and locations will not, as yet, be known.

If the kinds of work are listed, tenderers can be invited to quote unit prices for them, the result being called a **schedule of rates**. Work will normally be carried out on demand, measured on completion, and valued at the rates on the schedule. Such contracts are usually valid for a stipulated period of time, and are called **term contracts**.

COST REIMBURSEMENT CONTRACTS

Sometimes, building work is required urgently, e.g. following a fire, a civil disturbance, or an earthquake. On occasion, decisions on what works are required to an existing building can only be made after opening up some of its parts, e.g. when underpinning an external wall in a city centre.

In such circumstances, a simple building contract can suffice. Payment to the contractor can be based on the identification and reimbursement of the costs directly incurred on the site (called the **prime cost** in Chapter 16). In addition, a fee will be payable to the contractor as a contribution to their general business costs, or **overhead**. This can be either a fixed amount, or a variable amount calculated as a proportion (usually a percentage) of the prime cost. Different proportions can be applied to the direct costs of materials, labour, plant hire and subcontracts. Competition can be introduced by inviting bids based on the amount of this contribution or on these percentages.

Simple prime cost contracts offer no incentive to economy. Just the opposite, in fact, and the financial risks remain with the owner. To provide some constraint on costs, the contractor may guarantee the maximum cost; alternatively the fee may be based on an agreed best estimate and fluctuate in inverse proportion to the actual cost. Arrangements for making payments, dealing with unsatisfactory work, disputes, and the termination of the contract should be covered in the Conditions. A reliable construction management team is vital if this type of contract is to be successful.

Such agreements are sometimes called **daywork** contracts. This arrangement can also prove useful when incorporated into other types of contract. For instance, unforeseen work may be required on a contract for a lump (stipulated) sum. If a suitable price cannot be agreed in advance, or cannot be calculated by measurement and valuation, it can be paid for as daywork. The contract documents should stipulate how the prime cost is to be identified, and what the percentage additions are to be.

If a contractor is instructed to undertake work without any prior agreement on payment, their charges are likely to be based on daywork. Such payments are sometimes described by the Latin term *quantum meruit*, meaning **as much as is deserved**.

MANAGEMENT CONTRACTING IN THE UK

In a management contract, a contractor becomes a member of the professional team and makes preparations for the management of the construction Works while it is still being designed. The contractor's staff will comment on the buildability (constructability) of the design, and will contribute to and agree the cost plan. They will also make preparations to provide site facilities and to place Works contracts with those contractors who will carry out the actual construction.

Any of the types of contract previously discussed can be used for these Works contracts. As soon as work can start, the management contractor becomes responsible for ensuring that the Works are completed to time, cost and quality targets. Financial risks are thereby largely passed to the managing contractor. This procurement method is most appropriate for large, complex projects that are required in a hurry.

CONSTRUCTION MANAGEMENT CONTRACTS

Particularly in the USA, a professional construction manager, or **lead manager**, may be employed by the owner both to act as **owner's representative** (possibly in collaboration with the architect), and also to apply building economics, management and financial skills to the project. The construction manager, rather than the architect, will be responsible for cost estimating during the design stages, for obtaining bids, arranging and administering Works contracts, and generally ensuring that the cost, time and quality objectives of the project are met. The types of Works contracts will be chosen on their merits, and will, as usual, determine the risks being transferred.

DESIGN AND BUILD

In the USA, the AIA forms of agreement for design/build envisage the owner contracting with a design/builder. This organization then contracts with an architect for design services and with various construction organizations for the actual building work. The intention is to employ methods of parallel working to produce the completed Works in the shortest possible time. There is no set basis for compensation (payment). The agreement with the owner is sometimes called a **turnkey contract**.

In the UK, some design-and-build organizations themselves provide a complete service for an owner. This method of procurement can be beneficial where the organization specializes in a particular building type, e.g. housing, industrial, which can be visited by intending owners. Others may seek outside help with the design work. Some contractors may simply offer a design/build service as a matter of expediency. As with all methods of parallel working, the selection of a contractor with experienced and capable staff is vital.

Although an owner might prefer to contract with a single organization for both design and construction, they may doubt if this will provide them with what they need. One alternative is for the owner to appoint the designers, who proceed as far as completing the scheme design and, possibly, a descriptive specification (see Chapter 10) for the detail design.

Tenders are then invited from design-and-build contractors, the successful one taking over the owner's contract with the designers. The legal device is called **novation**. This enables the contractor to influence the designers' decisions on the detail design, the final choice of products, buildability etc., while constructing the building to the scheme design already approved by the owner.

SALE AND LEASEBACK

This variant of 'design and build' is where a development organization agrees to buy the site from the owner, and to design, finance and construct a building on it to the requirements of the previous owner, and then to lease it back to them.

SUBCONTRACTS

Main contractors are likely to adopt the same kinds of contract agreements with their subcontractors irrespective of the conditions of their main contracts. These can range from a simple lump sum offer and acceptance by correspondence to a formal contract based on a standard form. However, where the main contract contains conditions that affect subcontractors, especially those concerning payment, it is important that they match.

As most, if not all the Works may be constructed by subcontractors, it follows that they will be involved in most of the disputes and will carry most of the risks. Where payments by the owner to the main contractor are delayed, a 'pay when paid' policy by the contractor can be unjust to subcontractors. If either of the main parties becomes insolvent, subcontractors may never be paid. A possible improvement in contractual arrangements is where payments by the owner are made directly and promptly into a trust fund and from there to the various organizations engaged on the Works.

17.5 COMMENTARY

Although this chapter might appear to be about contracts, it is really about people. While the AIA has commented that 'The process of building is intrinsically adversarial', the causes of friction are, to some extent, avoidable. They can include poor definitions of responsibilities, restrictions on freedom of access to information, incorrect timing of appointments, lack of skill and experience and, of course, human failings such as pride, egotism, lack of confidence, overconfidence, jealousy, greed, and ignorance. The Report *Constructing the Team* by Sir Michael Latham (1994) calls for many changes in the working practices of the UK industry, and seeks to promote a 'win-win' culture, particularly between contractors and their subcontractors.

The Housing Grants, Construction and Regeneration Act 1996 will take effect sometime during 1997. Thereafter 'pay-when-paid' provisions will become ineffective and all participants in new contracts will be entitled to stage payments. Disputes are to be resolved by adjudication (see page 308). Where not provided for in the contract conditions, the government's Scheme for Construction Contracts will apply.

17.6 SUMMARY

The nature and purposes of contracts for building work were analysed. The rights and duties of the owner, the contractor, and the administrator under lump (stipulated) sum and bill of quantities based contracts were considered. The features of schedules of rates contracts, cost reimbursement contracts, management contracting, construction management contracts, and design/build contracts were outlined.

REFERENCES

Latham, Sir Michael (1994) *Constructing the Team*, Department of the Environment, HMSO, London.

Leaning, John (1906) The assimilation of the practice of quantity surveyors, *The Surveyors' Institution Transactions*, Paper 317, No 494, Library of the Royal Institution of Chartered Surveyors.

FURTHER READING

American Institute of Architects. *Architect's Handbook of Professional Practice* (vol. 2), AIA, Washington.

Chappell, David (1993) *Understanding JCT Standard Building Contracts*, E & FN Spon, London.

Jones, R. (1988) *A Practical Guide to Subcontracting*, E & FN Spon, London.

Latham, Sir Michael (1994) *Constructing the Team*, Department of the Environment, HMSO, London.

Murdock, J. R. and Hughes, W. (1992) *Construction Contracts, Law and Management*, E & FN Spon, London.

Whitfield, Jeff (1994) *Conflicts in Construction*, Macmillan Press Ltd, Basingstoke.

Standard forms of contract, available from the bookshops mentioned in the Introduction.

Estimating and tendering

18

18.1 INTRODUCTION

Our viewpoint in this chapter is that of a general contracting organization invited to bid for a fixed price lump sum contract to construct Works in accordance with the drawings, specification and other operative information. We shall assume that the design is complete, that the specification prescribes the products to be used (see Chapter 10), and that production drawings have been prepared.

Thus, all the information likely to be required for building will be available to the estimator, although this ideal situation is somewhat unlikely. The Commentary at the end of the chapter includes some observations on what to do when the design data is incomplete.

The process of estimating and tendering involves exercising judgement about what is likely to occur in the future, both on the site and commercially, as well as making mental and actual calculations. Construction estimating is neither analytical nor comparative estimating as practised in other industries. Neither is it based on any formal work measurement procedures. The more it can be based on offers from third parties, i.e. subcontractors, products suppliers, plant hirers etc., the more the risks are passed to others, and the more certain the cost estimate will become.

The procedure of preparing a bid will be considered in the following three stages:

1. Collect background information, or **Works appreciation**.
2. **Estimate the prime cost** to be incurred as a result of site activities. We shall base our predictions of expenditures on the quantities of construction products required and the durations of activities (see Chapters 11 to 15 and Appendix B). There are separate sections for (a)

general approach and preparation, (b) products quantities and costs, (c) labour and plant durations and costs, (d) Works overhead costs, and (e) estimating procedures.

3. **Adjudicate** on the estimate and decide how much the job should contribute to the general overhead costs.

In the next chapter, we shall be considering various arrangements and shortcuts used by the industry to minimize the costs of bidding. These include the provision of measured data by the design team, and the direct pricing of units of finished work.

TO BID OR NOT TO BID

Perhaps between four and six organizations may be invited to bid for a contract, depending on its size, although the number can be much higher. The income arising from each successful bid must pay the costs of preparing all the unsuccessful ones as well.

It is tempting to bid for every possible job, especially when work is scarce. Unfortunately, that is when many other contractors will be doing the same, and the likelihood of success is that much less. It is safer to bid only for those contracts that will suit the organization, and it may be more profitable to find or create your own opportunities for work, e.g. by negotiation with a long-standing customer.

If the professional team is following the procedure for selective tendering mentioned in Chapter 6, they will invite potential bidders to indicate their willingness to bid some weeks before the tender documents will be ready. Factors which will affect the decision to bid include:

- whether the estimators, planners and management can prepare a bid in the time allowed;
- the reputation and financial standing of the owner;
- the reputations of the professionals;
- when the planning and building control permits for the Works are expected;
- whether the contract will be on a standard or modified standard form, or has been drafted specially. Non-standard contracts should be viewed with suspicion, and legal advice sought;
- the location of the site;
- the starting date;
- the contract period;
- the spare capacity within the organization for undertaking work during that time.

18.2 STAGE 1: WORKS APPRECIATION

STUDY THE CONTRACT DOCUMENTS

The following are likely to be included in the contract documents:

1. the drawings;
2. the specification;
3. possibly, a ground exploration report;
4. a tender form to be completed and delivered to the specified address before the specified date and time;
5. a clearly identified tender envelope addressed to the owner or the designers and stating that it must not be opened until after the final delivery date and time;
6. either the title of the standard contract to be used, and details of the variables to be added to it, e.g. names of the owner, administrator and the professional team, Works start and completion dates, any phasing of completion, or
7. a full copy of the contract agreement and conditions.

A close study of the documents should reveal their completeness or otherwise, and the thought and care that has gone into their preparation. Consider the following questions.

1. Are the drawings merely those prepared during the design process, or are they specially prepared production drawings?
2. Are the drawings referenced to each other?
3. What is the quality of the details on the assembly drawings?
4. Is the specification a separate document, or have specification notes been written on the drawings? If the latter, are they legible?
5. If the specification is a separate document, are the details on assembly drawings referenced to it?
6. Does the specification prescribe what products to use, or is it descriptive, thus leaving the bidder to make decisions on products?
7. Do the sections of the specification document match the probable allocation of the Works to the various subcontractors?
8. Is there space on the site for temporary buildings, materials compounds, cranes and other fixed plant, and for temporary spoil heaps?
9. Where will the site access be, will it be adequate, and will there be a turning space for vehicles?
10. Will temporary roads be required?
11. What are the implications of the ground exploration report?
12. Are there any onerous contract conditions?

VISIT THE SITE

The following should be investigated, and, where appropriate, photographed or video recorded.

The locality:

1. local roads, and their suitability for heavy vehicles;
2. evidence of vandalism and the level of security advisable;
3. adjacent buildings, and their need for supports;
4. existing sewers and manholes;
5. existing water, gas and electricity supplies;
6. any work in progress where subsoils are exposed;
7. local tips, their capacities, distances and charges;
8. the availability of local labour and labour-only subcontractors;
9. local products suppliers, plant hirers, road hauliers;
10. construction activities in the locality and their possible effects on the employment of operatives, competition for subcontract work, and the supply of bulk materials and ready-mixed concrete;
11. addresses of local planning and building control officials, and public utility undertakings;
12. address of the local Job Centre or other employment agency;
13. opinions of local residents on possible problems, as voiced during conversations in bars etc.

The site itself:

1. the depth and quality of the top soil;
2. existing trial holes and borings;
3. the type of subsoil and the likely stability of the sides of voids;
4. the level of the water in the ground, its likely variability during the site preparation processes, and whether a water extraction system will be needed;
5. weather conditions to be expected during the contract period;
6. the possibility of surface water flooding and the need for pumps;
7. the possible effect of rain on the surface of the site;
8. buildings to be demolished;
9. trees and undergrowth to be cleared;
10. the presence of overhead power lines and telephone wires;
11. any right of way or right of light that must be preserved;
12. work to be carried out in existing buildings (these should be visited again when preparing the detailed estimate).

METHODS STATEMENT

Information gleaned from the above will indicate a likely strategy for carrying out the work. Initial decisions on matters such as the sequence of the main activities, the major items of plant, and the layout and management of the site will be included in a **methods statement.** This will be a partial expression of the full methods statement discussed in Chapter 12.

Tactical matters will include groundworks and the construction of structural elements. Their durations and costs will depend on the processes to be carried out and the resources being employed. Unless suitable methods have been evolved on previous jobs, or are traditional, a feasible method should be worked through and written down for each activity. Once the structure has been built, it will provide workplaces for those carrying out later processes, and there will be fewer uncertainties regarding their methods. The object is primarily to ensure that the costs allowed for in the estimate will fund the activity.

Even when the professional team do not ask for an outline plan to be prepared, it may still be prudent to produce one. However, when the time comes to construct the Works, all such decisions are likely to be thought through afresh by those who will be managing the construction.

18.3 STAGE 2 – ESTIMATING THE PRIME COST

(A) GENERAL APPROACH AND PREPARATION

Construction plan

A simple plan may consist of a bar chart of major activities. More complex Works might be analysed in some detail and expressed as a network (see Chapter 12). The durations of activities may be based on the planner's own calculations or, preferably, on information arising from the estimating process.

One object will be to identify factors that might cause uncontrollable delays, e.g. products with uncertain delivery dates; the provision or modification of services by public utilities and local authorities who are not contractually obliged to be helpful. The plan will also indicate the durations of Works overhead period costs and the overall duration of the Works.

Subcontractors schedule and enquiries

Information for the estimate will be supplied by many people and organizations, and they should be involved as early as possible. Clearly, it is vital to ensure that all the works called for by the contract are allowed for in the estimate. A study of the contract documents will indicate:

(a) the prime cost (PC) sums for work that must be carried out by:

1. subcontractors nominated by the design team, and
2. public utility companies.

(b) what activities could be carried out by either:

1. directly employed operatives (if any),
2. labour-only subcontractors, or
3. other subcontractors.

These last two are sometimes called **domestic subcontractors**. A list of organizations to be invited to quote for the various domestic subcontracts will constitute the subcontractors' schedule. Each must be supplied with a copy of every relevant document, so that they can identify and measure the quantities of products and materials to be provided and incorporated into the Works. Some may expect the contractor to provide such quantities. Within a UK contracting organization, the quantity surveyor is usually responsible for the measurement and valuation of subcontractors' work. The whole process may be called **enquiries**.

When the quotations are received, they should be checked to ensure that they accord with the enquiry. Entering their amounts on the schedule will identify those who have not yet responded and who need to be prompted. The late arrival of quotations makes the work of an estimator particularly difficult and stressful (or, for some, exciting). For instance, it is not unknown for a labour-only quotation to be delivered at the last minute when the invitation was for both labour and material!

Products quantities and process durations

We shall adopt the **inputs** approach to estimating. That is, we shall calculate and price:

1. the quantities of products to be delivered to the site, and
2. the times for processing them, as discussed in Chapter 14 and in Appendix B.

The procedure is outlined in Figure 18.1.

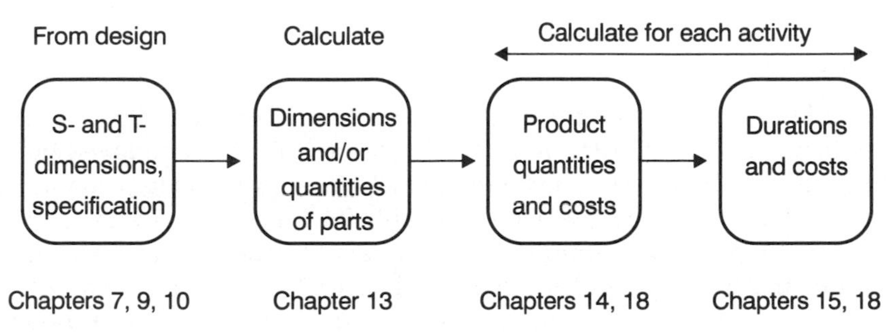

Fig. 18.1 The activity estimating procedure.

Obtaining the quantities of products is a straightforward matter of either counting or simple arithmetic, and a list of product quantities should be produced for each activity.

As described in Chapter 12, an activity is a set of processes that could be carried out by the same operatives or gang using a limited variety of main products to achieve a clearly identifiable result. One or more activities may be carried out by a particular skill-group, possibly employed by a subcontractor. Decisions on who to employ are made by the contracts manager.

On the site, an activity may be carried out as a series of sub-activities, bearing in mind the work to be carried out by other gangs. Planning, co-ordinating and control on the site, including maintaining continuity of employment for the various operatives, will be the responsibility of site management.

Estimating the durations of activities will be based on past performance. However, as we have already seen, if only the overall duration of an actual activity is known, the performance achieved can only be expressed as the rate at which the main products were processed. No data on the time spent on dependent products can be produced unless their individual times for fixing are also known. Yet, of course, many processes do involve a variety of dependent products. The two possible ways of dealing with this when estimating the total time for an activity are:

1. use an inclusive performance rate by selecting the point on the range of performance rates that will reflect the additional products to be fixed, or
2. adopt a two-stage approach of:
   (a) calculating the duration using a performance rate that would be suitable if there were no additional products, and
   (b) adding a suitable amount of time for the fixing of the additional products.

(B) PRODUCTS QUANTITIES AND COSTS

PRODUCTS ENQUIRIES

Quotations must be sought from manufacturers, suppliers or builders' merchants for the products to be worked on by directly employed operatives and labour-only subcontractors. The approximate quantities required should also be given as these will affect the unit price rates.

Calculating the net quantities or numbers of products in the various parts of the works was described in Chapter 14. Working with whole numbers of formed main products gives correct quantities even though some units will be reduced in size to make them fit. Where the material cut off cannot be used elsewhere, it can be called **unavoidable waste**. This also occurs when the contents of packages or containers of products (e.g. packages of blocks; cans of paint) are not fully used.

Avoidable waste can be seen on any building site, and reflects the quality of the Works management. Its causes include:

- the removal of defective work, or work that does not conform with the contract;
- over-construction, e.g. excess concrete in foundations due to digging too deeply or using too wide an excavator bucket;
- carelessness and neglect by operatives when unloading, storing, stacking, moving and processing materials and protecting completed work;
- pilfering (a euphemism for 'stealing a small quantity') by operatives;
- vandalism.

The quantity calculations of materials to be incorporated into dimensional parts should allow for unavoidable waste and, possibly, for some avoidable waste. With large quantities, contracts managers can exercise some control over wastage by authorizing the site management to arrange for delivery in instalments as needed and by stipulating maximum quantities.

Only the actual number of components, e.g. doors, locks, required for integral parts will be ordered. Loss or damage to these is always avoidable, and will involve re-ordering. Pilfering, vandalism and theft would be better considered during the adjudication stage.

Units for ordering and purchasing

Loose formless materials are ordered and purchased in units of mass, e.g. per tonne, per kg. Small quantities may be supplied in containers.
Liquids are ordered and paid for by giving the capacities of their containers in liquid measure, e.g. litres, gallons. Wet-mixes of concretes are ordered and purchased per m^3.
Class B formed products are often available in a variety of sizes, e.g. timbers, pipes and tubes, steel sections, profiled sheets. The supplier may simplify their pricing by 'quantifying' one or more dimensions and by charging per multiple unit of quantity, e.g. the total of all the different lengths of 100 × 50 mm timbers at a rate per 100 m.

All-in product rates

The following costs may be incurred in addition to the supplier's charges.

- Delivery to the site, when transported by the contractor's own vehicles.
- Unloading and storing. This can cause contention as the vehicle driver

may not wish to help. On large sites, one or more labourers may be employed on general duties, including unloading from vehicles, their costs being included in the Works overhead. Otherwise, operatives will have to interrupt their work to do this.

- Distributing products to the various workplaces around the site, possibly using vehicles allowed for in the Works overhead.
- Special lifting gear such as a mobile hydraulic crane for components that are too large or heavy for the facilities on the site.

Where possible, the above should be regarded as part of the cost of the product, and apportioned to each unit of purchase, e.g. per 1000 bricks.

(C) LABOUR AND PLANT DURATIONS AND COSTS

All-in hourly rates for operatives

Estimating the cost of an operative's time is simplified by the use of an all-in rate. This is an expression of all the costs of employing the operative for one productive hour. It is calculated by totalling all the costs likely to be incurred during a period of time and apportioning them over the number of productive hours in that period.

As we saw in Chapter 16, the all-in rate for an employee of a stable organization can include a proportion of all the expenditures of the organization, together with the planned surplus, or 'profit'. This approach is likely to apply to individuals, to specialist subcontractors – and to professionals.

General contractors and other organizations will have a fluctuating, or unstable, labour force, whose total productive hours are uncertain. Obviously, their overhead costs, etc. cannot be apportioned over an unknown number of productive hours and must be recovered separately. In such cases, the all-in rate will be based solely on the marginal cost of employing one extra employee. Because the individual costs change from time to time, it is convenient to put the calculations on a spreadsheet (see Figure 18.2).

The training levy is paid annually. In winter, there are fewer holidays, and daily working hours may be shorter. Thus, arithmetical precision can only be achieved by considering costs and hours worked over a whole year. However, such precision can give a false impression of 'correctness', and ignores the waywardness of humankind. Instead, we will base our calculations on the costs of a working week, plus an apportionment of the more obvious annual costs.

Employers may negotiate directly with employees regarding their rates of wages and hours and conditions of work. More likely, these will be established by negotiation and arbitration between their representatives. In general contracting in the UK, the authoritative document is the

Working Rule Agreement published by the National Joint Council for the Building Industry. This is reviewed annually in June. Many specialist industries have their own agreed rules.

Employees are paid provided they are available for work. Thus, wages paid during bad weather and other non-productive time must be allowed for in the all-in rate. They are also paid during public and annual holidays.

Weekly expenditure on employees

- Wages, based on the guaranteed minimum weekly earnings, various allowances, 'plus rates' and other extra payments for skill or responsibility, together with overtime and travelling time. The weekly minimum earnings consists of the basic rate plus (at least) the guaranteed minimum bonus.
- Employers' contribution to the costs of operatives' annual holidays, and to the Death Benefit and Retirement Benefit Scheme. This credits scheme is run by the Holidays Management Company, which sells stamps to employers for fixing each week to cards owned by the operatives. Thus, funds are accumulated from which the company will reimburse the costs of wages paid during annual holidays to whoever is the operative's employer at the time.
- National Insurance Contributions, calculated as a percentage of the employee's earnings.
- Insurance of the employer's liabilities arising out of their duty to care for the safety of their employees and members of the general public. This cost may also be based on employees' earnings.

Annual payments

- The Construction Industry Training Board levy should be apportioned over 49 working weeks.
- Wages paid for eight public holidays.

Previous annual costs for the following will be revealed from records, and can be expressed as a percentage of gross wages.

- Daily sick pay for illness in excess of three days.
- Severance (or redundancy) pay, if not regarded as a general overhead cost.
- Production lost during inclement weather. This can be allowed for by reducing the number of productive working hours in the week.

Hours worked

The total weekly cost of an employee will be divided by either the number of hours which the operative spends on the site, or the smaller number of hours actually spent in construction operations. Whichever method is chosen must also be used when deriving productivity data for use in estimating.

Figure 18.2 uses 1992 figures to show the factors to be included when calculating an all-in hourly rate for (say) a concrete finisher in the UK. It can be simplified by excluding items 15 and 17, and dividing by 34, rather than 36 hours, to give an all-in rate of £5.37. Of course, an all-in daily rate will be one-fifth of the weekly total.

ALL-IN GANG RATES

Although individual operatives do sometimes work on their own, most of them will work as members of integrated gangs. An all-in gang rate will be the sum of the rates of individual members, plus the extra costs of their supervisors. When plant is used solely and continuously on productive work, its costs and those of plant operators, fuel, lubricants, and

	A	B	C
1	*Weekly wages and expenses paid to operative*		
2		#	Notes
3	Basic rate for adult operator	118.37	
4	Guaranteed minimum bonus	14.24	
5	Payment for extra skill code C	7.80	
6	Payment for intermittent responsibility	0.00	
7	Overtime payments	0.00	
8	Daily travel in excess of 6 km: 10 km @ 8p	4.00	Conveyed by employer
9	Payment for work in difficult conditions	0.00	
10	Tool and clothing allowance	0.00	
11	Weekly total:	144.41	
12	Hourly equivalent:	3.70	39 hour week
13	*Employer's additional costs*		
14	Weekly stamp for annual holidays etc.	18.70	
15	Weekly equivalent of 8 public holidays	4.86	(7 * 8 + 7 hrs)/48 weeks
16	Employer's National Insurance etc.	13.00	9% of weekly total
17	Av. 5 days/year sickness and injury payments	2.50	@£59.00 per week + statutory sick pay
18	Construction Industry Training Board levy	0.83	£40 per year
19	Employer's Liability Insurance	2.89	Might be about 2%
20	Total of weekly additional costs	42.78	
21	Gross weekly total:	187.19	
22			
23	Normal working week = 39 hours, non-productive hours per week = 3		
24	All-in hourly rate #:	5.20	Weekly total / 36 hrs

Fig. 18.2 Calculating an all-in hourly rate for an operative in the UK.

daily plant maintenance can also be included. Otherwise, such costs should form part of the Works overhead.

(D) WORKS OVERHEAD DURATIONS AND COSTS

The Works overhead

The methods statement mentioned above can be extended to include tentative decisions about the Works overhead enabling items, e.g.

1. the site managers, supervisors, storekeepers, general operatives, crane operators, and other 'non-productive' staff to be employed on the site;
2. the staff concerned with safety on the site;
3. insurances;
4. site accommodation, fixed plant, stores, and areas secured by fences and gates, i.e. **compounds**. These will incur single fixed costs for delivery, preparing bases, erection, maintenance, dismantling and removal, plus period costs while on the site;
5. transport on the site;
6. daily transport of operatives to and from the site;
7. scaffolding and other temporary workplaces;
8. temporary water, electric and telephone service installations and charges;
9. services to be provided to the professional team as stipulated in the contract documents, e.g. furnished hut and telephone with fax machine for clerk of Works.

A schedule of Works overhead items can be listed at this time, and their fixed costs estimated. Those whose costs will depend on their time on site, i.e. the period costs, can be based on the estimated durations of the related activities when these are known.

(E) ESTIMATING PROCEDURES

Activity estimating model

The constituents of the estimated cost of an activity are illustrated in Figure 18.3 and are as follows.

1. The gross quantities of main and dependent products, calculated as described in Chapter 14, and allowing for avoidable and, perhaps, some unavoidable waste. Their all-in price rates should be based on the lowest of the net prices quoted for their supply, allowing for discounts, and should include for delivery, storing, and bringing to the workplace.
2. The total duration of the activity, calculated from the quantities of main products as described in Chapter 15 and Appendix B, and priced

Item details	Quantity	Unit	Rate #	Cost #
Activity: loadbearing masonry internal walls				
100 mm Conclite solid facing blocks	4800	nr	0.85	4080.00
140 mm Conclite solid facing blocks	1150	nr	1.20	1380.00
Cement, lime and sand mortar	5	cu m	65.00	325.00
100 mm polythene damp-proof course	7	rolls	3.00	21.00
140 mm polythene damp-proof course	2	rolls	4.00	8.00
Lintels over doorways	8	nr	12.00	96.00
Gang: 2 masons, 1 assistant , laying 100 mm blocks @ 22 per hour and 140 mm blocks @ 16 per hour, including dependent products	290	hours	20.00	5800.00
Total			#	11710.00

Fig. 18.3 An example of estimating the prime cost of an activity.

at suitable all-in daily or hourly rates. Durations must allow for interruptions between processes and sub-activities, the numbers and variety of dependent products to be built in, and the site factors that will affect performance.

Charges by labour-only subcontractors may be based on rates for processing each unit of quantity of main product, e.g. laying 1000 bricks.

Assembling the estimate

The estimated prime cost will include:

- the main contractor's and subcontractors' estimates for construction activities;
- selected quotations for direct plant hire;
- the Works overhead;
- costs of clearing rubbish and cleaning the building;
- sums included in the contract documents, e.g. the contingencies sum; prime cost (PC) sums for the costs of works by subcontractors nominated by the administrator;
- the contribution to the general overhead required in connection with the PC sums for subcontractors' work, although it is preferable to include this in the 'mark-up' (see below); this is usually called 'profit', and is a common example of the misuse of this term;
- costs of providing facilities required by such nominated subcontractors (known as **attendance**).

To the above must be added the estimator's evaluation of the risks inherent in undertaking the Works:

- the costs of avoidable wastage of materials;
- expenditure in correcting defects during construction and also at or after practical (substantial) completion, when it may be called **snagging**;
- expenditure in maintaining the works for the period stated in the contract;
- the additional costs of protection, space heating, and other measures necessary when building during the winter;
- the possible effects of price rises during the construction period. A **firm price** contract is one where the contractor is assumed to have allowed for all fluctuations in the costs of materials and labour. The alternative is for the contract to include conditions dealing with the reimbursement of such price fluctuations.

The above are reported to the managers who are responsible for marketing the services of the organization and maintaining its workload in accordance with its business plan. They exercise their commercial judgement in deciding how much the job should contribute to the general overhead (see Chapter 16), i.e. what the **mark-up** shall be, hence the name **adjudication** for this final stage.

18.4 STAGE 3: ADJUDICATION

As we saw in Chapter 16, the difference between the income from a contract and its prime cost is the contribution the contract makes to general overhead expenses. In deciding what contribution to aim for, internal considerations will include:

- the expenditures arising from being in business, that is, the **general overhead**;
- the expectations of the owners or shareholders for dividends;
- whether the Works can be undertaken with existing supervisory and management staff. If not, what suitable extra staff are available for employment on short-term contracts?
- the risks involved in fluctuations in the prime costs; while the contract for the Works is likely to be for a firm price and not subject to adjustment for fluctuations in the prices of materials or labour, offers from suppliers and subcontractors are not likely to be held firm for more than a few months; thus, allowances must be included in the tender (bid) for all likely increases or decreases in costs;
- the risks of delays due to lack of control over public utilities, specialist product manufacturers and other independent organizations who have no contractual duty to collaborate;
- the risk of liability for liquidated damages because the Works will not be completed within the contract period;

- the risks that dust and noise pollution arising from the Works will result in claims from neighbours, and legal proceedings;
- the expense and risk attached to the owner's requirements for performance or 'on demand' bonds;
- the maximum amount of working capital required to bridge any gap between interim payments by the owner and the amounts owed to their own employees and to suppliers, subcontractors etc.; legislation requiring employees' wages and taxes to be paid regularly is likely to be extended to payments to other organizations;
- the need for work to keep the resources of the organization employed in both the short and the long term.

External considerations include:

- the extent of competition within the local construction market;
- transport and access to the site;
- the availability of operatives, subcontractors and plant;
- the likelihood of being 'messed about' by the professionals and the owner, and consequential delays and costs;
- the possible extent of vandalism;
- the culture and ethics of the society where the Works are to be undertaken. Factors might include the power of unions and local politicians, the effectiveness or otherwise of local bureaucracies, and the expectations of individuals regarding payments for the exercise of influence.

As a result of considerations such as the above, a mark-up is added to the estimate to convert it to a tender (bid).

18.5 COMMENTARY

In preparing any estimate, gross products quantities and their costs can be calculated by arithmetic. It is the durations of the activities that are uncertain. As, in many cases, the same specialists are likely to be invited to tender by each bidder, every bid will be based to some extent on the same lowest sub-bids.

A bidder may be reluctant to submit a bid but may prefer not to reveal this. The estimator may then seek a **cover price** from an acquaintance employed by a rival bidder that will ensure that their bid is unsuccessful. Although this practice has been deprecated by everyone, it is not likely to disappear.

When an invitation to bid comes directly from an owner, and no design team is involved, the bidder must make the design decisions as well, i.e. it becomes a design and build project. This can occur when an owner requires a small improvement or extension, or even a small new building, but sees no advantage in engaging a professional design team.

The agenda outlined in Chapter 6 should still be followed. That is, the owner must be encouraged to provide both a project brief and a design brief. These will state in general terms what is required, e.g. the changes to be made to the existing building, or what the extension or new building must provide.

Someone employed by or instructed by the bidding organization must then decide on the scheme design, on suitable technical solutions and on those products that will provide the required attributes. The bidder or their consultants may also have to prepare drawings and obtain the approval of the Building Control and Planning authorities for the project. Once the owner has approved the proposal, the estimating procedure can begin.

18.6 SUMMARY

The viewpoint has been that of a general contracting organization preparing a bid based on the operative information on the Works. The procedure is outlined in Figure 18.4.

Works appreciation listed some general matters and identified significant aspects of the site and its locality. A general introduction led to a detailed consideration of the cost elements to be included in the **estimated prime cost** of the Works on site, and the procedure to be followed. This begins with making 'enquiries' to possible suppliers and subcontractors. The following is calculated for each activity:

1. the gross quantities and costs of main and dependent products, and;
2. their predicted durations, based on performance rates for processing the main products. The factors to be included in all-in labour rates were listed and the calculations illustrated. Those for employees of main contractors were contrasted with the totally inclusive rates for employees of more stable organizations, e.g. subcontractors, professionals.

The mix of single and period costs of the enabling processes is called the Works overhead. This is added to the estimated costs of the activities to give the total estimated prime cost.

Under **adjudication**, internal and external matters were discussed that would affect the feasibility, risks and financial success of the contract. These influence the decision by management on the amount the contract should contribute to the general running of the organization. Adding this 'mark-up' to the estimated prime cost converts it into a tender (bid).

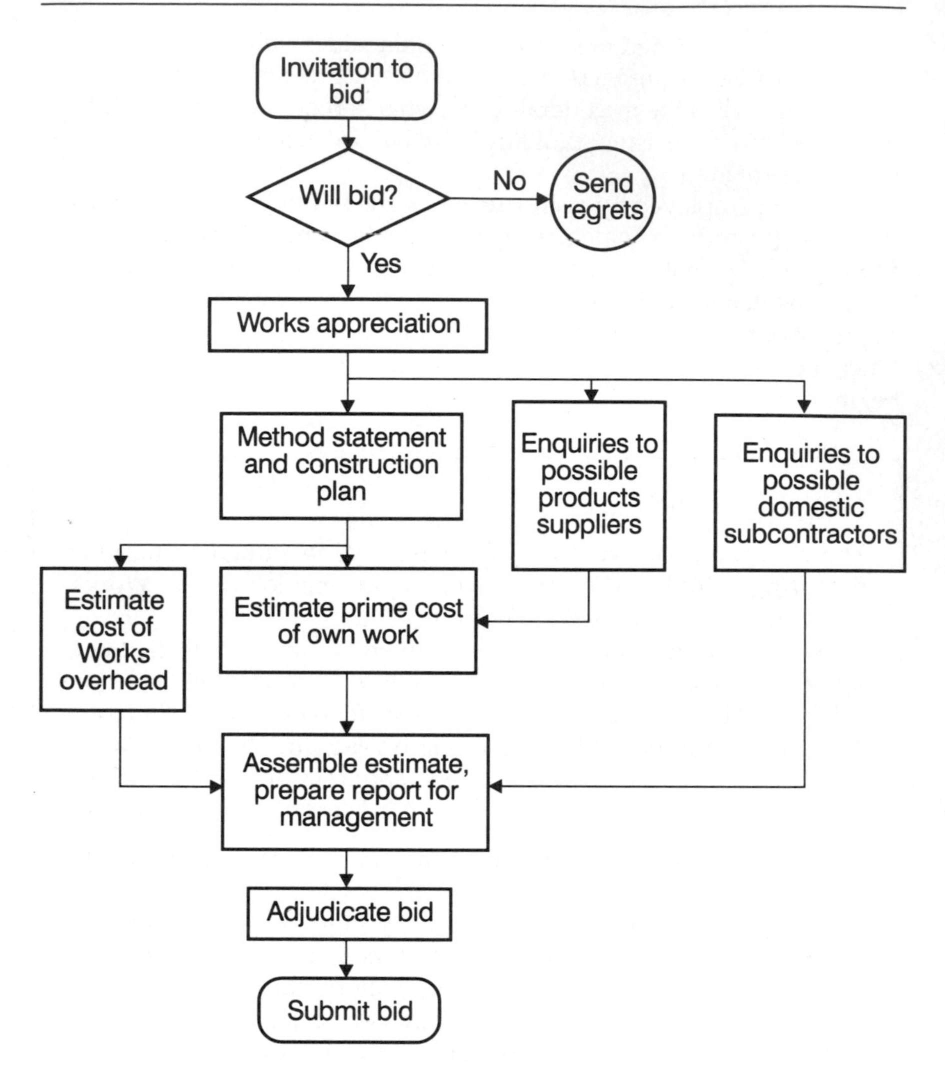

Fig. 18.4 A flow chart summarizing the bid preparation procedure.

REFERENCES

National Joint Council for the Building Industry (annually) *Working Rule Agreement*, 18 Mansfield Street, London W1M 9FG.

FURTHER READING

Chartered Institute of Building (1983) *Code of Estimating Practice*, CIOB, Englemere, Kings Ride, Ascot, Berkshire SL5 8BJ.
Illingworth, J. R. (1993) *Construction Methods and Planning*, E & FN Spon, London.

Pricing bills of quantities

19

19.1 UNIT PRICING

PRICES INSTEAD OF ESTIMATED COSTS

The procedure described in the last chapter for estimating the prime cost of an activity was based on predicting the quantities of the various **inputs** of materials, operatives' time etc. Each quantity was then multiplied by its estimated unit cost. Operatives' performance rates were based on the rate of processing the main product, while allowing for the amounts of dependent products built in at the same time. The total of these activity costs, the Works overhead costs, and a suitable contribution to the general overhead, became the bid price.

The alternative being considered in this chapter is outlined in Figure 19.1. This is to:

1. **itemize** and **measure** each of the different kinds of **outputs** of **in-place** or **finished** work constructed with a particular type of main product; putting this another way, it is to itemize and calculate the quantities of the parts of the completed building; we shall call these **main items**;
2. **itemize and measure additional items** for those dependent products, e.g. lintels, hip tiles, whose quantities do not vary with that of the main item;
3. **estimate** a money price per unit of quantity of each of the items. The income resulting from multiplying (1) and (2) by (3) must be enough to fund all the costs of achieving the Works. The prices of the additional items will be expected to include for some fixing time even though this cannot be separately identified during site processes.

Although all buildings are unique, many of their parts will be much the same. Thus, estimators in both general building and specialist enterprises find themselves dealing repeatedly with the same kinds of products and processes, although their amounts and costs will vary from job

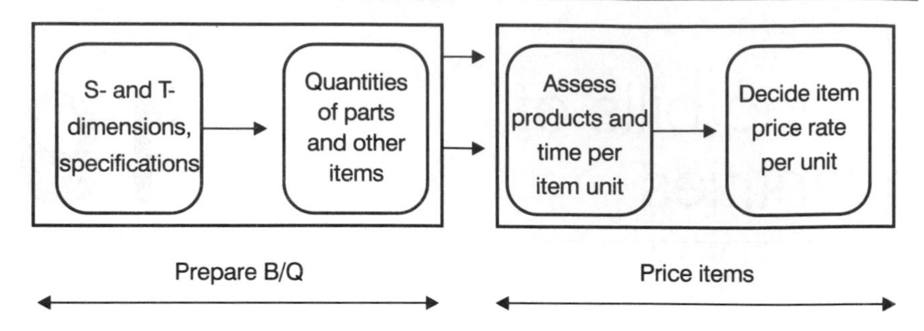

Fig. 19.1 The unit rate pricing procedure.

to job. In unit price estimating, this experience is quantified into money rates per unit of quantity of finished work and applied to the measured quantities of the job in hand.

Unit pricing is speedy when compared with activity estimating, but the rates can only be approximations of the realities of the activities. Also, the prices are often rounded. Unit pricing contributes no information on products quantities, activity durations, or the durations of Works overhead period costs, and there will be little or no project planning.

19.2 MEASUREMENT

QUANTIFYING DIMENSIONAL PARTS

We have been distinguishing between dimensional parts, e.g. walls, wall finishes, which are constructed to whatever size is required, and integral parts, e.g. doors, notice boards, which are made by fixing individual components. Integral parts can be counted and fully described. The same could be done with every dimensional part at every location, but then each would have to be priced separately.

Instead, dimensional parts that are the same except for their sizes are made into a single item by quantifying some or all of their dimensions and totalling the results. In considering which dimensions to quantify, we shall find that the quantification rule suggested in Chapter 13 is widely followed. The exceptions are identified later. The rule states:

- the quantity of a measured item should be calculated from its extensive S- dimensions, while the T-dimensions should remain in the description.

Figure 19.2 shows two rows of data on such a part. The dimensions columns are somewhat rearranged for clarity. The last row shows how this data can be made into a single item by partitioning it into:

1. the item quantity, which is calculated by multiplying the i and k dimensions and the number in each row and totalling the results, and

				Dimensions and other data on a part of an element			
i	*k*	*nr*	*j*	*Part name*	*Products*		*Extra processing*
6.40	3.20	2	0.14	Wall	140 mm conclite blocks	Cement, lime and sand mortar	Flush jointed one side
4.60	3.20	1	0.14	"	"	"	"
					Measured item		
Item quantity					*Item description* (see Figure 19.4)		
55.68	sq m	140 mm		Wall	140 mm conclite blocks	Cement, lime and sand mortar	Flush jointed one side

Fig. 19.2 The partition of data on an item and its aggregation into a single quantity and its description.

2. the *j* dimension and the rest of the item description, which is common to both rows of data.

Although this keeps the number of items to a minimum, certain information can no longer be carried by such **main item** descriptions. For example, in Figure 19.2,

1. the lengths of the individual quantified dimensions are lost; without these, the number of class B products cannot be calculated precisely (see Chapter 14);
2. information about the shape of the part is lost;
3. information on how much there is at any one location is lost;
4. the descriptor for the surface finish can be included because its quantity is the same as that of the main item.

Dependent products and those surfaces whose quantities do not vary with those of the main item (see Figure 19.3) have to be described in **additional items**.

19.3 BILLS OF QUANTITIES

STANDARD METHODS OF MEASUREMENT (SMMS)

An estimator who also measures and calculates the quantities will mentally decide what to measure and price, and what to allow for by increasing the unit prices. The process involves combining information from the drawings etc., with knowledge gained over the years about the industry. However, only the results of this thinking will be written down.

When measured data such as a bill of quantities (B/Q) is provided by others, but without the detailed drawings or even the specification, the estimator must be able to trust both the item descriptions and the

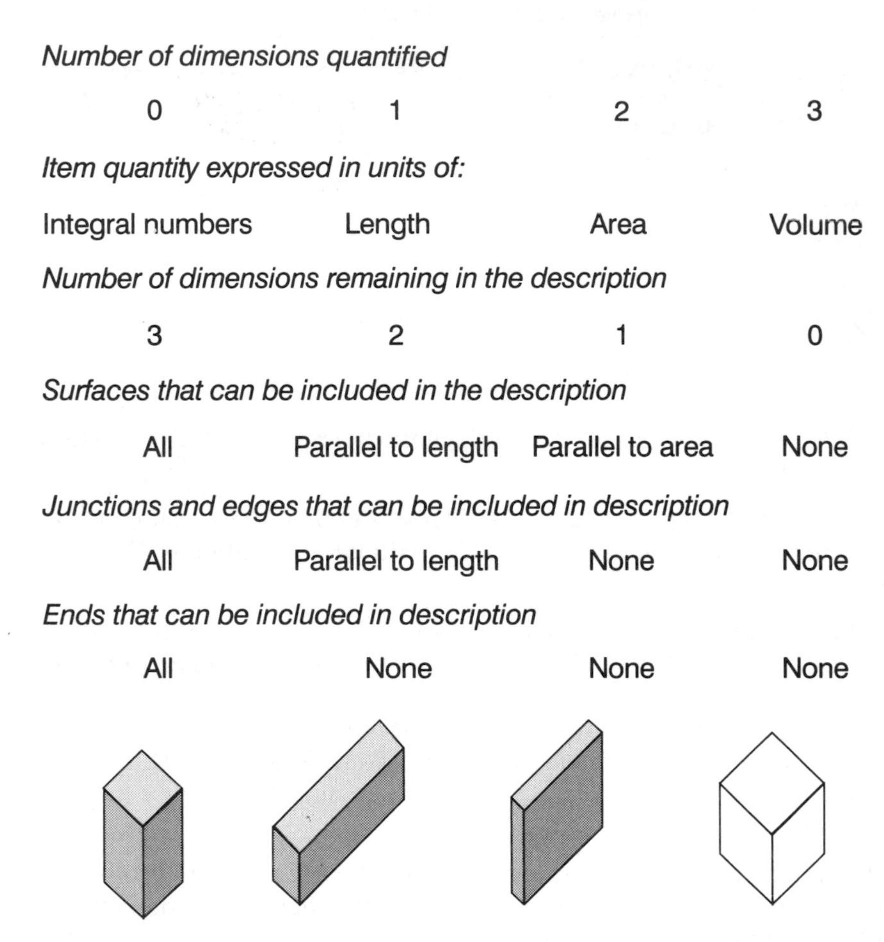

Fig. 19.3 The relations between the quantification of the dimensions of a part and the information that can be carried by an item description.

quantities. Standard Methods of Measurement are intended to provide such assurance.

Chapters 11 and 17 briefly reviewed the historical development of general contracting in the UK, including the evolution during the 1800s of bills of quantities for proposed Works. These were based on the already customary 'after-measurement' of 'in-place' work. This method of procuring both building and civil engineering works was exported to most parts of the then British Empire.

The codifying of measurement conventions into both national and international Standard Methods of Measurement (SMM) is now well established. They are important because they provide the 'rules of the game'. Their main object is to:

- identify the classes of construction entities that are to be separately itemized;
- state what information is to be given in their descriptions; and
- state what dimensions are to be used when calculating their quantities. Usually, these are not stated directly. Instead, they are implied by mentioning any T-dimensions, e.g. the thickness, that are to be included in the description and by giving the unit of quantity, e.g. m^2.

THE ARRANGEMENT WITHIN BILLS OF QUANTITIES

The first section in a B/Q is likely to be called the **Preliminaries**. This was the term used in the first (1922) SMM for what surveyors had been calling **general items**. In the 7th edition, the titles are merged into 'Preliminaries/General conditions', and this section is likely to contain the following:

- descriptions of the site, the Works, and the drawings and other operative information;
- either the full contract conditions; or
- the title and clause headings of the standard form of contract being used; also, any modifications to it, and the values that are being given to its variable features (contract period, time for honouring certificates etc.);
- the requirements of the owner regarding site matters;
- general items of Works overheads.

In addition, there will be various sums for inclusion in the bid to cover the costs of:

1. specific work not yet designed;
2. undefined work;
3. work by subcontractors nominated by the design team;
4. work by statutory undertakings;
5. products from nominated suppliers.

Sums for the costs of (1) and (2) are likely to be called **provisional sums**. The others will be called **prime cost,** or **PC sums**.

An item for the contribution to the general overhead is unlikely. Instead, it is intended that this contribution will be apportioned over the prices of the measured items. Such repricing is likely to be done only if the tender is successful, and a contract bill is being prepared.

B/Q in the UK are most likely to be arranged according to the Common Arrangement of Work Sections (CAWS) mentioned in Chapter

10. Thus, the work of each subcontractor should be identifiable. Where a separate specification is not being provided, the sections are likely to begin with some preamble clauses describing suitable products and related processes.

ITEM DESCRIPTIONS

The following descriptors must be included in a **main item** description if it is to communicate all it can about the qualities of the work.

1. the T-dimensions of solid parts or of the members of open-framed parts;
2. its shape, if not planar, e.g. circular on plan to 10 m radius;
3. the name of the part;
4. its inclination, if neither horizontal nor vertical;
5. the name and other descriptors relating to the main product(s), or a reference to where these can be found in the specification;
6. the descriptors of any dependent products that are in proportion to the item quantity, if not in the specification;
7. descriptors of any surface treatment carried out at the time which is in proportion to the item quantity.

Some of these will also be included in the descriptions of **additional items** for dependent products.

Three different ways of expressing this descriptive data are given in Figure 19.4. The individual descriptors in an item description should consist only of identifiers and references to related clauses in the specification, as in the second example in Figure 19.4.

(Ref)	(Description)	(Quantity)	(Rate)	(Total price)
d	140 mm thick walls built of conclite concrete blocks laid in cement, lime and sand mortar and flush jointed on one face as the work proceeds.	56 m²		
	OR			
d	140 mm conclite blockwork walls flush jointed on one face. Spec. F10/106	56 m²		
	OR			
	Conclite blockwork Spec. F10/106			
	Walls			
d	140 mm thick; flush jointed on one face	56 m²		

Fig. 19.4 Alternative arrangements of the descriptors in an item description.

Even so, descriptions can still be written as prose, as in the first example. Here, conjunctions and prepositions link the various descriptors into a single statement, although not necessarily into a complete sentence. It was normal practice until the late 1960s. The third example uses a standard phraseology. Within each work section, items may be classified under element, part or location headings.

STANDARD PHRASEOLOGIES

Individual items often have many descriptors in common. This has led to the decomposing of item descriptions into their individual descriptors, with those appearing in two or more items being extracted and treated as subsidiary headings, or classifiers. In such cases, only the descriptors that distinguish one item from another remain to be written against the item reference and quantity, as shown in the last example in Figure 19.4.

All the descriptors etc. likely to be given in descriptions can be listed in the form of a **standard phraseology**, and either used manually or put on a computer database. Writing a B/Q then becomes a matter of selecting classifiers appropriate for the various levels of a group of items, followed by item residual descriptors, sizes and quantities.

Of course, using a computer is a speedy way of producing this kind of B/Q and reducing costs to the professional team. But to some extent, the price is paid by estimators and others, who have to learn to read back up the page to find out what the items and the various headings are telling them.

THE QUANTIFICATION RULE AND ITS EXCEPTIONS IN A TYPICAL SMM

The majority of the measurement conventions in an SMM are likely to follow the Quantification Rule. However, work items made with concrete, i.e. (A1) products, present difficulties as their plastic nature enables them to be used for a variety of different shapes. For instance, we could visualize a concrete stepped strip foundation on a sloping site if its length were given as its quantity and its width and thickness remained in the description. However, foundations with different cross-sections would require separate items. Also, the additional concrete in the overlapping steps and in small projections for piers etc. must still be measured somehow.

Instead, the various sizes of foundations and other parts made of concrete are usually reduced to single items by quantifying their T- as well as their S-dimensions. Unfortunately, this **over-quantification** prevents estimators from visualizing the actual work, although the T-dimensions will be shown on the drawings. Some concrete work items may be given in thickness stages in an attempt to give some indication of their T-dimension(s).

In the following consideration of likely exceptions to the Rule, we shall follow the order of our main products classes.

- Parts made with **(A1) Concretes** and **(AB1) reinforced concretes** are likely to be expressed in units of volume. Such over-quantification means that the area of the **formwork** (actually, the area of the surfaces requiring support, the actual formwork being somewhat larger) has to be measured as an additional item. In any case, its manufacture, erection and dismantling are separate activities, and are carried out by different operatives.
- **(A2) Thick** and **(A3) Thin coatings** to narrow surfaces may be grouped into width stages and billed in units of length.
- **(A4) Fillings** All fillings, including those of constant thickness under ground slabs, may be measured as volumes. If so, the surface blinding to receive a ground slab must be measured as a separate item.

Most main items and additional items with class B products, including **(AB2) small structural units in mortar,** may be expected to follow the Quantification Rule.

Main items with class C components also follow the Rule, and are counted. In the UK, before 1963, all joinery articles were measured as areas or lengths as had been the case when, in the past, they were hand made by joiners who were paid for 'piecework'. Nowadays, door frames and linings may still be measured the old way, by giving the total length of each size and profile. The remainder are recognized as being 'things', and are counted.

(D) Soils and rocks Except for removing the layer of topsoil, the various excavations are usually given as volumes. This is probably the only way of dealing with the varying depths resulting from irregular natural and finished ground levels.

19.4 ESTIMATING AND TENDERING (BIDDING) PROCEDURE

The estimating and adjudication procedures generally will be much the same as that described in the previous chapter. However, the contribution to the general overhead will probably be apportioned to the various measured items. It is likely that quotations will be sought from suppliers and domestic subcontractors by sending copies of extracts from the B/Q.

UNIT PRICE RATES

Unit price rates may be calculated by dividing the costs incurred in producing similar work in the past by the quantity of that work. Alternatively, they can be built up using the amounts of product and time that can be apportioned to each measured unit of work (see Figure 19.5).

The average performance rate per measured unit of work can be predicted by using the estimating function described in Chapter 15. Choose a suitable performance rate from a range of possible rates such as those in Appendix B. If A is the amount of product per unit of measured work and P is the chosen performance rate with that product, then $A / P =$ the time to construct that unit of work D.

This is sometimes referred to as a **production standard**, or **constant**. As no performance rate is ever likely to be constantly maintained, this term has attracted some criticism. Even so, when calculating the estimate, the production standard will be the assumed arithmetical **constant of proportionality** between the measured quantity and the time and cost of the process.

It must be said, though, that many unit price rates are not based on any analysis of products and process costs at all. Instead, they will be what the estimator believes will be sufficient to cover the various costs. They may have been taken from, or be based on published 'price books', with or without modification to suit the circumstances of the Works. On the other hand, computer-aided estimating software may generate unit

Unit rate calculation for one sq m of 140 mm wall						
Cost element	*Conversion factors*		*Quantity*	*Unit*	*Rate #*	*Cost #*
Materials cost for 1 cu m of cement, lime and sand mortar (1–1–6)						
Cement	Proportion 1/6	0.17				
	Bulk density	1.44	0.24	tonne	100.00	24.00
Lime	Proportion 1/6	0.17				
	Bulk density	0.52	0.09	tonne	220.00	19.80
Sand	Bulking of damp sand	1.30	1.30	cu m	15.00	19.50
					Total	63.30
Price rate for one sq m of 140 mm wall of 450 × 225 mm conclite blocks						
Blocks	1/0.45/0.225	9.88				
	8% wastage	1.08	10.67	nr	1.20	12.80
Mortar	No. of blocks per cu m	1074.00				
	Wastage 6%	1.06				
	No. of blocks	9.88	0.01	cu m	63.30	0.63
2 + 1 gang	No. of blocks	9.88				
	Laying 16 per hour	0.06	0.59	hours	20.00	11.80
					Net unit rate	25.23

Fig. 19.5 Analysis of unit rate for the item illustrated in Figures 19.2 and 19.4.

prices from data on products quantities, gang times per unit and current cost rates, all of which can be readily modified and updated.

PRICING THE CONTRACTOR'S OWN MEASUREMENTS

Item descriptions in bills of quantities will prescribe main and other products. However, when preparing a bid for a contract based on drawings and a performance or descriptive specification, the bidding enterprise is left to decide what these products will be.

Also, someone will have to decide what items to measure and what to allow for in their price rates for the costs of unmeasured items. For instance, a specialist in the construction of buildings of a particular purpose group might begin by measuring the gross floor area and considering a suitable price per unit of floor area for the whole building enclosure. In such cases, the unit price would be based on analyses of the actual costs of previous contracts.

Another approach to predicting the cost of a building envelope is to:

1. multiply the area on plan by the sum of the rates per m^2 for the parts of all the **horizontal elements**; their total cost is then added to
2. the costs of the parts of **vertical elements**, i.e. internal and external walls, calculated by building up inclusive rates per linear metre and applying these to their girths. This is shown on Figure 19.6. Parts with vertical S-dimensions, such as walls and their finishes, will have areas per linear metre equivalent to their heights, as these heights are being multiplied by 1.00. Similarly, the amount of the volume of the concrete foundation will be numerically the same as the amount of its sectional area. Skirtings, copings, damp-proof courses, etc. will have quantities of one linear metre. The vertical section is, in effect, rotated round the building space, as indicated in Figure 19.6.

Any construction with a constant cross-section, e.g. drainage and other pipelines, can be priced by multiplying its length or girth by a price per lin m for the parts shown on its cross-section. Even so, this will still leave many parts to be priced separately.

Some estimators are uncomfortable with having to take off measurements and make decisions about products before they can get on with their proper job. In such cases, the bidder's quantity surveyor may be asked to prepare a bill of quantities for them to price, using the measurement conventions of the current Standard Method of Measurement.

In 1967, Ferry reported that, in the UK, 'The SMM appeared to dominate the whole field of measurement in many of the firms, even when it was not mandatory. . . . The principal reason for this . . . seemed to be that the firm's surveyors are trained in its use . . . and would find it difficult to change.'

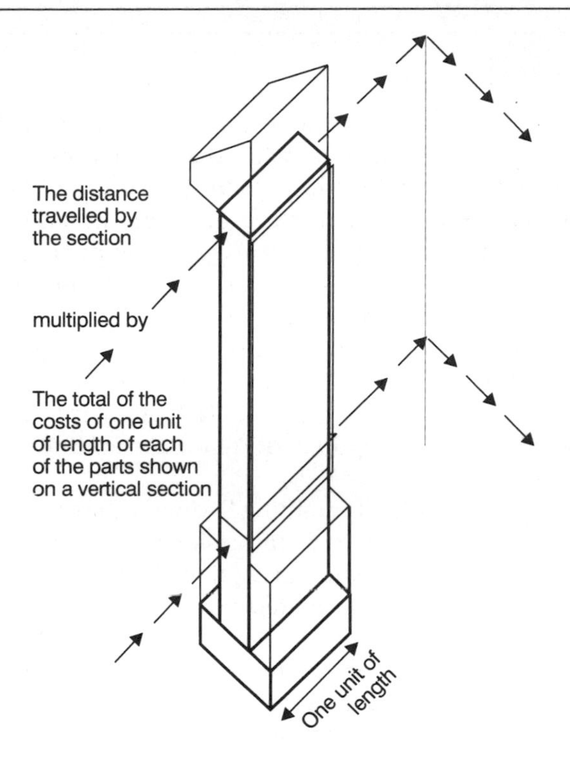

Fig. 19.6 Estimating the cost of the external walls element by multiplying the girth of its rotating section by the sum of the costs of one linear metre of its parts as shown on a vertical section.

19.5 COMMENTARY

The building contract will normally allow the owner to retain a small proportion of each interim payment in case, for example, the contractor becomes unwilling to make good defective work. This retention may be offset during the adjudication process by increasing the value of the earlier worksections, and making equivalent reductions in later ones. This is called **front end loading.**

Having one B/Q produced by the design team for all bidders to price is an economy for the competing enterprises. It also means that bids are prepared on the same basis, and so can be directly compared. As bill item descriptions are prescriptive, the design can be assumed to be complete, and the estimating process is straightforward. Having a priced B/Q as a contract document enables variations in the design of the Works to be readily measured and priced. It also provides cost data that the design team can apply to other projects. A B/Q has meaning for the owners, too, as it itemizes and describes what they may expect to receive.

19.6 SUMMARY

The detailed measurement and costing of inputs described in the previous chapter is contrasted with the alternative procedure of measuring and pricing the outputs from construction processes. The Works appreciation, estimating and adjudication procedures are much the same, although the process of unit pricing is quicker.

Estimators who measure for their own purposes will instinctively allow for all the implications of the measured items when deciding prices. However, when pricing a bill of quantities prepared by others, they need to have confidence that they understand the implications of the data. Standard Methods of Measurement for preparing bills of quantities are intended to give that assurance.

A Quantification Rule of retaining T-dimensions in item descriptions and quantifying their S-dimensions was suggested. Most items in a B/Q follow this Rule. Exceptions include items with class A products, and those of excavated voids, which are measured as volumes.

The inevitable reduction in information on a part when its dimensions are quantified was described. Those descriptors likely to be found in an item description were listed, and possible arrangements were illustrated, including an example of the application of a standard phraseology.

The advantages of bidding on a B/Q provided by the design team and the consequent uses of the priced B/Q as a contract document were outlined.

REFERENCES

Ferry, D. J. O. and Holes, L. G. (1967) *Rationalisation of Measurement*, Royal Institution of Chartered Surveyors, London.

FURTHER READING

Chartered Institute of Building (1983) *Code of Estimating Practice*, CIOB, Englemere, Kings Ride, Ascot, Berkshire SL5 8BJ.

Managing cost and value 20

Value is that amount of some commodity, medium of exchange etc. which is considered to be an equivalent for something else; the material or monetary **worth** of a thing (OED); what someone is prepared to pay for something.

Price is the amount of money for which anything is bought and sold; an estimate of the value of something.

Cost is what a purchaser paid for something.

20.1 MANAGING RISK

Any situation that is influenced by a large number of factors, and any artefact that consists of a large number of parts, can be described as being **complex**. This will apply to the commercial environment in which the construction industry operates as well as to the buildings themselves. While at any one time, we might think we know a great deal about our commercial environment, the various factors that should be taken into account when deciding to proceed with a project are changeable by nature.

The **risk** associated with a business (or any other) decision is the degree of probability that assumptions made regarding these factors will turn out to have been wrong, that expectations will not be realized, and that the outcome will adversely affect the organization.

FACTS AND PREDICTIONS

The quantity of some part of a building, the load-bearing capacity of a beam, and other information calculated using appropriate formulas and factors, can be no more accurate than the data from which they are derived.

Even more uncertain will be predictions based on assumptions about human behaviour and the weather. Examples include the performance rates of operatives, the wastage of materials, the availability, skills and reliability of individuals, and the achievement of quality, time and cost targets.

The cost determinants on which a bid for a building contract is based may be numbered in thousands. These determinants will also vary in proportion from one building to another and with the passage of time and the state of the economy. Thus, cost predictions based on the historical costs of similar buildings will also be uncertain, particularly when dealing with an incomplete design and its likely cost a year or more into the future.

Future movements in bank interest rates and international exchange rates are highly uncertain. These will reflect the complexity of world markets, the enormous number of commercial transactions which take place daily, decisions by governments on taxation and public expenditure, and the hopes and fears of millions of individuals.

UNCERTAINTIES

Thus, while information on such factors might be accurate at the time, their values are likely to change during the years that it usually takes to design and construct a building, and there will be **uncertainty** regarding their accuracy, both initially and later. Although the design process is intended to remove most of the uncertainty over what the building work will be, some will remain, particularly when dealing with an existing building.

Check lists of information to be gathered and things to do in various circumstances have been given in a number of chapters. Obviously, such lists cannot be comprehensive and, unless they are extended to match the project, the information gathered as a result is not likely to be either. Generally, matters that have been overlooked cause the most trouble, and none of us is ever free from such imperception. The information on which a business decision is based will be subject to at least three kinds of uncertainty:

1. all the relevant questions are not likely to have been asked and answered;
2. the information gathered is not relevant or not reliable, e.g. assumptions about soil conditions, promises by third parties, the financial status and reliability of the owner, the availability of the site;
3. the information is correct for the present, but will be different in the future, e.g. interest rates, market prices, weather conditions, available staff.

Uncertainties can be reduced by:

- not hurrying when making decisions;
- having information confirmed;
- negotiating changes in the circumstances;
- avoiding reliance on individuals from other organizations;
- timing, e.g. avoiding site preparation during the winter; staggering Works start dates;
- levelling demands on the resources of the organization, e.g. not undertaking contracts that are excessively larger or smaller than normal.

REDUCING THE RISKS

A construction organization is always at risk when undertaking work controlled by others:

1. the cost and time targets laid down may be unrealistic;
2. the organization may lack suitable financial and technical resources;
3. the technical and managerial staff of the organization may not be competent;
4. other organizations involved may not be reliable, e.g. specialist subcontractors, products suppliers, professionals, public utilities;
5. design information and other instructions may not be provided on time, may be incomplete or incorrect, or may be changed.

Risks can be modified by:

- suitable compensation for accepting them;
- avoiding them, e.g. by only undertaking commitments well within the competence of the organization;
- transferring them to others, e.g. by subcontracting, by insuring, by taking and acting on independent professional advice;
- sharing them with others, e.g. by having a fixed rate mortgage, the mortgagee accepting the risk that the rate might go up, and the mortgagor that it might go down;
- initiating formal internal policies, procedures and controls (see below);
- being security minded.

Making a business decision implies having an objective and choosing between alternative actions. The circumstances of each alternative action and the factors that might influence its outcome can only be discovered by hard thinking. The implementation of a decision must be monitored vigilantly. Readers who have not yet experienced what is sometimes called **Murphy's** or **Sod's law** will do so soon enough. This promises that **if something can go wrong, it will**.

INTERNAL AUDIT

In addition to the risks associated with undertaking construction work there are also the normal business risks of loss or damage due to fraud, fire, theft, incompetence, vandalism and social unrest. The purpose of an internal audit service is to provide the organization with an independent review of the adequacy and effectiveness of its procedures and internal controls. This is particularly important where the organization is decentralized. Matters reviewed should include:

- compliance with approved policies and procedures and the law;
- the wise use of human and financial resources;
- the competence of individuals;
- the exposure to fraud;
- the suitability of information available to management;
- the preservation of management and commercial data, e.g. computer equipment and records, and their duplication and dispersal;
- the security of physical assets;
- insuring of liabilities as employers, etc, and their liability limits;
- money management.

20.2 WORK-VIEWS

THE SOCIETAL VIEW

The local planning authority will, on behalf of society in general, seek to ensure that development takes place only where the surrounding area is ready to receive it. This will be where water, electricity and gas supplies, roads, sewers, telephone lines etc. are already in place and, in the case of a new housing development, where public facilities such as schools, public transport, shops and garages are available for residents.

It is the creation of such an **infrastructure** around a 'green field' site that transforms it into a potential building site. The increase in value of such a site compared with its previous value as, say, grassland, is called betterment, although the amount will depend on supply and demand. (There is a saying that 'the three factors which determine value are: (1) location, (2) location, and (3) location', or, in North America, (1) position, (2) position, and (3) position.) Although individual landowners will profit from expenditure by others on the infrastructure, they will, in the UK, pay Capital Gains tax on any increased value when such land is sold.

Once completed, a building will make continuing demands on the community for energy, water, waste treatment and disposal, and for care and maintenance. In return, it will benefit those who live, work, learn, shop, worship or play in it, and who pay taxes on it. A project can also bring disbenefits for those in the neighbourhood, e.g. noise, loss of privacy, traffic congestion, air pollution.

In the UK, the policies of planning authorities regarding individual development proposals, and what might and might not be permitted, will be expressed in local plans. Policies concerned with strategic changes in the use of larger areas of land for some social purpose will be stated in the structure plan. This plan will indicate to highways authorities and public utility undertakings where to extend the infrastructure, and will identify possible locations for development. A structure plan will also be concerned with conserving and improving the environment and with mineral working, waste disposal and land reclamation.

THE VIEWS OF INDIVIDUALS AND ORGANIZATIONS

Those who work as members of the design team, or in the building materials industry, or on construction sites, will be concerned primarily with 'doing their job' and earning their livings. Not many of them will be interested in the qualities of the completed building. Similarly, no single organization involved in a new project will be concerned with the whole of its costs and benefits to society. Instead, they will all have what might be called **limited work-views**. That is, they will operate in the interest of one individual or organization for a limited time.

A developer may be concerned simply with maximizing the margin between the total expenditure on a project and its selling price when completed. An owner who intends to occupy a project when completed may only be interested in its initial costs, although a more thoughtful owner-occupier will also be concerned with the continuing costs of occupation for the whole of its life.

Where some or all the development funding is borrowed, the lender may place certain constraints on the procedures to be followed and the type of development. Thus, the professional team should adopt a work-view that embraces those of the owner and their advisers.

Those involved in the general management of a construction organization will strive to keep it busy and to maximize the contributions from its various contracts to the general overhead costs. Site and contracts managers will be concerned with keeping construction costs within the estimated prime costs. Subcontractors tend to have very limited work-views that may or may not include meeting the general contractor's completion dates.

An owner who is promoting the development of a site will wish to ensure that the value of what is built is greater than its cost. In the next but one section on project appraisal, our work-view will be that of the potential owner-occupier of a proposed project. But first, we must consider the costs of owning and occupying buildings. Not only are these important in themselves; they will also influence the market prices of buildings, and help to set prudent cost limits for new building projects.

20.3 THE COSTS AND VALUE OF A BUILDING

In this final chapter, we return to the first sentence in the book, 'Buildings are built for people to use.' Where a property is available for lease or purchase, every potential occupier will have their own opinion of the benefits to be gained from being able to use the building. Although we shall continue to distinguish between the owner and the occupier of a property, they are quite likely to be the same individual or organization. The chief benefit of owning the building you occupy is the freedom and security it gives.

Once a building or other asset has been acquired, its cost is a historical fact. But when selling that asset, 'bygones are bygones', and the price the present owner paid for it will be irrelevant. Instead, its selling price will reflect its value to the new buyer, who will be able to compare its asking price with those of similar new and old buildings on the market at the time, and to negotiate accordingly.

BUILDING OCCUPATION

An occupier is primarily interested in the quality and usefulness of the accommodation provided for people, goods, vehicles and plant. These were considered in Chapter 1. There may be disbenefits, e.g. 'I can't get the BMW into the garage.' Some features may completely disqualify the property, e.g. loading bays unsuited to modern transport vehicles.

Other aspects will relate to the location. These include:

- whether town, suburb or country;
- distance from neighbours;
- access to the road and railway networks;
- local public transport;
- local facilities, e.g. schools, shops, churches, entertainment, sports centres, ski resorts;
- local employment or trading opportunities;
- availability of suitable labour.

Expenditures arising from the use of the property (sometimes called **costs-in-use**) will be either occasional or continual, although payments for those will usually be made periodically. They will include the costs of the following.

Occasional

- buying, maintaining, and renewing furnishings and equipment;
- periodic maintenance of building services, e.g. boilers, cookers.

Continual

- if the property is not owned, the rent to the owner;
- space heating and cooling, hot water and other domestic services; obviously, an old building with a poor standard of insulation is going to cost more to heat than a modern one of equivalent floor area; other pointers to high heating costs will be a relatively high external walls girth index (see Chapter 8) and large windows;
- electric lighting and power;
- water and water treatment charges;
- caretaking, window and other cleaning, gardening;
- security;
- central and local government taxes;
- insurance of furnishings and equipment against damage or loss due to fire, theft etc.;
- insurance against claims by visitors to the buildings, i.e. 'third party claims'.

BUILDING OWNERSHIP

Buildings must be cared for if their values as investments are to be maintained. The costs-in-use associated with owning a property include the following.

Occasional

- Renewing applied decorations.
- Renewing worn mechanisms, e.g. door locks and handles, electric light switches, boilers, lifts.
- Renewing parts affected by the climate, e.g. rotting woodwork, degraded plastic gutters, pointing to exposed brickwork, decayed flat roof coverings.
- Renewing worn parts, e.g. floor coverings.
- Overcoming errors in detail design, e.g. sound transmission, condensation on thermal bridges, cracking, settlement.
- Periodic refurbishment. As time passes, public expectations regarding building standards tend to increase, and life styles change. Also, some buildings have to be adapted to meet changes in manufacturing methods and increasing technology. Modifications may also be required because of changes in the law, e.g. fire and smoke alarms in houses let to more than one tenant, i.e. in **multiple occupation**. Thus, it is prudent to expect to refurbish a property from time to time during its life. Some so-called **loose fit** buildings are especially designed to be adaptable to changes in use.
- The organization and administration of the above.

Continual

- Interest charges on the capital invested, whether these funds are borrowed from someone else, or belong to the owner. Investing your own funds means you have lost the opportunity of obtaining a return by investing them elsewhere.
- Replacing the capital invested in the building by accumulating a sinking fund over its expected life. The building may have been bought with funds provided under a mortgage agreement with, say, a bank or a building society (savings and loans). Both interest and capital repayments may be combined into periodic payments (see formula (16.c) in Chapter 16).
- Service contracts, e.g. lift maintenance.
- Insurance to cover reinstatement costs resulting from a fire, flood, tempest, subsidence, public disorder etc. and the consequential loss of income or benefits.
- Insurance to cover liabilities arising from having a duty of care to visitors and other members of the public.
- The organization and administration of the above.

The benefit (if such it is) of leasing your property to someone else is the regular income it provides and the possibility that rents and the market value of the property might increase at least at the current rate of inflation. At one time, the security offered by this kind of investment was summed up in the phrase 'as safe as houses'. More recently, though, economic activity has stopped expanding and has even declined. In consequence, some rents and the equivalent selling prices have been declining also.

When a property you own is let to one or more tenants, the following costs may be incurred in addition to the normal costs of ownership:

- having to manage the lettings, including advertising vacancies, selecting and dealing with tenants, and rent collection;
- late payment of rent;
- unpaid rents and taxes, i.e. bad debts;
- difficulties and costs of debt collection;
- legal and other costs of regaining possession;
- irrecoverable costs of cleaning and/or repairing damage after a tenant has left;
- income lost while the property is untenanted, i.e. losses from avoidable **voids**;
- income lost when the tenants are moved elsewhere for the duration of maintenance works, i.e. losses from **unavoidable voids**;
- (where more than one tenant) the costs of services to common parts, e.g. caretaking, cleaning, heating, lighting, security.

CHOOSING BETWEEN ALTERNATIVES

Intending owners or occupiers will wish to provide themselves with a selection of alternatives (see Chapter 6). For example, they may seek to choose between:

- alternative buildings or parts of buildings available on lease;
- purchasing and adapting an existing building;
- designing and constructing a new one.

Although the final choice is likely to be made subjectively, and be based on personal judgement, the economic implications of the alternatives can be considered more objectively. We have already identified many kinds of ownership and occupancy costs, but their amounts and their timing will be different for different properties. How then, can the different options be compared?

One approach is to calculate what sum invested at compound interest at the start of a new project or occupancy would meet all its demands for cash throughout its life. In other words, we could calculate the total present value (PV) of the various single and periodic future expenditures.

These should be costed at current values. The rate of interest should not include an inflation element, and could be 3–4% per annum. This approach can be called a **single cost appraisal** and uses the equivalence formulas described in Chapter 16. Formula (16.1/a) will discount future occasional expenditures and (16.1/c) will calculate the present value of continual expenditures if they are treated as series of future payments. The total costs of the alternatives can then be compared.

20.4 PROJECT FINANCIAL APPRAISAL

The financial appraisal of a project can take the form of a balance sheet similar to those in Figures 20.1 and 20.2. Costs are shown on the left, and their total must be less than the cost limit based on the value put on the project by the owner. The constituents of each set of costs will be considered next.

DEVELOPMENT COSTS

1. **Site acquisition** If the site is owned already, its value will be either its cost or its likely selling price. But if it has yet to be acquired, the asking price should not be assumed to be affordable. There may be little connection between what others might appear willing to pay and what it is worth to you. Rather, its **residual value** as a site for the project should be tested by deducting all the other likely costs from the cost limit as shown in Figure 20.2. Any higher price can be afforded only if the other costs can be reduced.

Costs	Value
The development costs include those of: 1. site acquisition, 2. readying the site, 3. financing the project, 4. fees and administration, 5. Works contracts costs and taxes, 6. furnishings and equipment, 7. initial letting.	**The cost limit for the project** is based on one of the following: ● the funds available, ● the approved budget, or ● the estimated market value of the completed project.

Fig. 20.1 Sections of a project appraisal balance sheet.

From this residual value, legal charges, taxes, finance charges and professional fees for valuation advice must be deducted, and the remainder compared with the asking price to see if it can be afforded. This approach can be used when deciding what limit to adopt when bidding at an auction. Of course, the more financially efficient a development is, the greater the site value to its developer. Even so, one would still seek to negotiate as low a purchase price as possible.

2. **Readying the site** Some of the following will be necessary before a site will be ready to receive new Works, no matter what these might be:

- demolishing and clearing existing buildings and sealing off the service connections;
- clearing the site of rubbish, undergrowth, trees etc.;
- pollution treatment or removal, and pollution insurance;
- re-routing sewers and electrical transmission and telephone lines;
- re-routing public footpaths;
- installing fencing and gates;
- extending public utilities to the site;
- constructing access roads.

These extend the matters discussed in Chapter 6.

3. **Financing the project** We shall assume that the owner will have to borrow the finance for:

1. carrying out the project, funds being drawn as and when required and, at the end,
2. investing in the completed project and repaying (1).

Until the Works have been completed, the building cannot be occupied and earn rent, and so offers little security to an investor.

Project balance sheet		
Cost		*Value*
1. Site acquisition	??????	Cost limit 500000
2. Readying the site	15000	
3. Financing the project	25000	
4. Fees and administration	40000	
5. Works costs and taxes	320000	
6. Furnishings etc.	25000	
7. Initial letting	15000	
Subtotal	440000	
1. Site acquisition	60000	
	500000	500000

Fig. 20.2 The residual approach to valuing the site.

Consequently, providing loans for construction is a high-risk specialist business activity, and their relatively high interest rates reflect these risks. The rate can be less if the owner has other properties that can be used as security. Also, the providers of long-term finance may be willing to fund the construction as well.

The lender will charge for negotiating and arranging the loan, and will charge interest on the amounts borrowed for interim payments etc. These charges will be included in the project balance sheet, and may be capitalized by being added to the amount of the loan.

After completion, the project is likely to be refinanced using a long-term loan from, say, a pension fund or insurance company seeking a secure investment and charging a lower rate of interest. This loan will be used to repay the moneys borrowed to finance the construction, and to pay the interest charges. It will be secured by a mortgage agreement with periodic repayments of capital and interest.

4. **Fees and administration** Besides the fees and charges of the professional team and the site inspectors, costs may be incurred for surveying the site and producing site plans, and in soil investigations. Fees will also be charged by whatever public bodies are responsible for Building Control and for giving planning approval.

The owner will also incur costs, e.g. those of the commissioning team and their advisers, when preparing the project brief (see Chapter 6).

5. **Variable development costs** These will be mainly for the Works contracts and the amounts of VAT or other taxes payable on them. Obviously, they will depend on the design, and are considered in some detail in the next section.

6. **Furnishings and equipment** Some of these might be included in the building costs. Otherwise, the owner's estimated cost for these items should be included. If the building is to be wholly or partially let to tenants, their leases may stipulate that the owner will finish and furnish the spaces to the tenants' requirements.

7. **Initial letting** The costs of advertising for and negotiating with prospective tenants for, say, a new block of shops and offices, and any financial incentives such as an initial reduction in the rent, must all be included in the project cost. So, too, must possible rent losses due to delays perhaps extending to years before finding suitable tenants.

THE COST LIMIT

The professional team must say clearly if funds are insufficient as, once they have accepted a cost limit, any overrun will cause trouble both for themselves and for the owner.

The construction of Blenheim Palace, near Oxford, illustrates what can go wrong when funding and lines of responsibility are unclear. It was to have been a gift from Queen Anne to her victorious general, the first Duke of Marlborough. An inscription states that 'this house was built ... by Sir J. Vanbrugh'. He had been appointed surveyor in 1705, and had both designed it and organized its construction. However, the supply of cash ceased when the Queen fell out with the Duchess in 1710 and much grief followed.

Work stopped, and tradesmen, materials suppliers and the surveyor went unpaid. Although, after Queen Anne died, the now ailing Duke proceeded with the work at his own expense, he refused to pay for earlier work. Grinling Gibbons and others were unwilling to reduce their 'customary charges' from Crown rates to lesser ones more appropriate for a duke, and abandoned their work. Later, the Duchess fell out with Vanbrugh, and took over and completed the project herself.

The cost limit may be either:

1. pre-set by the owner; sometimes, the funding will have been raised or promised already, e.g. for a Scout hut; a corporate owner-occupier is likely to include funds for its building programme in its annual capital budget.
2. built up by the professional team from the balance sheet cost elements shown in Figure 20.1; this is considered in the next section;
3. set by reference to the financial return that can be achieved from the investment. The capital equivalent of this return is sometimes described as the **gross development value**, or GDV. The following **residual** approach will apply whether the property will be occupied by the developer-owner or by a subsequent purchaser. It is illustrated in Figure 20.3.

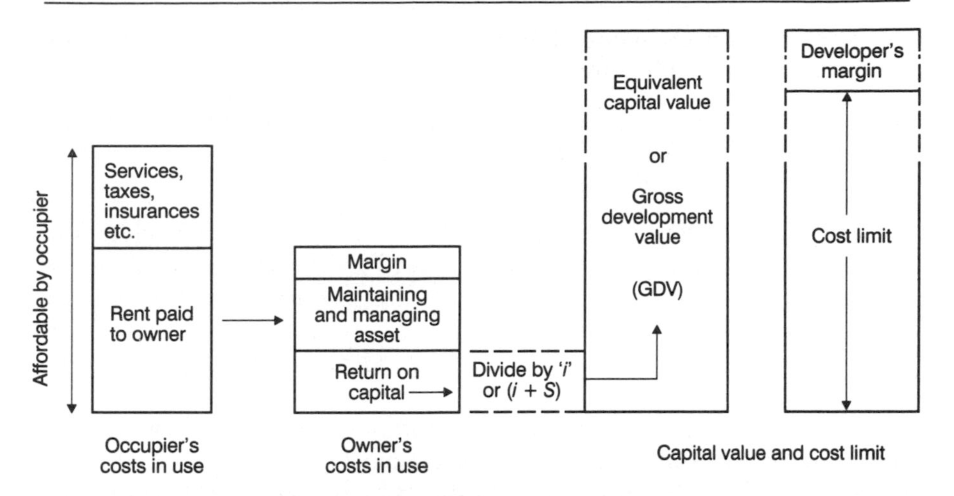

Fig. 20.3 A residual approach to the relations between the expected rent and other occupancy costs of a building, the owner's costs-in-use and management costs, its capital value, and the project cost limit.

A RESIDUAL VALUATION OF THE COST LIMIT

The market value of a property will depend on the benefits that flow from occupation, and the amount an occupier is prepared to value them at, and pay as rent. An estimate of what the rent might be for a project as yet unbuilt will be based on the rents now being paid for similar buildings, on predictions of what these might be in the future, and on an assessment of the risks.

However, an occupier will have many other expenses, such as heating and cleaning, as well as the rent. These were considered earlier. It might be better to start with what an intending occupier might regard as a suitable and affordable amount to pay in total for the accommodation. The amount left after deducting all the other expenses from this total will be the maximum available for the rent.

The owner will have expenses to meet out of the rental income, e.g. maintenance costs and insurance. These were also considered earlier. There ought also to be an owner's margin as payment for accepting the risks of investing in an asset of uncertain value. The residual amount will be available to pay interest and to repay the capital invested, period by period. This capital sum, the **gross development value** (GDV), can be calculated by dividing the residual portion of the rent by one of the following:

1. the mortgage repayment function $(i + S)$, i.e. formula (16.1/c) in Chapter 16. This assumes a known life expectancy, but table (c) in Appendix C shows that, the more the periods, the closer this gets to just i. Also, the higher the rate of interest, the earlier this happens.

Thus, for projects whose life expectancy is long but indefinite, the alternative is to divide by

2. a suitable interest rate *i*. The reciprocal of this rate of interest is sometimes called the **year's purchase in perpetuity**. It dates back to when it was easier to multiply than to divide.

The residual difference between the GDV and the margin which the project should contribute to the developer's overhead costs will be the cost limit for the project. Alternatively, if the cost limit has already been set by either (1) or (2) above, the difference between it and the GDV will be the developer's margin. Its amount will indicate if the project is worth doing.

Where a developer is in business to promote building projects for sale, the contributions from successful projects must, in addition, pay for the costs incurred on the unsuccessful ones. The residual surplus between the year by year sum of these contributions and the developer's total overhead costs will be the developer's reward for initiative and taking risks.

Clearly, the more the cost of the Works to the owner, the less the residual development margin. But as the Works cost will be many times this margin, a small increase in the former will reduce the latter by a relatively larger proportion, and vice versa. For instance, if the desired margin is 20% of the GDV, an increase of 15% on the Works cost (80% * 1.15 = 92%) will more than halve the margin to 8%.

Thus, whatever the method used for setting the cost limit, the professional team must strive to keep within it.

20.5 PLANNING AND CONTROLLING THE PRICE OF THE WORKS

COST DATA

The quantities and prices of the items in contract bills of quantities for past jobs provide a professional team in the UK with an important source of cost information. This information may be summarized and made available to others through the Building Cost Information Service or by being published in a building periodical. Although the approach relies on the pooling of data from many projects, some private owners may, understandably, be unwilling to contribute. However, the data will be based on accepted bids, and their final costs may be quite different. For instance, the bid may be subject to materials and labour price fluctuations.

Even so, each B/Q will relate to a particular project, location and time, and its prices must be suitably modified before they can be applied to a different project at a different location at some future time. For example:

1. **Changes in price levels** These can be expressed as a series of quarterly indexes calculated from either changes in wage rates and materials prices, i.e. in input costs, or changes in bid prices. They will be found in Table 2.1, the Construction cost and price indices in Part 2 of the quarterly *Housing and Construction Statistics*, which should be in all UK public reference libraries. The professional team should state the dates of its price predictions.
2. **The effects of the morphology of the building** The effects of size and shape on the relative quantities of the building elements were considered in Chapter 8. Obviously, these can only be considered when there is a design to measure. Until that time, any modification to single rates used for estimating can only be intuitive.
3. **Variations in the qualities of buildings** Differences in qualities are difficult to quantify, especially where an element has more than one technical solution.
4. **Main contractor's pricing policy** regarding 'front end loading', the apportionment of Works overhead costs and the basis of calculating the contribution to the general overhead. These factors and their uncertainty were considered in Chapters 18 and 19.

COST PLANNING UNITS

Cost planning is a procedure for setting cost and spatial standards before starting to design and for maintaining control of costs during that process. This contrasts with the measurement of design efficiency indexes during or at the end of the outline proposals or during the scheme design stage (see Chapter 8).

When analysing the historical costs of a building, the total cost of either an element or the whole building is divided by its quantity to give its unit price rate. The units used in cost planning will depend on the stage reached in the design process and the quantitative design data available.

1. The **functional unit rate** was introduced in Chapter 6. This is the result of dividing the total cost of a building by the number of units of accommodation it provides, e.g. number of two-person flats, lettable office floor area, school places.
2. An **element unit rate** R_e is the price of an enclosing element divided by its **element unit quantity** in m^2. The professionals in each country will have their own rules for obtaining these quantities.
3. The **element rate per m^2 of gross floor area** R_f is the price of an element divided by the sum of the horizontal areas of all enclosed floors. This is measured within the internal faces of the external walls and over stairwells, internal walls etc. This rate enables the element costs on different jobs to be compared.

4. The heating and cooling element cost can be analysed by dividing it by the steady state energy demand to give a **cost per kW**.
5. The **rate per m²** of gross floor area is the price of the whole building divided by the gross floor area measured as above, and is widely used when making cost comparisons.

However, it is advisable to consider ranges of unit price rates, not single ones, as the published rates for a particular class of building can deviate widely from their mean values.

THE VARIABLES IN THE COST EQUATION

Leaving aside the external works, we are always concerned with three aspects of a proposed building:

1. its size (or quantity) Q. This might be either its useful or its gross floor area or a number of units of accommodation;
2. its specification (or quality), expressed as a price rate per unit P; and
3. its total cost C.

In each situation, only two of these will be known, and can be regarded as the independent variables in the **cost equation**. The third will always be the unknown dependent resultant. For instance:

1. where the amount of the Works Q and desired quality price rate P are known, its cost is calculated by $Q * P = C$, (20.1)
2. where the cost limit for the Works C and its maximum size Q are known, the affordable maximum quality price is calculated by $C / Q = P$, (20.2)
3. where the cost limit C and a suitable quality price rate P have been established, its maximum size is $C / P = Q$. (20.3)

If the owner's requirements for accommodation and for the maximum expenditure are incompatible, then the design team should indicate in their report the effects of reducing the one or increasing the other. Once realistic proposals have been agreed, the professional team can move to the next stage.

20.6 COST PLANNING STAGES

STAGE A INCEPTION

Information will be sought on:
1. expenditures on the acquisition of the site;
2. the fees to be charged by members of the professional team;
3. the GDV or the cost limit being placed on the project by the owner;
4. the number of functional units of accommodation to be provided.

STAGE B FEASIBILITY

At this stage, the only quantitative data is that contained in the owner's design brief. This must include the **quantity of functional units** required by the owner, e.g. the number of motel rooms, school places, or other units of accommodation, or the useful floor area. Either cost equation (20.1) or (20.2) will apply.

(20.1) The cost limit for the Works will be the product of either:

- (a) the functional quantity and a rate selected from a band of functional unit rates, or
- (b) the required useful floor area, the likely design efficiency index (to give the likely gross floor area), and a rate selected from a band of rates per m^2 of gross floor area, plus
- the likely cost of the external works.

This cost limit can be incorporated into the balance sheet to check if it is affordable.

(20.2) If all the items on the development balance sheet except the Works cost have been identified and costed, the residual item will be the Works cost limit. This limit should be reduced to allow for the costs of the external works. When the remainder is divided by the functional quantity, the result will be the affordable functional unit rate. A comparison of this rate with those of other projects (after modification) will indicate whether the project is feasible or not.

While an owner may be keen to know what the Works will cost, a specific figure casually offered at this stage is likely to return thereafter to haunt the professional team. Instead, they should advise what the upper and lower bounds of the Works price are likely to be at the time of bidding.

STAGES C AND D OUTLINE AND SCHEME DESIGN STAGES

Having previously set a cost limit for the building and considered its quality, we can use cost equation (20.3) to indicate a maximum gross floor area. If the owner requires a certain useful floor area, this can be divided into the gross area to check that the design efficiency index is realistic (see Chapter 8). If the gross floor area of the design cannot be kept within the indicated limits, the factors in the equation must be reappraised.

A list of all the elements, with the cost limit for the Works apportioned to them, together with their element rates per m^2 of gross floor area will constitute a **cost plan**. Neither the brief nor the scheme design should be changed after being approved at the end of this stage.

STAGE E DETAIL DESIGN STAGE

To be economic, the hidden parts of elements should not have T-dimensions that are larger than they have to be for technical reasons. Examples will include foundations, frame members, solid floors, insulation.

On the other hand, the occupiers of a property may require that some elements, e.g. some internal finishes, the outside faces of the envelope, the main entrance, are to have certain visual attributes (see Chapter 7). This enhancement can be stated as a **cost/worth ratio**. This is the cost of the chosen technical solution divided by the minimum cost of achieving the same standards of safety, health and energy efficiency, i.e. its worth. Judgements on appropriate ratios, and on the amounts of the extra costs, can be made in collaboration with the owner.

Cost/worth ratios feature in **value engineering** (see 'Project audit', below). This is a body of procedures whose objective is to optimize value for money. They include (see Chapter 7):

1. investigating the regulative, supportive and visual functions and attributes of each element; and
2. comparing the economic attributes of alternative technical solutions.

The following is taken from Appendix J of the *Housing Manual* (1949).

> The aim should be to avoid or minimise difficult and costly work on the site by the simplification of construction and method, and by the fullest possible use of the most economical materials and components. Nevertheless, the initial cost of materials and components is not necessarily indicative of economy, and a careful evaluation of alternatives should always be made. . . .
>
> The endeavour should be to adopt constructional designs and methods requiring the minimum number of separate operations to complete a component [*sic*] such as a first floor or roof. Investigation has confirmed that a multiplicity of operations is one of the major causes of high cost, due to the time lost in preparing and clearing up each section of the work.

Thus, simplicity is an aid to buildability and so to economy.

SAVE NOW AND SPEND LATER, AND VICE VERSA

Technical solutions may be chosen because of their low initial costs, although this may be at the expense of occupiers later on. The opposite will also be true. For example, the initial quality of external parts, internal finishes and applied treatments, hardware, and mechanical and electrical services will determine the frequency of their renewal costs. Also, the more efficient (and probably more expensive) heating or air conditioning and lighting installations, and the thermal insulation of the building, the lower the future energy consumption. As decisions are made, the cost plan can be reviewed and amended and the cost limit either confirmed or, with the owner's approval, modified.

As an alternative to a cost plan, a **cost model** may be prepared. In essence, this is a dimensional model of the enclosing elements as described in Chapter 8, together with services, installations, external works etc. and with element unit rates attached, extended and totalled. As always, unit rates are enhanced to allow for the parts and dependent products that are not measured. Heating and cooling costs can be estimated by calculating the steady state energy requirements and pricing this per kilowatt.

THE BID AS THE FINAL COST CHECK

The lowest bid is the final check on the cost plan. However, the construction market can be very uncertain, and when the lowest bid for the Works exceeds the cost limit, this may not indicate incompetence. Rather, it may be prudent for the professional team to aim high but to have a predetermined policy on how savings might be achieved without harming the design too much.

Simply reducing the amount of the lowest bid by making arbitrary reductions to the cost/worth ratios of finishes etc. can turn out to be ineffective. When the time comes for the work to be carried out, their reduced quality may just not be acceptable.

OWNER'S COST CONTROL DURING CONSTRUCTION OPERATIONS

Although a contract might be for a lump sum, careful cost control of the Works will still be required. The contract conditions will provide many reasons for adjustments to the contract sum, including those considered in Chapter 17. Uncertainties should be avoided where possible. In particular, the financial effect of variations (or **architects' instructions**, or **change orders**) should be agreed before being put into effect.

20.7 PROJECT AUDIT

In any audit, the actions of one party are examined by another party. The object of a project audit will be to consider whether the owner's requirements for a new building are being met, and if the design team have been doing what they should. A formal approach called **value engineering** and using **cost/worth ratios** is practised by specialists in the USA at the scheme design stage.

Alternatively, the following list of questions could be asked during an audit. The related chapter references are given in brackets.

By the end of the scheme design stage

1. (1, 7, 9) To what extent have the requirements for each room or other space not been achieved? Can any over- or under-provision of floor area be corrected?
2. (1, 2) What features of the building or the external works will users or visitors find unsatisfactory?
3. (1, 3) What alternative systems might be more economical for (a) the structure, and (b) each of the services?
4. (6) Which members of the design team have not yet been appointed?
5. (7) Where has the fire escape and fire containment strategy not been achieved?
6. (8) To what extent are the design efficiency indexes better or worse than average?
7. (20) Is the value of the building likely to exceed its cost by a suitable margin?
8. (6, 17) Will the chosen procurement method provide a satisfactory building within the owner's time and cost limits?
9. (6, 7) Has a planning supervisor of proven competence been appointed and has the project been notified to the Health and Safety Executive?
10. (7, 9) Have the proposals of the design team been approved by the owner?

At the end of the detail design stage

1. (1) What are occupiers and visitors likely to think of each room and the building as a whole? Consider the lighting and sound levels, the fittings and their positions, the finishes, the circulation and means of escape, the direction signs, and all those other qualities of the internal environment.
2. (4, 9) Have the fire protection requirements been achieved?
3. (1–4, 7, 9) Have any of the desired attributes of each of the functional elements not been achieved?
4. (7, 8, 20) Are the cost/worth ratios of the various elements appropriate?
5. (4, 6, 9) Can the building as a whole be constructed safely and easily, and has the pre-tender stage of the health and safety plan been completed?

6. (6, 12, 17) Is the building buildable by its planned completion date?
7. (6, 7, 20) Have the appropriate regulatory authorities approved the design proposals?

20.8 SUMMARY

The wide variety of uncertainties and the risks involved when undertaking construction projects were considered. The UK's legal framework for the control of development and the differing viewpoints of the individuals concerned were discussed. The costs of ownership and occupation of buildings and the relations between rent and capital value were outlined.

A **project appraisal balance sheet** for setting and controlling costs at the design stage and a **residual value** approach to budgeting were described. Methods of cost planning and control that might be used at the various stages in the design process were outlined. A two-stage audit of the design proposals was suggested, and related to the contents of earlier chapters.

REFERENCES

Building Cost Information Service. *BCIS Subscription Service*, BCIS, Kingston upon Thames.

Department of the Environment. *Quarterly Housing and Construction Statistics*, HMSO, London.

Technical Appendices of the *Housing Manual 1949* (1949) HMSO, London.

FURTHER READING

Cadman, D. and Austin-Crowe, L. (1991) *Property Development*, E & FN Spon, London.

Flanagan, Roger and Norman, George (1993) *Risk Management and Construction*, Blackwell Scientific Publications Ltd, Oxford.

Jardine Insurance Brokers Ltd (1987) *Risk Management*, Kogan Page Ltd, London.

Johnson, Stuart *et al.* (1993) *Greener Buildings: The Environmental Impact of Property*, Macmillan Press Ltd, Basingstoke, Hampshire.

Manser, J. E. (1994) *Economics – A Foundation Course for the Built Environment*, E & FN Spon, London.

Murdock, J. R. and Hughes, W. (1992) *Construction Contracts, Law and Management*, E & FN Spon, London.

Scarrett, Douglas (1991) *Property Valuations*, E & FN Spon, London.

Stone, P. A. (1988) *Development and Planning Economy*, E & FN Spon, London.

Local and structure plans of the appropriate planning authorities.

APPENDIX A
Glossary

Only those important terms that are used in chapters after the one in which they are introduced are given here. They are also included in the Index. Cross references are italicized.

activity A sequence of processes by operatives employing the same set of skills on a set of products, and resulting in the construction of one or more clearly defined parts of the Works at one or more locations on or in the building or on the site. The construction or delay in construction of parts or portions of parts of an element regarded by management as an entity for the purposes of planning, organizing and controlling the Works.

adjudication The final stage in the preparation of a bid by a general contractor. The process of deciding what contribution the contract should aim to make to the general overhead costs of a construction organization.

attribute A quality of a part of a functional element arising from the properties of its substances. The contribution a part makes to the qualities of the element as a whole, so that the element can fulfil its functions, e.g. being weathertight, fire resistant, and strong and stable. While properties can be measured, attributes can only be presumed and are seldom tested. Cf. *property*, *substance*.

avoidable waste Materials that have become useless while on the site. Causes may include the carelessness, incompetence or negligence of operatives, and vandalism.

bid An offer, or **tender**, by one party to provide another party with a service in return for a stipulated payment, or vice versa.

building enclosure The roof, the walls, the floors and those other functional elements that enclose and subdivide the building space and create the various rooms and other internal spaces.

building envelope The roof, the external walls and the lowest floor.

building space The overall space within the building envelope consisting of the internal spaces and the upper floors and internal walls between them.

category A substance A substance whose important properties develop as a result of chemical changes to some of its constituents after they have been incorporated into the building, e.g. concretes, mortars, plasters, bitumens, asphalts, paints and other liquid treatments, adhesives.

category B substance A substance that acquired its important properties before being incorporated into the building, e.g. fired clay, natural stones, natural wood, metals, plastics, glass.

client The customer of the professional team; the initiator of a construction project and, ultimately, the owner of the completed building.

component A complex formed product that becomes a part when fixed in position, e.g. a door, a washbasin. A prefabricated part with its own attributes, and contributing particular functions.

construction method An analysis of the way a building was constructed, including the content of the various activities, the operative skills employed, the kinds of plant used (if any), and the extent to which similar processes were carried out either simultaneously or as a sequence. Alternatively, a plan for doing so.

construction process Those actions performed by either a single operative or a gang with the intention of progressing the Works in some way.

contribution The difference between the income arising from a Works contract and the costs directly incurred on the site. The amount of money a contract contributes towards the running of the organization, i.e. to the general overhead.

co-ordinating size The effective size of a formed product when incorporated into a building. Its visible contribution to the Works. This can be more or less than its manufactured, or **work** size.

dependent product A product used solely in conjunction with a particular kind of main product where this is technically inadequate, e.g. masonry reinforcement, pipe fittings. Cf. *main product*.

dimension A measurement of linear extension in a particular direction. A vector quantity.

dimensional part A part of a building that is constructed in place to the dimensions required by the design. Cf. *integral part*.

dominant product See *main product*.

duration The hours or days when operatives are effectively employed on an activity. Cf. *elapsed time*.

economic attribute A characteristic of the technical solution of a part of a building whereby physical, human, and financial resources are wisely used, both initially and during its lifetime.

elapsed time The hours or days between two events, usually the start and end of an activity, although operatives may not be working all the time. Cf. *duration*.

element See *functional element*.

enabling process Setting up a facility that will enable the construction processes to take place, e.g. the site accommodation, or removing it afterwards. The costs of the enabling, or **indirect** items, constitute the *Works overhead*.

estimate As a noun, an estimate is an offer or bid.

estimating The process of predicting the costs of the enabling, site preparation and construction processes to be incurred on the site, i.e. the prime cost. It can also mean the whole process of bid preparation.

formed product A product whose substance, size, shape and rigidity will have been established during its manufacture, i.e. one with a category B substance.

formless product A liquid or granular product which, after site processing, placing in position, and chemical transformations, becomes what we are calling a category A building substance.

function One of the purposes of a functional element, as expressed by what it does for observers, for other elements, and for the internal spaces and their occupiers.

functional element An essential, distinct, and well-defined construction, being a constituent of a building which is always in the same position with respect to the other elements, and which always has the same functions, no matter what it is made with. Usually abbreviated to *element*.

functional unit A unit of measurement of the owner's requirements, usually either a unit of useful floor area or a unit of a particular kind of accommodation.

general overhead The sum of those costs that will be incurred by an organization in the short term, irrespective of the amount of construction work done. The costs of being in business.

integral part A part of an element made by fixing a component.

main product The class of product chosen to provide a part with its required attributes. Cf. *dependent product*.

managing system Those employees of an organization who regulate and maintain the operative systems on the various sites by controlling the movement of operatives, plant and construction products across the boundaries between sites and their environments. Cf. *operating system*.

materials A general name for the liquids and solids with which a building is made. More specifically, we call them **products,** as their manufacturers will be offering them for sale.

measure To ascertain how much of a thing there is, i.e. its extent, using the ratio scale of numbers and a particular unit.

operation See *activity*. May refer to a sub-activity, i.e. that portion of an activity that takes place at one of a number of locations.

operative system All the operatives, management personnel, plant and other resources involved on the site in the management and execution of site preparation, enabling and construction processes.

operative information Production information on the Works to be constructed on the site which the operatives must follow.

part of an element Materials in an appropriate position within an element which contribute to its functions by having suitable attributes.

performance rate The average rate at which operatives can process products to the required quality standard.

practical completion The term used in the UK to describe when the building is ready for occupation except for some minor items; in the USA this is called **substantial completion**.

prime cost The sum of a group of direct costs, particularly those relating to one production cost centre.

products The liquids and solids, i.e. the *materials* offered for sale, from which a building is made.

project The whole of the activities for which the owner will be paying in order to achieve the desired outcome. Alternatively, the subset of those activities directly connected with or occurring on the site.

project planning A management procedure of analysing the total building process into activities, cataloguing their features, identifying their interrelations, resource requirements and likely durations, and deciding on their sequence to achieve cost and time targets. The graphical process of achieving and expressing this.

property A physical characteristic of a substance that can be studied and, perhaps, measured in a laboratory, e.g. thermal resistance, coefficient of linear expansion, mass density. Cf. *attribute*.

quantification rule The idea that the measured quantity of a dimensional part should be based solely on its S-dimensions.

quantify The process of carrying out arithmetical operations on certain of the dimensions of a part to obtain its quantity.

regulative attribute A quality of a part of an element that enables it to regulate the conditions between one space and another, and between the inside of a building and its environment, e.g. thermal resistance, being weathertight.

resources The people, the plant and the working capital that continue to be available to an organization. Also, when planning a project, the materials used in the building process.

risk The maximum adverse effect a decision could have on the fortunes of the organization.

schedule (noun) A two-dimensional array of data. Usually, the data will be about related parts that have different technical solutions at different locations; (verb) The action of assigning resources to tasks during the construction planning process.

site preparation processes Removing unwanted buildings and other features, clearing the site, and changing its surface configuration by excavating or filling so that it will accommodate the works to be constructed.

spatial or S-dimension The distance between pairs of opposite walls, or between the floor and the ceiling of a space, thus indicating the length, or the breadth, or the height of the space. Also, the dimensions of constructions that are based on those of the spaces, e.g. the lengths and widths of ceilings and floors, and the lengths and heights of walls. Cf. *technical* or *T-dimensions*.

specification A statement describing some of the qualities of one or more existing or proposed objects.

stable organization An organization having a predictable number of future employees which can recover all its costs by charging an all-in hourly or daily rate for its services. Cf. *unstable organization.*

substance The physical matter in which certain qualities, or properties, exist. In our case, the properties of the solid substances to be found in buildings which give its parts their attributes.

substantial completion See *practical completion.*

supportive attribute A quality such as being strong and stable, and being dimensionally stable, that enables an element to support itself and provide support for other elements as necessary.

task A simple sequence of actions by one or more operatives using those kinds of materials regarded as the province of their group.

technical solution A set of decisions on the construction of the parts of an element, including its location, the main and dependent products to use, their positions within the element, and the widths and/or thicknesses of the parts.

technical or **T-dimension** One of the widths or thicknesses of a part of an element decided as part of the technical solution. Cf. *spatial* or *S-dimension.*

total building process The totality of those processes that take place on the site of the Works, including enabling, site preparation and construction processes.

unavoidable waste Excess material that, while having to be purchased and delivered to the site, was not incorporated into the building, and cannot be reused elsewhere, e.g. the unused contents of packages of blocks and cans of paint or wood offcuts.

uncertainty The extent to which some information might be incorrect, either now or in the future, and should not be entirely trusted.

unstable organization A construction organization with a future labour force of uncertain size and types of skills.

visual attribute Those qualities of the exposed surface of an element of a building that contribute to an *attractive appearance.*

work Something made or done. A convenient way of referring to a variety of different constructions, e.g. external works.

Works The results of the various site activities intended to achieve the owner's objectives.

Works overhead The costs incurred on a construction site that cannot be assigned to any particular construction, e.g. site huts, site staff and other enabling items. Also called **indirect items** or **Preliminaries.**

work size The actual size (usually with related tolerances) to which a formed product is manufactured.

SOME OF THE RELATIONS BETWEEN TERMS

1. Element **functions** and **attributes** are either:

(a) **regulative,**
(b) **supportive,**
(c) **visual,** or
(d) **economic.**

2. Building **substances** acquire their properties either:

(a) **after the construction process** (category A), or
(b) **beforehand** (category B).

3. **Category A substances** arise from the use of **formless products.**

4. **Category B substances** are found in **formed products.**

5. A product is either a

(a) **main product** independently chosen to provide a part with its attributes or
(b) **dependent** on that choice, and needed where the main product is technically inadequate.

6. The size of a formed product may mean either:

(a) its actual, or **work size,** or
(b) its effective, or **co-ordinating size.**

7. The process of **building** consists of **site preparation, enabling, and construction processes.**

8. The parts of a building are either:

(a) **dimensional parts** constructed to whatever size is required, or
(b) **integral parts** made by fixing individual **components.**

9. Public knowledge includes statements describing generally applicable **concepts** and **conventions** and information published by, e.g. products manufacturers. Other information used by participants in a project will be **private** and specific.

10. **Drawings** are either **pictorial** or **orthographic.**

11. **Specifications** state either:

- the attributes of the elements (**performance** specification),
- the properties of their substances (**descriptive** specification), or
- the descriptors of the products to be used (**prescriptive** specification).

12. **Dimensions** are either:

(a) the **spatial** or **S-dimensions** of the building spaces decided during the scheme design stage, or they are
(b) **technical** or **T-dimensions** of the building fabric decided when considering the element technical solutions during the detail design stage.

13. The stages of **preparing a tender** (**bid**) are:

(a) collect background information;
(b) make enquiries and obtain prices for products and subcontracts;
(c) estimate and total the prime cost; and
(d) adjudicate on the estimate and decide the contribution.

14. **Measured items** are either:

- **main items** giving the quantity of the part and its main product; or
- **additional items** giving details of dependent products or surfaces which cannot be included in the main item.

15. The **quantity** of a measured item should be calculated from its extensive S-dimensions, while its T-dimensions should remain in the description. Main exceptions are items of concrete work and excavated voids which are measured as volumes.

16. **Site management** will be concerned with the **processes** that can be assigned to one or more workpeople so as to complete an **activity** at a particular **location.** In contrast, **contracts managers** will be more concerned with the aggregation of those **activities** that can be assigned to a group of workpeople (who may work for a subcontractor) and which will constitute a **work package**.

17. An organization is either **stable** or **unstable** depending on whether its labour force is established or fluctuating.

APPENDIX B
Operative processes and performance rates

B.1 PERFORMANCE RATES

INTRODUCTION

This appendix extends the discussions in Chapters 11, 12 and 15. It is intended to encourage a thoughful approach to estimating the durations of activities when planning and cost estimating.

Each of the following sections relates to one or other of our classes of construction products (see Chapter 5). Each begins with an introduction to the skills of those who work with those kinds of products. A process which is likely to achieve the highest performance rate is then described, and followed by a discussion on the technical factors that will reduce this rate.

Figures are given showing ranges of process performance rates that might be achieved with some main products, depending on the constraints. They are not intended to be comprehensive. Unless the members of a gang are given, the ranges of performance rates will be for one tradesman with suitable assistance. The scales are approximately logarithmic and indicate the performance that might be achieved in an eight-hour day and the average amount per hour. They assume that the same process continues throughout the day.

Although estimators in the UK tend to use hourly performance rates, the daily output has the advantage of being more easily derived from progress on the site. It is also a more convenient unit to use when comparing the estimated with the actual progress on the site.

SELECTING AND APPLYING PERFORMANCE RATES

The figures indicate the likely bounds to performance with particular main products. They are not based on any formal site research but will be assumed to include for both effective and ineffective employment. To select a rate:

1. identify the correct figure and the main product or process;
2. consider the constraints that will reduce the performance rate; the more the constraints, the less the rate;
3. starting with the highest rate on the right, move back to a point in the range that seems to represent the circumstances of your job; check this by considering what other constraints would bring you back to the lowest rate;
4. visually extend this point upwards or downwards to the appropriate scale and read it, interpolating as necessary.

The estimated duration of the activity can then be calculated using formula (15.2), the 'estimating function'. This is (see Chapter 15):

amount of main product / performance rate = duration, or

$$A / R = D$$

B.2 PERFORMANCE RATES WITH FORMLESS MATERIALS

Dimensional parts made with formless materials The input and output quantities will be the same. For example, one m^3 of wet-mix concrete becomes one m^3 of hardened concrete in place. The material applied to one m^2 of a surface becomes one m^2 of coating. Performance rates for concreting and filling will be expressed as m^3 per day and per hour, and coatings of all sorts as m^2 per day and per hour.

(A1) CONCRETES

Unreinforced (or **mass**) concrete is only used where it will be in compression.

Highest performance rate: continuously discharging ready-mixed concrete directly into surface trenches and other voids at ground level, with little or no spreading and surface finishing.

Constraints: concreting processes are discussed under class AB1, reinforced concrete.

Typical performance rates in m^3 for a gang of four operatives are shown in Figure B.5.

(A2) THICK COATINGS (E.G. TYPE 5 BASE)

Tradesmen include plasterers, asphalters, floor layers. Tasks include receiving and moving the constituent materials, mixing, moving, applying and finishing. The most commonly used hand tool is the plasterer's trowel.

Highest performance rate: applying a lightweight plaster undercoat to a slightly porous surface such as that of common brickwork.

Constraints

1. **Coarsely textured material**, e.g. cement mortar used for renderings.
2. **Surface inclination** Gravity will encourage the material to flow or fall off any sloping or vertical surface, or from an underside, or **soffit**.
3. **Smooth base** offering little adhesion for the coating, e.g. engineering brickwork, precast concrete.
4. **Narrow base**, e.g. coatings to roof curbs, skirtings, window reveals.
5. **Surface finish**, e.g. trowelling, scratching (to provide a 'key' for the next coat).
6. **Accessibility** Working above shoulder-height, or low down, or on the far side from where the operative is standing, i.e. **overhand**, or from scaffolding.
7. **Temperature** Bitumen-based materials are more easily worked at higher temperatures when they are less viscous.
8. **Incorporating dependent products** such as plaster angle and edge beads, expanded metal reinforcement for asphalt.

Some likely performance rates for one tradesman are suggested in Figure B.1. To convert a number of square metres to their equivalent number of square yards, add one-fifth.

(A3) THIN COATINGS (E.G. TYPE A SURFACE FINISH)

Operatives are painters and their assistants, and general operatives. Their tasks include preparing the materials, preparing the background for treatment, using a brush, roller or spray-gun to apply the materials, and, possibly, texturing the finished surfaces. The larger the brush, the more quickly a surface can be coated, and the less frequently it has to be re-loaded.

Highest performance rate: using a wide brush to apply a free-flowing material to the top of a smooth, slightly absorbent, horizontal surface, e.g. applying a coat of hardener to a concrete floor.

Constraints

1. **High viscosity of the material**.
2. **Surface inclination** See (A2) above.
3. **Porous or rough base**, e.g. insulation board, the 'hairy' surface of sawn wood, the slightly porous and absorbent surfaces of planed woods and untreated plaster.

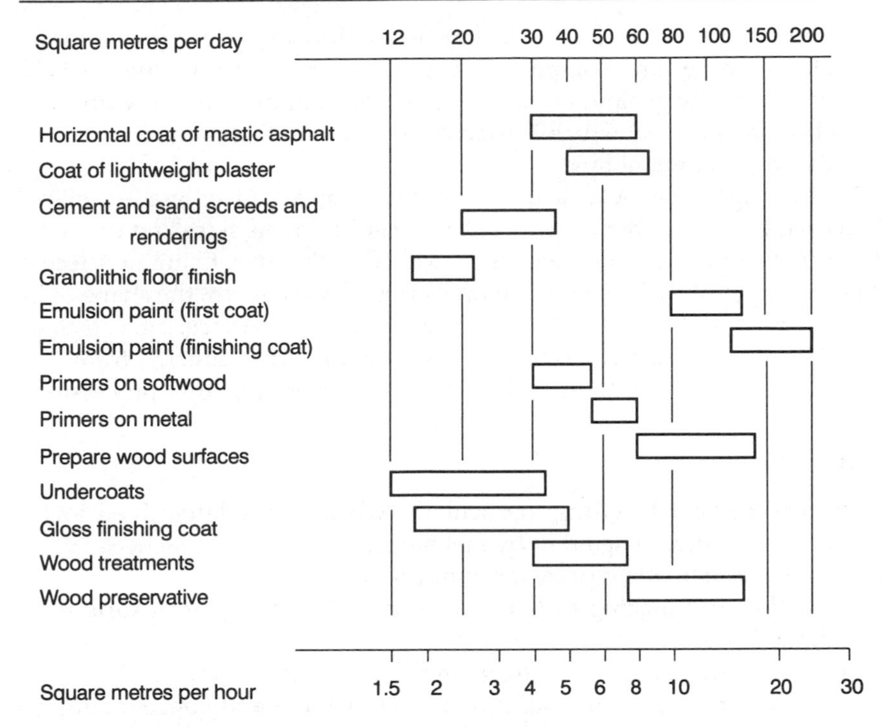

Fig. B.1 Some likely performance rates with class A2 and A3 products.

4. **Narrow bases** such as window frames and casements, where the edges of coatings have to be **cut in** at junctions with the glass etc. and where smaller brushes will be necessary.
5. **Surface treatments** such as stippling.
6. **Accessibility of surface**, e.g. above head level, low down, or working from a temporary workplace.

Some likely performance rates for one tradesman are suggested in Figure B.1.

(A4) FILLINGS (I.E. GRAVEL TYPE)

A **graded** filling material consists of a mixture of differently sized particles of hard material in predetermined proportions. The processes of forming beds for ground floors, beds or surrounds for drainage pipelines, and filling to make up levels, are usually carried out by general operatives. The mechanical plant may include dumpers, hydraulic backacters, compactors and rollers.

Initially, the material is likely to be deposited in heaps, from where it will be dug, loaded, moved and deposited in its final position. It may then be spread and either levelled or finished to a gradient (that process is called **grading**) and compacted. The top surfaces of filled beds below floors next to the ground are likely to be covered, or **blinded**, with sand or other fine material which is then compacted and smoothed to receive the damp-proof membrane.

A drainage pipeline will usually be laid on a bed of graded filling. Afterwards, more of the material will be filled round its sides and up to the level of its top. The trench will be backfilled with either filling or selected soil, or a layer of each, depending on its position, its depth, the shape of its cross-section, and the loads that will be applied at the surface, e.g. vehicles.

Highest performance rate: tipping the material directly from the transporting vehicle into the void to be filled without further processing.

Constraints

1. **Spreading and levelling,** particularly when to predetermined levels or slopes, either by hand or by mechanical plant.
2. **Compacting** by hand or by mechanical plant.
3. **Selecting and placing individual units by hand**, e.g. at unsupported edges of beds.
4. **Loading from a temporary heap**, moving and depositing on site.
5. **Blinding** the top of the filling with fine material and consolidating to form a smooth surface at the desired level.

A range of performance rates using mechanical plant where possible is shown on Figure B.8.

B.3 PERFORMANCE RATES WITH FORMED PRODUCTS

Thing-thinking These notes are concerned solely with factors that affect the number of units or pieces of the formed material, i.e. the number of **things** likely to be processed in one unit of time either by one craftsman with assistance or by a suitably sized gang. Of course, using larger units would enable a portion of work to be built in less time.

The most significant factors are the shape and size of the unit and its weight, which will depend on its substance. The types of fixings will depend on both the units themselves and the background to which they are being secured.

(B1) ASSEMBLAGE OF UNIFORM-SIZED RIGID FORMED PRODUCTS WITH ONE MINOR DIMENSION, THEIR FITTINGS, ACCESSORIES AND FIXINGS (I.E. TILE TYPE)

These will become surfaces and will be fixed to some compatible kind of structural background. Alternative methods are:

1. **secured to an open framework** (e.g. roof covering type A1); tasks include bringing the units and fixings to the workplace, setting out, reducing dimensions of some units to fit the length of the row or column, placing and securing; operatives are slaters, tilers, roofing workers;
2. **secured to a continuous base** (e.g. internal finish type C); tasks as before, except that securing will be by adhesive or a mortar bed to a continuous base; operatives will be wall tilers, plasterers, specialist floor layers.

Highest performance rate: placing plain roofing tiles (work size 265 × 165 mm) on battens without nailing.

Constraints

1. **Size, shape and weight of units** The smaller units can be manipulated with one hand. Others require the operative to use both hands. Obviously, the heavier the units, the more tiring they are to handle. Bituminous felt square butt slates (called asphalt strip shingles in North America) are about 1 m long and have an effective width of about 130 mm.
2. **Substance of units** The substance of tiles, i.e. ceramic, concrete, natural stone, does not directly affect the rate at which they can be laid. However, it may influence the sizes and shapes that can be manufactured and whether they can be cut to size.
3. **Type of fixings** Roofing tiles have self-supporting nibs and, where not exposed to high winds, whole courses may be left unsecured. Natural slates are not self-supporting. Instead, each must first be holed and then held while being nailed. All units that form vertical surfaces will be individually secured. Where units are to be continuously attached to a background, the substances of both the units and the background will determine both a suitable adhesive, e.g. mortar, tile adhesive, and the process. Asphalt shingles are fixed with large-headed clout nails and their exposed faces are continuously bonded to the ones below.
4. **Nature of the work generally** The co-ordinating dimensions of overlapping units and any supporting battens can be adjusted to fit within the overall dimensions of the surface being created. However, where units are laid edge to edge, their pattern, e.g. checkerboard, herringbone, has to be set out and maintained during laying, and units at the margins must be cut to fit. The smaller the surface being covered, the greater the proportion of edge length to area.
5. **Surface finish** Although the edges of floor and wall tiles might appear to be straight and square, the tiles will be spaced slightly apart

and have their joints **grouted** to ensure a continuous surface and to avoid capilliarity.

6. **Unusual arrangement of units** Units are sometimes bedded in mortar to form window sills, decorative features etc.
7. **Incorporating dependent products** Tile fittings at junctions and edges require special attention, and interrupt the operative's rhythm. Similarly, soakers, lead slates for service pipes, flue outlets etc. will have to be positioned in a roof covering to maintain the continuity of the weathershield.

A range of likely performance rates for one tradesman is shown in Figure B.2.

(B2) LARGE UNITS USED IN COMBINATION AND THEIR FIXINGS

Units of this class are delivered to the site ready for fixing. A part will be made by joining a number of B2 units. (This contrasts with class C1 components where each individual unit becomes a part when fixed in its final position.) The construction process includes unloading, moving, lifting, positioning, and jointing or fixing.

Where possible, units will be lifted from delivery vehicles and immediately placed in position. This has the advantage of eliminating storage, double handling, and the risk of damage. A minimum of two operatives will be employed, together with a lifting device and operators. The units are too diverse for a highest output to be offered.

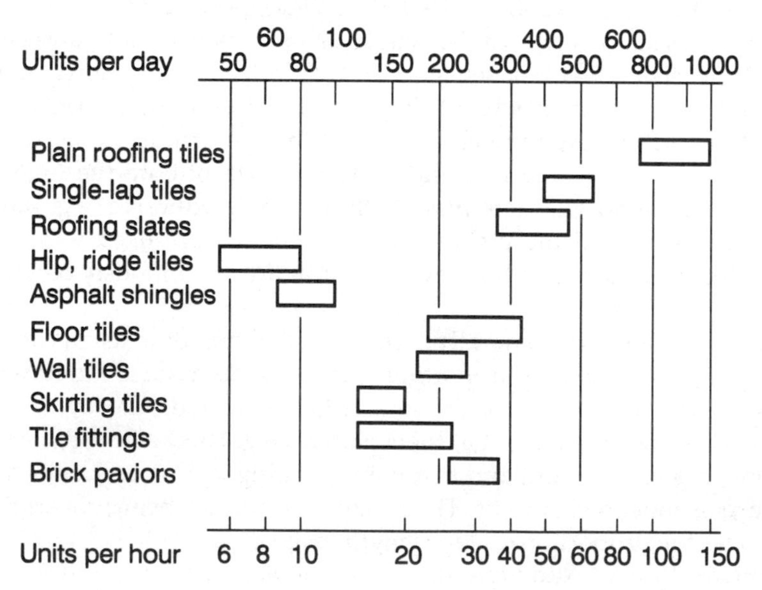

Fig. B.2 Some likely performance rates with class B1 products.

Constraints

1. **Size, shape and weight of units** If the lifting device already on the site, e.g. a tower crane, will be inadequate or unavailable, a special item of plant must be brought to the site, installed, dismantled and removed, e.g. a mobile crane with an extending jib.
2. **Substance of units** This will affect the robustness of the units and their requirements for storage and protection.
3. **Types of fixings** Spanning units will be fixed both to each other and to supporting elements. Fixings can be either 'wet' or 'dry', i.e. either with mortar or concrete, or with nails, bolts, clips, dowels, straps etc.
4. **Nature of the work generally** Some kinds of units will be placed directly on top of their supports, e.g. beams in a steel frame, timber trusses, wall panels. In such cases, they need only to be steadied while fixing takes place. Others, e.g. ashlar walling units and precast floor beams, will be secured by being bedded in mortar or some other class A material. Their weight must be supported from above while this bedding takes place and the material hardens.
5. **Surface finish** The exposed edges of the mortar joints of dressed stones may need pointing.
6. **Unusual arrangement of units** Trussed rafters to a complex pitched roof may include half-trusses and special shapes as well as diagonal bracing, purlins, binders and ridge, hip and valley members.
7. **Incorporating dependent materials** Precast units may be joined and/or finished with wet-mix concrete, possibly steel reinforced. This may delay progress.
8. **Interdependent operations** Mortices for the holding-down bolts of steel columns will be formed during the casting of their concrete foundations. After the steel frame has been erected, levelled and plumbed, and when the bases have been wedged and the connections secured, the spaces around the holding-down bolts and between the foundation tops and the column bases will be grouted with a suitable category A material.

Performance rates The processes of preparing, hoisting, placing and securing each kind of product will be different for each site. As a starting point, ask yourself if either two or three can be fixed in each hour.

(B3) RIGID FLAT SLABS AND SHEETS AND FIXINGS (I.E. PLYWOOD TYPE)

(B4) PROFILED SHEETS, THEIR FITTINGS, ACCESSORIES AND FIXINGS (I.E. CORRUGATED ROOFING TYPE)

B3 and B4 products have two major dimensions and are used to make surfaces. Tasks will include getting from store, moving, setting out and cutting to size, and fixing. Operatives include carpenters, plasterers, roofers, specialist fixers, glaziers and their assistants.

Highest performance rate: securing 1200 × 450 mm cavity wall insulation units with retainers clipped to wall ties.

Constraints

1. **Size, shape and weight of units** Some units are too large to be handled by a single operative.
2. **Substance of units** The substance will indicate how easy it will be to cut to size, and what tools will be suitable. Also, it will indicate whether it is tough enough to withstand directly driven nailing, or whether holes for unit fixings should be pre-bored. A substance may be so brittle, e.g. glass, that the units have to be secured indirectly, e.g. with beads and ribbons of mastic.
3. **Types of fixings** Fixings must be compatible with the substances of the units and their backgrounds and with the loads to be resisted.
4. **Nature of the work generally** Some flat sheet units are cut from larger sheets to suit the design, and are fixed individually to supporting backgrounds. Where flat boards or sheets are assembled to form larger continuous surfaces, the operatives are likely to determine the positions of intermediate supports and how the sheets will be cut.
5. **Surface finish** The heads of nails should be punched below the surface to ensure a smooth surface.

Some likely performance rates with products of classes B3 and B4 are given on Figure B.3. These are for one or two operatives, depending on the size of the unit.

(B5) FLEXIBLE SHEETS AND QUILTS (I.E. FELT TYPE)

Tasks may include moving, unrolling, cutting to size, folding or lapping at joins, positioning and fixing. Operatives include slaters, tilers, general operatives, bricklayers, carpenters, painters and decorators, floor layers, and their assistants.

Highest performance rate: laying insulation between rafters.

Constraints

1. **Size, shape and weight of units** Most products are supplied in rolls that can be handled by one operative. Polyethylene damp-proof membrane is supplied in 4 m wide rolls, and is joined by folding.

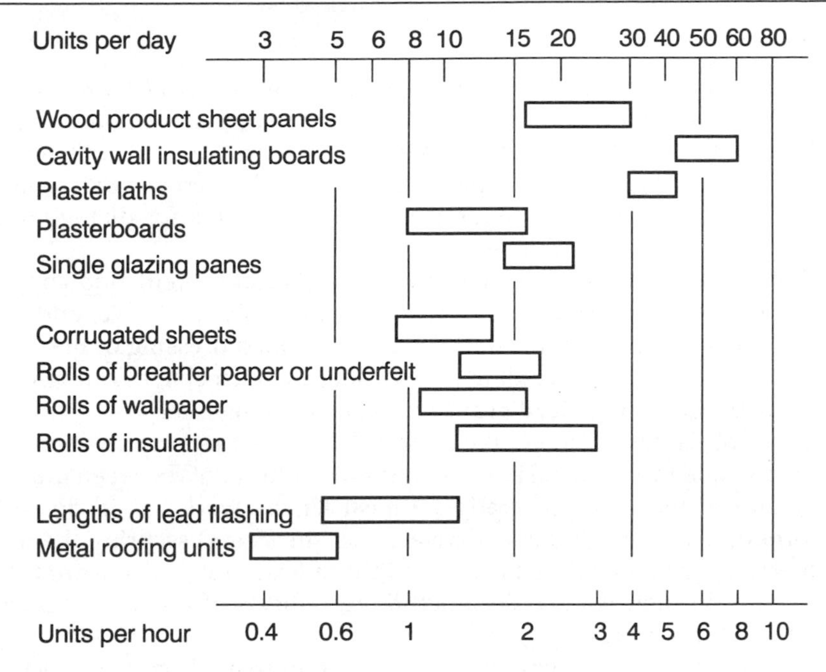

Fig. B.3 Some likely performance rates with products from classes B3–B6.

2. **Substance of units** Protective clothing and masks may be required when handling insulating materials.
3. **Types of fixings** Sheets requiring continuous support are likely to be bonded to a continuous base; those supported intermittently may be fixed with special nails.
4. **Nature of the work generally** Most sheets will be unrolled and cut to length before being either placed horizontally in position, or fixed to sloping or vertical supports. Some materials will be fixed above the operatives' heads, e.g. decorative ceiling papers.

Some likely performance rates for one operative with class B5 products are given on Figure B.3.

(B6) MALLEABLE SHEETS (I.E. SHEET LEAD TYPE)

Tasks undertaken by plumbers or specialist metal workers and their assistants include setting out and cutting to size, moving, shaping, placing in position and fixing.

Highest performance rate: straight lead flashing cut from a roll of suitable width, with one edge bent and secured with lead wedges in a joint between masonry units.

Constraints

1. **Size, shape and weight of units** To limit the effect of its relatively high coefficient of thermal expansion, the size of a unit will be limited to an effective area of not much more than $1m^2$. A lead unit might weigh 35–40 kg. One of copper or zinc would be perhaps only a fifth of this. They will be cut from rectangular sheets or strips that are normally supplied flat or in rolls.
2. **Substance of units** Copper and zinc sheets can be bent, and copper can be welded or soldered. Lead is exceptionally malleable, and the substance can be hammered (the process is called **dressing** or **bossing**) into quite intricate three-dimensional shapes. Such shapes can also be made by welding, or **lead burning** small pieces together.
3. **Types of fixings** Strips of the same substance are secured to the background and incorporated into the upstands of the joints between units.
4. **Nature of the work generally** Almost all the work has to do with forming weathertight joints between the sheets and securing them to their background. While the self-weight of lead units will ensure their stability in high winds, the much lighter sheets of other metals will rely on their fixings.
5. **Surface finish** Surfaces are usually left untreated, as a protective coating of oxides or carbonates is likely to form naturally from exposure to the atmosphere.
6. **Unusual arrangement of units** Specially shaped units can be used to cover awkward junctions between roof slopes and to weatherproof perforations in roof coverings for pipes and flues.
7. **Incorporating dependent products** An underlay will be required to isolate the sheets from their (probably nailed) continuous support and to ensure that all moisture is excluded.
8. **Interdependent operations** Inserting soakers between roofing units and fixing flashings to masonry walls.

Some likely performance rates with class B6 products are given in Figure B.3. These are for one or two operatives, depending on the size of unit.

(B7) SOLID SECTIONS AND FIXINGS (I.E. TIMBERS TYPE)

In the UK, all lengths of softwood intended for structural use or **carcassing** are likely to be referred to as **timbers**. In North America, this term seems to be reserved for wood sections exceeding 114 mm thick. Smaller sizes are called **lumber**.

Tasks may include moving, setting out, cutting to length, preparing ends, and fixing by carpenters and joiners.

Highest performance rate: lengths of tiling battens nailed to rafters.

Constraints

1. **Size, shape and weight of units** Carcassing and joinery timbers and other smaller sections can be handled by one operative. Where members are fixed at their ends, their length has little effect on the time per unit. Ends of structural members may be notched, e.g. to steel joists, splayed, e.g. at both ends of rafters, or birdsmouthed, e.g. wood rafters to wall plates. Joinery items may be mitred, e.g. external angles of skirtings and architraves, or scribed, e.g. internal angles of skirtings.
2. **Substance of units** Hardwoods are generally more difficult to cut, shape and nail than softwoods. Special care must be taken with hardwoods chosen for their appearance.
3. **Types of fixings** End fixings may be by directly nailing to other members or by nailing to straps or hangers themselves secured to other members or to masonry. Skirtings, architraves, fascias and other trims (sometimes called **mill** items, as their profiles are produced by rotating cutters) must be secured at frequent intervals to a suitable background.
4. **Nature of the work generally** Much of the work calls for precise measurement on the site, a high level of manual dexterity and, in the case of pitched roof members, an appreciation of the three-dimensional geometry involved.
5. **Surface finish** Where wood surfaces are to be painted, nail heads should be punched below the surface and the holes filled. The surfaces of woods chosen for their colour and the figure of their grain will be protected with clear or translucent finishes and their fixings should be secret. One method is **pellating** or **pelleting**, where screws are deeply countersunk, and the spaces above the screw heads are filled with matching wood disks, or **pellets**.
6. **Unusual arrangement of units** The ease with which woods can be cut and nailed make them suitable for many unusual arrangements.
7. **Incorporating dependent materials** Open frameworks may be stiffened by noggings or herringbone strutting between the members and connected to other elements with metal straps.
8. **Interdependent operations.** Plumbing, internal drainage and electrical services.

Some likely performance rates with class B7 products are given in Figure B.4. These are for one or two operatives, depending on the size of unit.

(B8) HOLLOW SECTIONS, THEIR FITTINGS AND FIXINGS (I.E. PIPE TYPE)

Tasks may include moving, setting out and cutting to length, preparing ends, making joints, fixing. They may be carried out by general operatives, drainlayers, groundworkers, plumbers, pipe fitters, steel fixers.

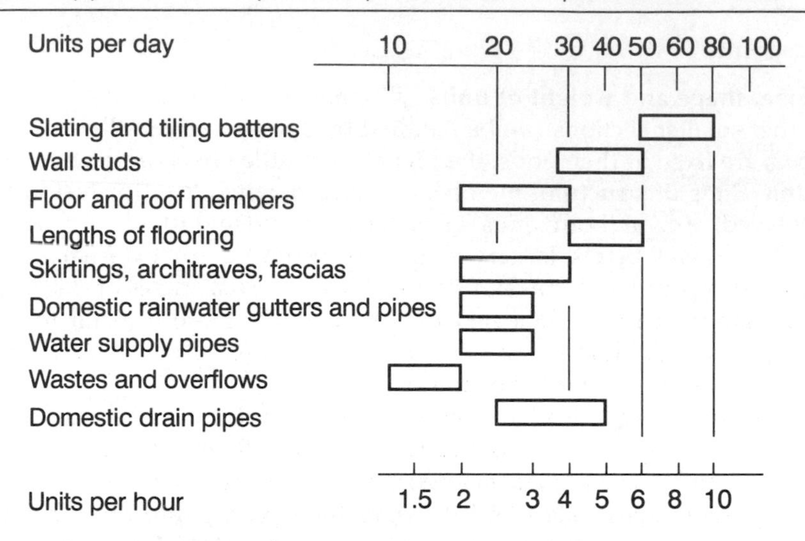

Fig. B.4 Some likely performance rates with class B7 and B8 products.

Highest performance rate: 100 mm clay field drain pipes in shallow trenches.

Constraints

1. **Size, shape and weight of units** Cranes may be required to lift, move and place large sewer pipes.
2. **Substance of units** The substance will determine how the units are jointed.
3. **Types of fixings** Units will be supported either continuously, e.g. drain pipes on filled beds, or intermittently by brackets or clips.
4. **Nature of the work generally** Most of the time goes in making joints in the running length, at changes in direction, or when connecting to another pipeline.
5. **Unusual arrangement of units** All pipelines consist of unique combinations of straight units and fittings of various kinds.
6. **Incorporating dependent products** such as insulation.
7. **Interdependent operations** Trenching, forming ducts, perforating floors and walls for pipelines, making good to finishes. Fixing components of various kinds, e.g. radiators, sanitary fittings, plastic inspection chambers.

Some likely performance rates with class B8 products are given in Figure B.4. They are for one tradesman with some assistance.

B.4 PERFORMANCE RATES WITH COMPOSITE CONSTRUCTIONS

(AB1) REINFORCED CONCRETE

This type of construction was introduced in Chapter 3. Figure B.5 gives some ranges of performance rates for erecting and dismantling forms, which will depend on their complexity. The time for making wooden forms can be estimated using data from Figures B.3 and B.4.

Operatives include reinforcement benders and fixers, formwork carpenters and their labourers, and general operatives. Plant may include ready-mix vehicles, mixers, cement silos, power shovels, dumpers, cranes and skips, concrete pumps, vibrators and power floats. Metal formwork, clamps and supports can be regarded as plant as they are designed to be reused very many times.

Highest performance rate: large, lightly reinforced bases at ground level.

Constraints on cycle time

1. intervals between deliveries of ready-mixed concrete, or the time for site-mixing a batch of concrete;
2. volume of one delivery or batch of wet-mix;
3. volume of wet-mix transported in one cycle, e.g. skip size;
4. extent of horizontal and vertical movement of wet-mix;
5. workability of wet-mix concrete and its need for compaction;
6. small T-dimensions of constructions, e.g. thin floors, narrow columns, staircases;
7. intensity and arrangement of reinforcement;
8. extent of surface finishing.

Other constraints on progress

1. height restrictions to limit the hydrostatic pressure of unset concrete on formwork;
2. quantity restrictions to limit the build-up of heat in the concrete as it sets;
3. need to support the structure until it is sufficiently strong to support itself;
4. surface treatment to restrict evaporation, e.g. sprayed film, plastic sheeting.

Performance rates are outlined in Figure B.5. The rates for erecting and dismantling wooden formwork are for one carpenter with assistance. The rate for fixing bar reinforcement is for one operative. The concreting rates are for a gang of four operatives with lifting plant if required.

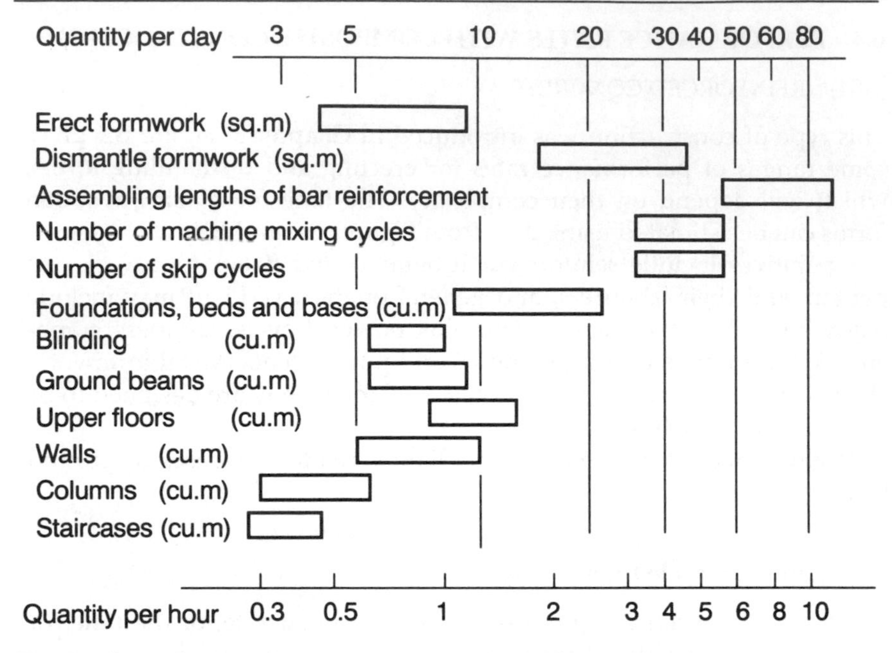

Fig. B.5 Some likely performance rates with class AB1 products.

(AB2) SMALL STRUCTURAL UNITS IN MORTAR, DEPENDENT PRODUCTS AND COMPONENTS

Tasks carried out by masons, bricklayers, and their labourers may include:

1. erecting scaffolding to provide a temporary workplace;
2. mixing and transporting mortar;
3. bringing units to the workplace;
4. building corners or erecting profiles;
5. laying units;
6. forming arches and cills;
7. building in dependent products such as damp-proof courses, ties, reinforcement, lintels and cavity trays;
8. building in window and door frames, or their profiles;
9. finishing exposed edges of mortar joints;
10. moving, dismantling and removing scaffolding and other temporary workplaces.

Highest performance rate: long, straight, thick walling with no openings and a rough finish, built at waist level from a ground floor with common bricks laid in English bond in a **plasticized** mortar, i.e. one with an additive to improve its plasticity.

Constraints on the rate of laying individual units

(It is likely that a portion of work could be built in less time when using larger units.)

1. **Size, shape and weight of units** When laying bricks, an operative will pick up, manipulate and place the bricks with one hand while keeping a trowel in the other for picking up and spreading mortar. In contrast, both hands are required to lift and place most blocks and walling stones, the trowel being taken up as required. Discrepancies in size and shape have to be accommodated by varying the thicknesses of mortar beds and joints, and/or by rejecting and selecting units, thus breaking the operative's rhythm. Operatives get tired more quickly when laying relatively heavy units such as engineering bricks, ballast concrete blocks and large stones.
2. **Substance of units** Units which only absorb moisture slowly, if at all, e.g. engineering bricks, walling stones and glass blocks, are not well held by their mortar beds. Consequently, they are more troublesome to lay. Also, the number of courses that will hold their position while the mortar remains unset, or 'green', is limited.
3. **Type of mortar** Mortars consisting solely of cement and sand are stiff and less easy to work with than those containing lime or plasticizer (or made with dirty sand).
4. **Nature of the work generally** Corners and ends are normally built before the rest of the walling. These are carefully plumbed and levelled (a painstaking process) to ensure that the courses in between will be horizontal, and that the walls will be vertical. Intersections with other walls, attached piers, and openings for windows and doorways interfere with the process of filling in between corners, and call for a careful setting out of the units and the preservation of vertical joints, or **perpends**. The same applies to the construction of detached piers. Where the dimensions of walls, including those between openings, are not multiples of the co-ordinating dimensions of the units, some units must be cut to fit. Cutting and fitting may also be necessary against the parts of other elements, such as a structural frame or precast floor beams.
5. **Surface finish** Extra care is required when aligning the exposed faces of units, and when finishing the faces of the joints. This is even more time-consuming where no workplace is available on that side of the wall and the work has to be done **overhand**. Performance rates may be further reduced where the faces of half- or one-brick thick walls have to be finished fair on both sides. The mortar at the exposed edges of the joints may be raked out and pointed with mortar of a different kind.
6. **Unusual arrangement of units** Bricks laid in facing bonds, e.g. Flemish, English garden wall, lack the regularity of English or

Stretcher (running) bond. Arches, brick-on-edge copings and sills, brick-on-end or **soldier** courses etc. also call for a more complex manipulation of the units, their careful jointing, and possibly some cutting. Curved work involves working to a template, the plumbing of the whole face, and cutting units to suit the radius. Battered walls, with courses which slope down from front to back, are also built to templates.

7. **Incorporating dependent products and components** Pausing to build in lintels, reinforcement, wall ties, damp-proof courses, soakers, wall insulation, window and door frames or their profiles, etc, all interrupt the operative's work-rhythm. Some cutting of units may also be involved.
8. **Interdependent operations** Forming openings for services, and having to build portions of work in advance, e.g. walls to support beams and floors, or later, e.g. filling between the feet of rafters, interrupt and delay the works.

Some likely performance rates for a single tradesman with suitable assistance are shown on Figure B.6.

B.5 PERFORMANCE RATES WITH COMPONENTS

(C1) COMPONENTS AND FITTINGS

Tasks might include moving, preparing, assembling, modifying and fitting, e.g. to another component, as well as fixing. Operatives include joiners, plumbers, heating and ventilating engineers, drainlayers, with their assistants.

Highest performance rate: fixing hat and coat hooks to softwood with screws.

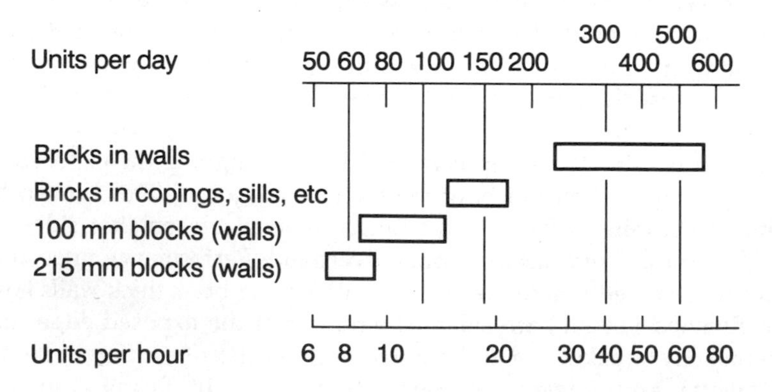

Fig. B.6 Some likely performance rates with class AB2 products.

Constraints

1. **Size, shape and weight of units** will indicate the number of operatives and the type of lifting or other plant required.
2. **Substance of units** Components are usually made of more than one substance and are likely to need protection. Extra care must be taken with hardwood joinery as their surfaces are intended to be exposed, and will not be obscured by paint.
3. **Types of fixings** usually depend on the type of component.
4. **Nature of the work generally** Components are self-contained parts that are fixed to other differing parts at different locations. The fixing process is usually simple.
5. **Interdependent operations** Hot water radiators may be removed temporarily to allow the walls to be decorated. Some fittings, e.g. washing machines, baths, will be connected to services.

Some likely performance rates for a single operative with assistance are given on Figure B.7.

(C2) FURNISHINGS

Tasks include moving, assembling, and placing or fixing in position.

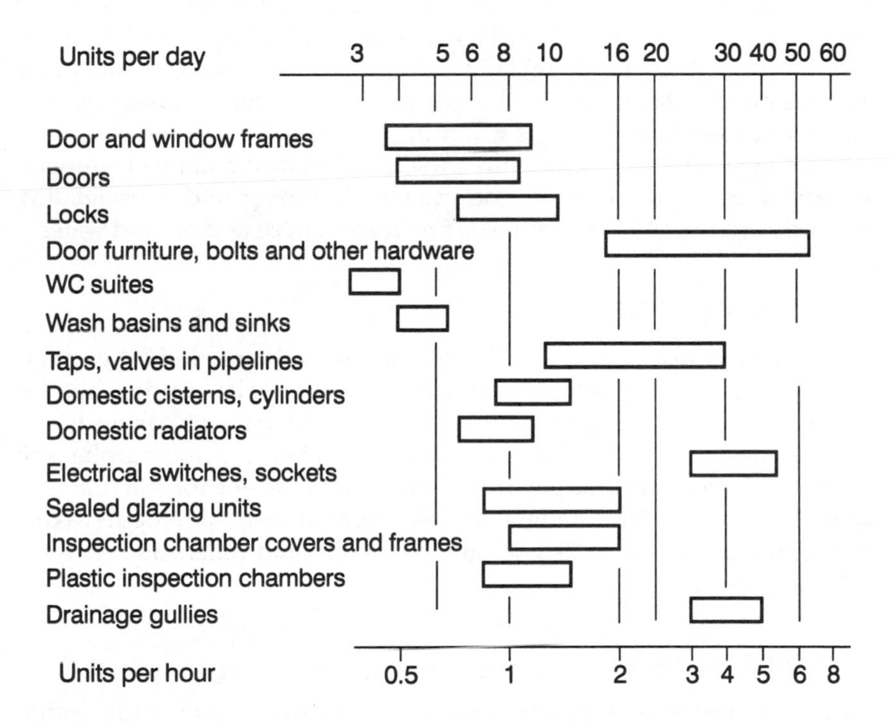

Fig. B.7 Some likely performance rates with class C1 components.

B.6 EXCAVATION AND FILLING

EXCAVATION

Excavating by hand or machine will involve some of the following tasks:

- breaking out the soil from its natural state and lifting it above the surface;
- depositing on site or into a vehicle;
- moving;
- either backfilling, or placing, or spreading; and
- consolidating.

Excavation processes, and the machines used, will depend on whether the excavation will take place over the surface or below it.

(a) **Excavation over the surface** results in the general lowering of the levels of the site. It can be carried out freely from the surface by machines able to dig down to just below their own wheels or tracks and move and deposit the soil. Access for machines and for vehicles to remove the soil is straightforward.

(b) **Excavating below the surface** will produce voids for specific constructions such as foundations, pipelines, and basements. The general level of the surface is left unchanged. Excavating machines will normally remain at ground level, although where the void is big enough, a ramp may be formed to enable machines to get in and out. The excavated material must be lifted up to ground level or into a vehicle, and this may call for additional machinery, e.g. a crane and bucket.

Associated tasks will include the provision and dismantling of supports to vertical or sloping faces of the ground, trimming and consolidating exposed surfaces, and keeping voids free from surface and ground water.

SELECTING MACHINES

Scrapers, face shovels, tractor bulldozers and shovels are used for excavating over the surface. Backacters and drag-lines can also do this, but are designed to excavate deeply below the surface, and to lift and deposit.

Taxed and insured vehicles are necessary when moving surplus soil on the highway, but dumper trucks are more efficient for moving and depositing soil on site. Graders will only spread or grade, roadbreakers and compressors with tools will only break out, and rollers and tampers will only consolidate.

MACHINE DIGGING

Excavating provides a good example of interdependent cyclic gang-working. The maximum output will be determined by the operators of the excavating plant. To maintain this output, the excavated material must be removed immediately, perhaps by the same machine, or in dumpers, road vehicles or barrows. This process is also cyclic. Either process can delay the other one.

Other operatives will direct the plant operators, maintain any exposed service pipes, shape and trim the sides and bottoms of the voids, and construct and remove the temporary supports to the sides. Compressed air or other powered tools may be necessary to break out rock or existing concrete and other hard constructions. Pumps will be needed when water is present. If work is being carried out on the highway, roadsigns, some form of traffic control, supervision, barriers and lighting will be necessary.

HAND-DIGGING

When excavating by hand down from the surface of the ground, it soon becomes impossible to dig while standing at ground level. The operative must then get into the void, and use a spade or shovel to lift the soil and deposit it on the surface for removal later.

Lifting by hand is limited to shoulder-height, and for greater depths, mechanical lifting (e.g. buckets and hoisting plant) may be adopted. However, if this is impracticable or not available, intermediate shoulder-height stages of re-digging, lifting and depositing must be introduced, using either steps cut in the soil or temporary staging. The health and safety of operatives is paramount.

DISPOSAL OF SURPLUS SOIL

Except for small amounts that will be backfilled after the construction work is in position, the excavated material will normally be moved away from the void. If it is to remain on the site, it will be placed either in its permanent location as required by the design, or in one or more temporary heaps designated by the constructor.

Surplus soil is often called **spoil**. If the design calls for the surplus soil to be removed, the constructor must decide its destination. Usually, this will be to a properly authorized tip for which a charge will be payable. There can be a local demand for vegetable soil.

IMPORTING FILLING MATERIALS

When the soil arising from excavations is insufficient to make up the finished levels of the Works, the rest must be imported. This is a normal feature of new road construction, where, to maintain suitable gradients, some stretches will be on embankments. Of course, other stretches may be in cuttings, and an economically efficient design will attempt to keep these volumes in balance.

However, any surplus at one location will have to be transported to where it is required, and still may not be enough. In such cases, the rest may be obtained by negotiating for, creating and excavating from a local **borrow pit**.

SUPPORTING THE SIDES OF VOIDS

In stiff clays and rock, the sides of quite deep voids will remain stable without external support. However, supports must be provided where operatives might otherwise be buried by a collapse. These can vary from the restraint offered by occasional poling boards and struts to continuous support from steel sheet piling.

It is mainly the mass of the triangular section of soil between the side of the void and the natural angle of repose of the soil that has to be restrained. This will vary with the square of the depth. Water present in the soil, additional loads from building materials, spoil heaps and existing buildings, and the effects of road traffic, will increase the need for support.

FACTORS AFFECTING EXCAVATING PERFORMANCE RATES

Among the factors affecting the productivity of excavating processes are the following:

- the nature of the soil or rock: if soil, whether cohesive or non-cohesive; if rock, whether explosives or breaking tools will be required; whether wet, moist or dry; the extent of bulking; the presence of tree roots;
- the presence of ground water;
- the state of the site and haul roads: whether dry and hard, moist, wet or water-logged;
- the extent of underground services and if they are to be preserved.

Machine digging

- Type of machine.
- Effective work done in each cycle, e.g. working capacity of bucket or blade, reduced to solid measure.
- Average cycle time. Where another plant item is used for removal, the process will include breaking out, filling the bucket, slewing, depositing and returning. If this average is based on observation, it will allow for delays due to manoeuvring and changing direction, waiting for

levels to be set and sides and bottoms to be trimmed, working round underground services, and moving to new work-positions. To some extent, the rate of working will depend on the accuracy required, including the avoidance of over-excavating.

Hand digging

- Digging by hand will normally be employed where it is not practicable to use a machine, e.g. small pier holes; trenches in very unstable soil where the sides are close-sheeted and strutted across the trench; small tunnels. Performance will depend on the space available for working, the depth (particularly when the surface is above shoulder-height), and the extent of obstacles such as trench supports and existing service pipes and cables.

Figure B.8 shows the ranges of outputs that might be achieved by one operative excavating by hand or by one machine with suitable assistance. To convert a number of cubic metres to the equivalent number of cubic yards, add one-third.

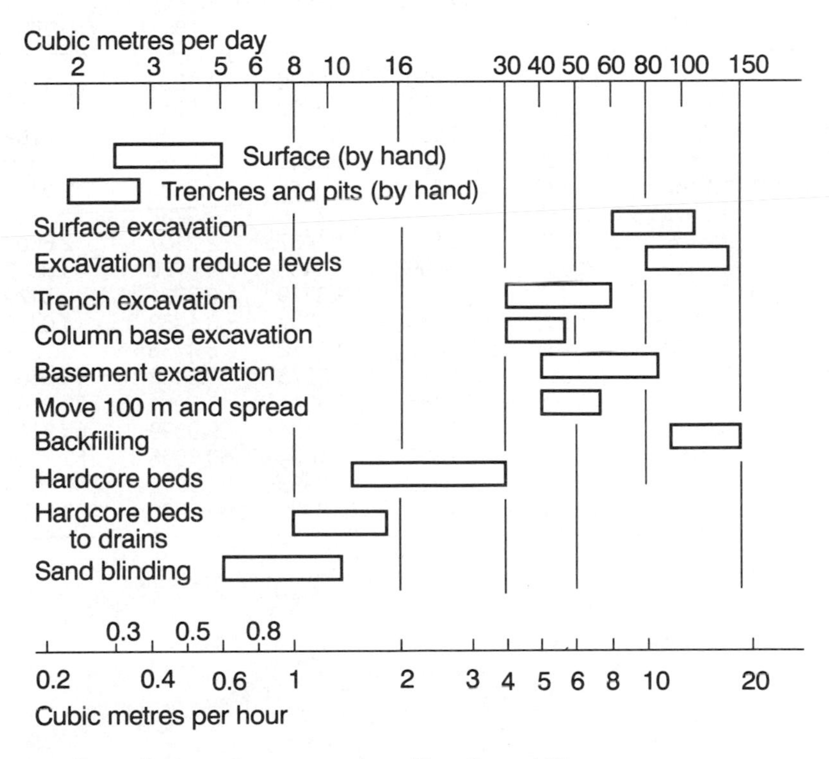

Fig. B.8 Some likely performance rates with soils and fillings.

APPENDIX C
Compound interest tables

C.1 (A) COMPOUND AMOUNT OF 1 AFTER *N* PERIODS

Present value multiplied by (a) = future worth
Future worth divided by (a) = present value

No. of periods	*Interest rate per period*					
	4%	*6%*	*8%*	*10%*	*12%*	*15%*
1	1.0400	1.0600	1.0800	1.1000	1.1200	1.1500
2	1.0816	1.1236	1.1664	1.2100	1.2544	1.3225
3	1.1249	1.1910	1.2597	1.3310	1.4049	1.5209
4	1.1699	1.2625	1.3605	1.4641	1.5735	1.7490
5	1.2167	1.3382	1.4693	1.6105	1.7623	2.0114
6	1.2653	1.4185	1.5869	1.7716	1.9738	2.3131
7	1.3159	1.5036	1.7138	1.9487	2.2107	2.6600
8	1.3686	1.5938	1.8509	2.1436	2.4760	3.0590
9	1.4233	1.6895	1.9990	2.3579	2.7731	3.5179
10	1.4802	1.7908	2.1589	2.5937	3.1058	4.0456
15	1.8009	2.3966	3.1722	4.1772	5.4736	8.1371
20	2.1911	3.2071	4.6610	6.7275	9.6463	16.3665
25	2.6658	4.2919	6.8485	10.8347	17.0001	32.9190
30	3.2434	5.7435	10.0627	17.4494	29.9599	66.2118
35	3.9461	7.6861	14.7853	28.1024	52.7996	133.1755
40	4.8010	10.2857	21.7245	45.2593	93.0510	267.8635
50	7.1067	18.4202	46.9016	117.3909	289.0022	1083.6574

C.2 (B) COMPOUND AMOUNT OF 1 PER PERIOD

Payment per period multiplied by (b) = future worth
Future worth divided by (b) = payment per period

No. of	*Interest rate per period*					
periods	*4%*	*6%*	*8%*	*10%*	*12%*	*15%*
1	1.0000	1.0000	1.0000	1.0000	1.0000	1.0000
2	2.0400	2.0600	2.0800	2.1000	2.1200	2.1500
3	3.1216	3.1836	3.2464	3.3100	3.3744	3.4725
4	4.2465	4.3746	4.5061	4.6410	4.7793	4.9934
5	5.4163	5.6371	5.8666	6.1051	6.3528	6.7424
6	6.6330	6.9753	7.3359	7.7156	8.1152	8.7537
7	7.8983	8.3938	8.9228	9.4872	10.0890	11.0668
8	9.2142	9.8975	10.6366	11.4359	12.2997	13.7268
9	10.5828	11.4913	12.4876	13.5795	14.7757	16.7858
10	12.0061	13.1808	14.4866	15.9374	17.5487	20.3037
15	20.0236	23.2760	27.1521	31.7725	37.2797	47.5804
20	29.7781	36.7856	45.7620	57.2750	72.0524	102.4436
25	41.6459	54.8645	73.1059	98.3471	133.3339	212.7930
30	56.0849	79.0582	113.2832	164.4940	241.3327	434.7451
35	73.6522	111.4348	172.3168	271.0244	431.6635	881.1702
40	95.0255	154.7620	259.0565	442.5926	767.0914	1779.0903
50	152.6671	290.3359	573.7702	1163.9085	2400.0182	7217.7163

C.3 (C) UNIFORM SERIES THAT 1 WILL BUY

Present value multiplied by (c) = payment per period
Payment per period divided by (c) = present value

No. of periods	*Interest rate per period*					
	4%	*6%*	*8%*	*10%*	*12%*	*15%*
1	1.0400	1.0600	1.0800	1.1000	1.1200	1.1500
2	0.5302	0.5454	0.5608	0.5762	0.5917	0.6151
3	0.3603	0.3741	0.3880	0.4021	0.4163	0.4380
4	0.2755	0.2886	0.3019	0.3155	0.3292	0.3503
5	0.2246	0.2374	0.2505	0.2638	0.2774	0.2983
6	0.1908	0.2034	0.2163	0.2296	0.2432	0.2642
7	0.1666	0.1791	0.1921	0.2054	0.2191	0.2404
8	0.1485	0.1610	0.1740	0.1874	0.2013	0.2229
9	0.1345	0.1470	0.1601	0.1736	0.1877	0.2096
10	0.1233	0.1359	0.1490	0.1627	0.1770	0.1993
15	0.0899	0.1030	0.1168	0.1315	0.1468	0.1710
20	0.0736	0.0872	0.1019	0.1175	0.1339	0.1598
25	0.0640	0.0782	0.0937	0.1102	0.1275	0.1547
30	0.0578	0.0726	0.0888	0.1061	0.1241	0.1523
35	0.0536	0.0690	0.0858	0.1037	0.1223	0.1511
40	0.0505	0.0665	0.0839	0.1023	0.1213	0.1506
50	0.0466	0.0634	0.0817	0.1009	0.1204	0.1501

C.4 TABLE SHOWING THE REDUCTION IN A LOAN AS REGULAR PAYMENTS ARE MADE

See Chapter 16, Figure 16.5
Repaying a loan of #100 000 over 30 years @ 8% interest per year

Period	*Payment*	*Principal at start of period*	*Interest*	*Loan reduction*	*Principal at end of period*
0					100000.00
1	8882.74	100000.00	8000.00	882.74	99117.26
2	8882.74	99117.26	7929.38	953.36	98163.90
3	8882.74	98163.90	7853.11	1029.63	97134.27
4	8882.74	97134.27	7770.74	1112.00	96022.27
5	8882.74	96022.27	7681.78	1200.96	94821.32
6	8882.74	94821.32	7585.71	1297.03	93524.28
7	8882.74	93524.28	7481.94	1400.80	92123.48
8	8882.74	92123.48	7369.88	1512.86	90610.62
9	8882.74	90610.62	7248.85	1633.89	88976.73
10	8882.74	88976.73	7118.14	1764.60	87212.13
11	8882.74	87212.13	6976.97	1905.77	85306.36
12	8882.74	85306.36	6824.51	2058.23	83248.13
13	8882.74	83248.13	6659.85	2222.89	81025.24
14	8882.74	81025.24	6482.02	2400.72	78624.52
15	8882.74	78624.52	6289.96	2592.78	76031.74
16	8882.74	76031.74	6082.54	2800.20	73231.54
17	8882.74	73231.54	5858.52	3024.22	70207.32
18	8882.74	70207.32	5616.59	3266.15	66941.17
19	8882.74	66941.17	5355.29	3527.45	63413.72
20	8882.74	63413.72	5073.10	3809.64	59604.08
21	8882.74	59604.08	4768.33	4114.41	55489.67
22	8882.74	55489.67	4439.17	4443.57	51046.10
23	8882.74	51046.10	4083.69	4799.05	46247.05
24	8882.74	46247.05	3699.76	5182.98	41064.07
25	8882.74	41064.07	3285.13	5597.61	35466.46
26	8882.74	35466.46	2837.32	6045.42	29421.03
27	8882.74	29421.03	2353.68	6529.06	22891.98
28	8882.74	22891.98	1831.36	7051.38	15840.60
29	8882.74	15840.60	1267.25	7615.49	8225.10
30	8882.74	8225.10	658.01	8224.73	0.37

Index